NEUROMETHODS

Series Editor
Wolfgang Walz
University of Saskatchewan
Saskatoon, SK, Canada

For further volumes:
http://www.springer.com/series/7657

Neurovascular Coupling Methods

Edited by

Mingrui Zhao

Department of Neurological Surgery, Brain and Mind Research Institute, Weill Cornell Medical College, New York, NY, USA

Hongtao Ma

Department of Neurological Surgery, Brain and Mind Research Institute, Weill Cornell Medical College, New York, NY, USA

Theodore H. Schwartz

Department of Neurological Surgery, Brain and Mind Research Institute, Weill Cornell Medical College, New York Presbyterian Hospital, New York, NY, USA

Humana Press

Editors
Mingrui Zhao
Department of Neurological Surgery
Brain and Mind Research Institute
Weill Cornell Medical College
New York, NY, USA

Hongtao Ma
Department of Neurological Surgery
Brain and Mind Research Institute
Weill Cornell Medical College
New York, NY, USA

Theodore H. Schwartz
Department of Neurological Surgery
Brain and Mind Research Institute
Weill Cornell Medical College
New York Presbyterian Hospital
New York, NY, USA

ISSN 0893-2336 ISSN 1940-6045 (electronic)
ISBN 978-1-4939-4465-1 ISBN 978-1-4939-0724-3 (eBook)
DOI 10.1007/978-1-4939-0724-3
Springer New York Heidelberg Dordrecht London

Printed on acid-free paper

Humana Press is a brand of Springer
Springer is part of Springer Science+Business Media (www.springer.com)

Dedication

This book is dedicated to my various scientific mentors whose wisdom and guidance have helped me maintain my curiosity and passion for neuroscience research: George Ojemann who taught me that the operating room can be used as a laboratory, Rafa Yuste who taught me the power of optical techniques and never to settle for anything less than your very best, and Tobias Bonhoeffer who taught me that good leaders must sometimes get their hands dirty to craft a scientific manuscript.

Theodore H. Schwartz

Preface to the Series

Under the guidance of its founders Alan Boulton and Glen Baker, the Neuromethods series by Humana Press has been very successful since the first volume appeared in 1985. In about 17 years, 37 volumes have been published. In 2006, Springer Science+Business Media made a renewed commitment to this series. The new program will focus on methods that are either unique to the nervous system and excitable cells or which need special consideration to be applied to the neurosciences. The program will strike a balance between recent and exciting developments like those concerning new animal models of disease, imaging, in vivo methods, and more established techniques. These include immunocytochemistry and electrophysiological technologies. New trainees in neurosciences still need a sound footing in these older methods in order to apply a critical approach to their results. The careful application of methods is probably the most important step in the process of scientific inquiry. In the past, new methodologies led the way in developing new disciplines in the biological and medical sciences. For example, Physiology emerged out of Anatomy in the nineteenth century by harnessing new methods based on the newly discovered phenomenon of electricity. Nowadays, the relationships between disciplines and methods are more complex. Methods are now widely shared between disciplines and research areas. New developments in electronic publishing also make it possible for scientists to download chapters or protocols selectively within a very short time of encountering them. This new approach has been taken into account in the design of individual volumes and chapters in this series.

Saskatoon, SK, Canada ***Wolfgang Walz***

Preface

The link between cerebrovascular autoregulation and brain function, also known as "functional hyperemia," is a phenomenon with a long tradition in neuroscientific laboratory investigation. The earliest hypothesis, from the late 1770s, was called the Monro-Kelly doctrine in which CBF was thought to be constant under both physiological and pathological conditions. The later work of Roy and Sherrington in the 1890s proposed that '*its vascular supply can be varied locally in correspondence with local variations of functional activity*', now named neurovascular coupling. In 1945, Kety and Schmidt first described a method of quantifying CBF in humans and brought brain blood flow research into a new exciting era. Through the middle to late twentieth century, advances in functional imaging techniques including fMRI, PET, and SPECT have improved our understanding of the relationship between brain activity and brain energy supply in awake humans. Neurovascular and neurometabolic coupling are critical to supplying the energy demands of brain tissue during both normal physiological function and pathological conditions. Nevertheless, most leaps in our understanding of neurovascular coupling have come from the laboratory, where high spatial and temporal resolution can be recorded using techniques that would be considered too invasive to use in humans. This book will bring the reader up-to-date with the current state-of-the-art techniques in measuring blood flow in the brain in the ongoing investigation of neurovascular coupling.

Each chapter in this book describes a different technique, or combination of techniques, applied to a specific species in either a healthy or an abnormal brain. It is important that most of these techniques can be applied to a variety of species to measure different aspects of neurovascular coupling in both normal and pathological brain states. Hence, the examples provided represent the interest of the specific laboratory reporting on the technique but not the sole application of this technique. This book thus provides a framework from which a multitude of additional experiments could be performed as these techniques are applied to a variety of other species and other regions of the cortex or pathological conditions. What is apparent from an overview of these chapters is the increasing importance and power of optical techniques in neurovascular coupling research. Likewise, the noninvasive nature of optical techniques render them useful in the neurosurgical operating room for use in humans, another common theme among these chapters. Likewise, in almost every chapter, multiple techniques are combined in order to measure signals from multiple sources, not just hemodynamic but also neuronal, metabolic, or glial. The combination of multiple techniques allows investigators to render conclusions on the coupling dynamics between these various sources of the signals.

In the opening chapter, Kennerley, Boorman, Harris, and Berwick use simultaneous fMRI and intrinsic optical spectroscopy to measure hemoglobin-based signals in an anesthetized rodent during whisker stimulation. While IOS provides high resolution 2-dimensional hemodynamic data, fMRI provides broader spatial sampling of the blood oxygen level dependent (BOLD) signal from the whole brain. By combining the two techniques, the sources of the BOLD signals are examined in more detail. In the next chapter, Radhakrishnan, Franceschini, and Srinivasan then use optical coherence tomography

(OCT) to measure neuronal and hemodynamic activity in rat somatosensory cortex in multiple cortical layers correlated with electrical recordings. OCT examines cortical hemodynamic activity in a laminar fashion providing layer-specific information at depths not available to IOS by using higher wavelengths with greater penetration depths. Continuing the exploration of neurovascular coupling mechanisms during normal somatosensory stimulation, Winn and Ko examine the role of adenosine and other neuropharmacological interventions in the regulation of cerebral blood flow during sciatic nerve stimulation. In order to obtain higher resolution and eliminate the influence of anesthesia, which has been shown to temper hemodynamic reactivity, Shih, Drew, and Kleinfeld use in vivo two-photon laser scanning microscopy in the awake mouse to measure RBC velocity and lumen diameter of small vessels using injection of fluorescent-conjugated dextrans during spontaneous activity and in response to somatosensory stimulation.

In the second section of the book, authors employ additional techniques to examine neurovascular coupling in the visual system. In this section, a particular emphasis on techniques for sampling deep brain structures is provided. Bélanger, Souza, Pouliot, Casanova, and Lesage measure neurovascular coupling in deep brain structures using confocal fiber-optic endomicroscopy to measure both calcium dye fluctuations in neurons and hemodynamic activity in the superior colliculus of the anesthetized rat. Likewise, Li and Freeman use the APOX, a deep probe sensor to simultaneously measure tissue oxygenation and neuronal activity in the central visual pathway of anesthetized cats. They find increased spatial localization of oxygen-based measurements of the initial dip compared with CBF or CBV. Vanzetta, Deneux, and Grinvald use widefield CCD- and CMOS-based imaging of intrinsic absorption of light to measure the RBC velocity in the cortex in the awake macaque and the anesthetized rodent to measure normal visual responses and spreading depression. This group also explores the utility of this technique in cat retina. Finally, Sato and Tanifuji use intrinsic optical signals to measure deoxyhemoglobin in the higher visual cortex of monkeys and correlate their findings with electrical recordings.

In the next section of the book, the investigators use several different techniques to examine neurovascular coupling during epilepsy. First, Xu, Paisansathan, and Pelligrino demonstrate the critical role of glia in vasodilation in anesthetized rats using video microscopy following both sciatic nerve stimulation and epileptic events. Zhao, Ma, Harris, and Schwartz show that IOS can be coupled with flavoprotein autofluorescence to measure oxygen metabolism along with hemoglobin signals during epileptic events in anesthetized rat. Then Kim, Hyder, and Blumenfeld use fMRI and LFP measurements to examine the etiology of the negative BOLD signal and its relationship to neuronal activity in anesthetized rodent models of epilepsy. In their data, it is clear that neurovascular uncoupling may be specific to pathological brain states. Ma, Zhao, Harris, and Schwartz use simultaneous IOS and either voltage sensitive dye or wide-field calcium imaging to examine neurovascular coupling in epileptic events in the anesthetized rat cortex. Again, uncoupling in pathological states is demonstrated. Finally, Jiang employs photoacoustic tomography to explore hemodynamic events associated with rodent seizures through an intact skull. The value of this technique, which will be further explored in later chapters, is the ability to measure hemodynamics through an intact skull, which provides translational techniques for noninvasive measurements in humans for clinical diagnostic purposes.

In the final section of the book, the investigators introduce a variety of additional neurovascular measurements which are used to examine cerebrovascular disease states such as ischemia, hemorrhage, and spreading depression. Piilgaard and Lauritzen correlate measurements of tissue oxygenation measured with polaragraphic electrodes, cerebral blood flow with laser Doppler flowmetry, and local field potential during spreading depression in anesthetized rats. Kazmi, Richards, and Dunn use laser speckle contrast and multi-exposure speckle imaging of cerebral blood flow during strokes in animal models and show the adaptability of the technique for use in humans during neurosurgical procedures. Finally, Ugliaro, Pfeil Barbour, and others describe the use of diffuse optical tomography, which is based on near-infrared spectroscopy, to measure hemoglobin-based signals during ischemia and subarachnoid hemorrhage transcranially in Bonnet Macaque.

This book provides an overview of a variety of techniques currently available to examine neurovascular coupling in both health and disease. It is impossible to cover all modern neurovascular methods in one single volume. We hope that this book can serve as the handbook for researchers to study neurovascular coupling in normal and pathological brain states. Understanding normal neurovascular coupling during sensory, visual, and ultimately higher order functions such as language and memory will be critical in the development of novel brain mapping techniques that can be employed in humans. However, the propagation of these techniques to diagnose disease will require a complete understanding of how neurovascular coupling mechanisms break down in disease states to permit accurate interpretation of vascular signals as a surrogate for neuronal activity. We hope that readers will enjoy this book and will help in further developing new methods for the next edition.

New York, NY

Mingrui Zhao
Hongtao Ma
Theodore H. Schwartz

Acknowledgements

We would like to thank all authors of this book for their outstanding contributions. Special thanks are due to Dr. Wolfgang Walz for making this book possible and the Springer staff for bringing it all together. Finally, we would like to thank our families for their continuous support and encouragement.

Mingrui Zhao
Hongtao Ma
Theodore H. Schwartz

Contents

Contributors

RANDALL L. BARBOUR • *Department of Pathology, State University of New York, Downstate Medical Center, Brooklyn, NY, USA; NIRx Medical Technologies, LLC, Glen Head, NY, USA*

FRANK C. BARONE • *Departments of Neurology, and Physiology and Pharmacology, State University of New York, Downsate Medical Center, Brooklyn, NY, USA*

SAMUEL BÉLANGER • *Département de Génie Électrique, École Polytechnique de Montréal, Montréal, QC, Canada*

JASON BERWICK • *Department of Psychology, The University of Sheffield, Sheffield, UK*

HAL BLUMENFELD • *Departments of Neurology, Neurobiology, Neurosurgery, Yale University School of Medicine, New Haven, CT, USA; Core Center for Quantitative Neuroscience with Magnetic Resonance (QNMR), Yale University School of Medicine, New Haven, CT, USA*

LUKE BOORMAN • *Department of Psychology, The University of Sheffield, Sheffield, UK*

CHRISTIAN CASANOVA • *Laboratoire des Neurosciences de la Vision, École d'Optométrie, Université de Montréal, Montréal, QC, Canada*

JEAN CHARCHAFLIEH • *Department of Anesthesiology, Yale School of Medicine, New Haven, CT, USA*

THOMAS DENEUX • *Institut de Neurosciences de la Timone, UMR7289, CNRS & Aix-Marseille Université, Marseille, France; Neurobiology Department, Weizmann Institute of Science, Rehovot, Israel; Unité de Neuroscience, Information et Complexité (UNIC), UPR CNRS 3293, Centre National de la Recherche Scientifique, Gif-sur-Yvette, France*

PATRICK J. DREW • *Department of Engineering Science and Mechanics, Pennsylvania State University, University Park, PA, USA; Department of Neurosurgery, Pennsylvania State University, University Park, PA, USA*

ANDREW K. DUNN • *Department of Biomedical Engineering, The University of Texas at Austin, Austin, TX, USA*

MARIA ANGELA FRANCESCHINI • *Martinos Center for Biomedical Imaging, Massachusetts General Hospital, Charlestown, MA, USA*

RALPH D. FREEMAN • *Group in Vision Science, School of Optometry, Helen Wills Neuroscience Institute, University of California, Berkeley, CA, USA*

TIGRAN GEVORGYAN • *Department of Surgery, The Brooklyn Hospital Center, Brooklyn, NY, USA*

HARRY L. GRABER • *Department of Pathology, State University of New York, Downstate Medical Center, Brooklyn, NY, USA; NIRx Medical Technologies, LLC, Glen Head, NY, USA*

AMIRAM GRINVALD • *Neurobiology Department, Weizmann Institute of Science, Rehovot, Israel*

SAMUEL HARRIS • *Department of Psychology, The University of Sheffield, Sheffield, UK; Department of Neurological Surgery, Weill Cornell Medical College, New York, NY, USA*

FAHMEED HYDER • *Department of Diagnostic Radiology, Yale University School of Medicine, New Haven, CT, USA; Core Center for Quantitative Neuroscience with Magnetic Resonance (QNMR), Yale University School of Medicine, New Haven, CT, USA*

HUABEI JIANG • *Department of Biomedical Engineering, University of Florida, Gainesville, FL, USA*

S.M. SHAMS KAZMI • *Department of Biomedical Engineering, The University of Texas at Austin, Austin, TX, USA*

ANEURIN J. KENNERLEY • *Department of Psychology, The University of Sheffield, Sheffield, UK*

ROBERT KIM • *Department of Neurology, Yale University School of Medicine, New Haven, CT, USA*

DAVID KLEINFELD • *Department of Physics, University of California at San Diego, La Jolla, CA, USA; Section of Neurobiology, University of California at San Diego, La Jolla, CA, USA*

KATHRYN KO • *Departments of Neurological Surgery, Weill Medical College of Cornell University, New York, NY, USA*

JOHN G. KRAL • *Department of Surgery, State University of New York, Downstate Medical Center, Brooklyn, NY, USA*

MARTIN LAURITZEN • *Department of Clinical Neurophysiology, Glostrup Hospital, Glostrup, Denmark; Center for Healthy Aging and Department of Neuroscience and Pharmacology, University of Copenhagen, Copenhagen, Denmark*

DANIEL C. LEE • *Department of Surgery, State University of New York, Downstate Medical Center, Brooklyn, NY, USA; Department of Medicine, Empire Clinical Research Investigator Program, Interfaith Medical Center, Brooklyn, NY, USA*

FRÉDÉRIC LESAGE • *Département de Génie Électrique, École Polytechnique de Montréal, and Research Center, Montreal Heart Institute, Montréal, QC, Canada*

BAOWANG LI • *Group in Vision Science, School of Optometry, Helen Wills Neuroscience Institute, University of California, Berkeley, CA, USA*

JENNY LIBIEN • *Departments of Pathology and Neurology, State University of New York, Downstate Medical Center, Brooklyn, NY, USA*

HONGTAO MA • *Department of Neurological Surgery, Brain and Mind Research Institute, Weill Cornell Medical College, New York, NY, USA*

SUNDEEP MANGLA • *Departments of Radiology, Neurology, and Neurosurgery, State University of New York, Downstate Medical Center, Brooklyn, NY, USA*

CHANANNAIT PAISANSATHAN • *Neuroanesthesia Research Laboratory, University of Illinois at Chicago, Chicago, IL, USA*

DALE A. PELLIGRINO • *Neuroanesthesia Research Laboratory, University of Illinois at Chicago, Chicago, IL, USA*

DOUGLAS S. PFEIL • *Department of Pathology, State University of New York, Downstate Medical Center, Brooklyn, NY, USA*

HENNING PIILGAARD • *Department of Clinical Neurophysiology, Glostrup Hospital, Glostrup, Denmark*

PHILIPPE POULIOT • *Département de Génie Électrique, École Polytechnique de Montréal, and Research Center, Montreal Heart Institute, Montréal, QC, Canada*

HARSHA RADHAKRISHNAN • *Department of Biomedical Engineering, University of California at Davis, Davis, CA, USA*

SERGIO A. RAMIREZ • *Department of Surgery, State University of New York, Downstate Medical Center, Brooklyn, NY, USA*

Lisa M. Richards • *Department of Biomedical Engineering, The University of Texas at Austin, Austin, TX, USA*

Takayuki Sato • *Laboratory for Integrative Neural Systems, RIKEN Brain Science Institute, Wako, Saitama, Japan*

Theodore H. Schwartz • *Department of Neurological Surgery, Brain and Mind Research Institute, Weill Cornell Medical College, New York Presbyterian Hospital, New York, NY, USA*

Andy Y. Shih • *Department of Neurosciences, Medical University of South Carolina, Charleston, SC, USA*

LeRone Simpson • *Department of Surgery, Texas Tech University Health Science Center, Lubbock, TX, USA*

Bruno Oliveira Ferreira de Souza • *Laboratoire des Neurosciences de la Vision, École d'Optométrie, Université de Montréal, Montréal, QC, Canada*

Vivek J. Srinivasan • *Department of Biomedical Engineering, University of California at Davis, Davis, CA, USA*

Manabu Tanifuji • *Laboratory for Integrative Neural Systems, RIKEN Brain Science Institute, Wako, Saitama, Japan*

Anthony D. Uglialoro • *Departments of Surgery and Neurology, Empire Clinical Research Investigator Program (ECRIP), State University of New York, Downstate Medical Center, Brooklyn, NY, USA*

Ivo Vanzetta • *Institut de Neurosciences de la Timone, UMR7289, CNRS & Aix-Marseille Université, Marseille, France*

H. Richard Winn • *Mount Sinai School of Medicine, New York, NY, USA; University of Iowa, Iowa City, IA, USA*

Haoliang Xu • *Neuroanesthesia Research Laboratory, University of Illinois at Chicago, Chicago, IL, USA*

Yong Xu • *Department of Pathology, State University of New York, Downstate Medical Center, Brooklyn, NY, USA; NIRx Medical Technologies, LLC, Glen Head, NY, USA*

Mingrui Zhao • *Department of Neurological Surgery, Brain and Mind Research Institute, Weill Cornell Medical College, New York, NY, USA*

Part I

Neurovascular Coupling Techniques in Normal Somatosensory Processing

Chapter 1

Simultaneous Functional Magnetic Resonance and Two-Dimensional Optical Imaging Spectroscopy

Aneurin J. Kennerley, Luke Boorman, Samuel Harris, and Jason Berwick

Abstract

The blood oxygenation level-dependent (BOLD) functional magnetic resonance imaging (*f*MRI) signal is a biophysical consequence of the hemodynamic response to neuronal activity. Intrinsic optical imaging spectroscopy (2D-OIS) is an invasive imaging method that can provide detailed two-dimensional information of this underlying hemodynamic response. Here, methods for the combination of *f*MRI and 2D-OIS in the rodent model are described. One of the numerous advantages of this simultaneous methodology is that 2D-OIS provides an independent hemodynamic measure to help interpret evoked or spontaneous BOLD signals. This is important when imaging animal models of disease where abnormal BOLD responses could result and thus be difficult to interpret without such an independent measure. From a more mathematical perspective the concurrent hemodynamic data can be used to help refine and/or test biophysical models of the BOLD *f*MRI signal—such models are important for the interpretation of human BOLD data in terms of the underlying neuronal activity. This has obvious benefits for calibrated BOLD studies heavily reliant on such models. The reverse is also true: using biophysical models the concurrent BOLD data can be used as an indicator of the hemodynamic response. This would allow modification of algorithms of light transport through tissue in optical imaging spectroscopy to better predict the hemodynamic response and which in turn could be used in the analysis of the BOLD *f*MRI signal. With all the above in mind, a final (but equally important) benefit of collecting multimodal data is that the number of animals needed for a specific study can be substantially reduced.

Key words *f*MRI, BOLD signal, Two-dimensional optical imaging spectroscopy, Hemodynamic response, Multimodal imaging, Whisker barrel cortex, Biophysical modeling, Neurovascular coupling

1 Introduction

Functional magnetic resonance imaging has revolutionized the field of cognitive neuroscience. It has allowed researchers unprecedented access into the workings of the human brain. However, there are limitations to the technique; the most commonly applied *f*MRI method, the blood oxygenation level-dependent (BOLD) signal, does not measure neural activation directly. It represents only a secondary hemodynamic marker of neuronal function, specifically the level of deoxyhemoglobin in the active brain region [1–3].

Mingrui Zhao et al. (eds.), *Neurovascular Coupling Methods*, Neuromethods, vol. 88,
DOI 10.1007/978-1-4939-0724-3_1,

The mechanism linking neural activation to the hemodynamic response is called neurovascular coupling. The existence of neurovascular coupling has been known for over a century [4]. However, a complete understanding of how neurons and blood vessels communicate is still lacking. Although the BOLD response can exquisitely localize active regions of the brain, it remains very difficult to interpret the underlying neuronal response that drives the BOLD signal, and, if *f*MRI is performed in isolation, only qualitative interpretations can be made. Complex biophysical models are required for the quantification of the neuronal response underlying the BOLD fMRI signal. Increasingly, researchers are combining imaging technologies to gain as much information as possible from each experiment.

In order to accurately interpret the BOLD signal in terms of underlying neural and hemodynamic responses the best solution would be to measure all three components simultaneously. This is a difficult technological challenge especially within the confines of the bore of an MRI scanner. For human based studies, research groups often use concurrent fMRI and Electroencephalography (EEG), the former used to measure both BOLD and blood volume/flow and the latter neuronal activity [5–9]. However, EEG offers very poor spatial resolution that is limited to the electrical activity at the cortical surface. Furthermore, using fMRI techniques such as arterial spin labeling and vascular space occupancy to measure both flow/volume and the BOLD response inherently limits the temporal resolution of the fMRI data. Thus, using such data in biophysical models is limited.

As a result, animal models, where more invasive high-resolution imaging techniques can be applied, have been used for investigation of neurovascular coupling mechanisms and preclinical parameterization of biophysical models. Many studies have measured neural activation, with multichannel electrodes, and hemodynamic responses with two-dimensional optical imaging spectroscopy (2D-OIS) and/or laser doppler flowmetry (LDF) [10–18]. Although these experiments have told us a great deal about the types of neural activity that drive the hemodynamic response, without fMRI one can only make predictions about the resultant BOLD signal. Expanding these animal studies to include fMRI as part of the multimodal experimental protocol would be of great benefit. Several groups now perform simultaneous neural measurements and BOLD fMRI in both rodent [19] and primate [20–25] animal models, providing more detailed information on how different types of neural activity may result in both positive and negative BOLD responses. However, these experiments critically lack a measurement of hemodynamic changes that is needed to fully understand and model how the BOLD signal, either positive or negative, is generated.

To that end we have developed a technique to measure hemoglobin changes within the cortex using 2D-OIS and the

BOLD fMRI signal simultaneously in the anesthetized rodent preparation [26–28]. Data can be collated with existing simultaneous electrophysiology and 2D-OIS data (using hemodynamics as a common measure) to give a more complete picture of neurovascular coupling and the resultant BOLD response. Data from this technique were used to refine complex tissue models for optical imaging spectroscopy analysis by constraining the light transport through tissue algorithms with the three-dimensional *f*MRI information [26–28]. The technology can also be used for a variety of scientific reasons:

1. To build, refine, and test current mathematical models of the BOLD fMRI signal. Biophysical models are often used for interpretation of human BOLD fMRI data in terms of neuronal activity. Current models are used in a technique known as calibrated *f*MRI [29–31] to estimate changes in the cerebral metabolic rate of oxygen consumption. However, such models contain many free floating, untested calibration constants and thus any inference to neuronal activity is qualitative at best. High-resolution concurrent fMRI and 2D-OIS data can be used to test/refine and anchor some of these parameters, leading to better estimates of the underlying neuronal activity [32].
2. To provide an independent measure of the hemodynamic response in the cortex allowing calibration between experiments in the same subject. For example, over the time course of a study several different fMRI sequences or scan parameters might be used. However, there is no guarantee that an evoked response, to the same stimulus, in an individual subject will remain the same over time (e.g., differences in anesthetic depth and neuronal state can lead to differing hemodynamics). Any changes in the measured *f*MRI signal could be an effect of the baseline physiology or the different sequence/scan parameters. An independent measure of the underlying hemodynamics can eliminate the possibility of a causal physiological link.
3. Increasingly, the BOLD signal is being used as a biomarker of disease progression [33–36]. However, if the BOLD signal is changing as a function of disease itself (i.e., by a breakdown in neurovascular coupling), without an independent measure of the hemodynamics, it would be difficult to understand the underlying hemodynamic cause of the BOLD signal change.
4. Multimodal imaging helps researchers to problem solve their experiment more expediently, which are often more difficult within the bore constraints of an MR system. For example, the lack of measurable BOLD response can be explained by either poor physiological state (leading to a reduced hemodynamic response) or a problem with the MR scanner/sequence parameters (e.g., slice in the wrong position and MR shimming artifacts). Although this may seem trivial,

it can amount to saving many weeks or even months in the laboratory changing scanning parameters when in reality it is the physiological state of the subject that caused the nonmeasureable BOLD condition (or indeed vice versa).

5. By collecting data simultaneously it greatly reduces the number of subjects needed if the experiments were done in isolation.

2 Materials and Surgical Tools

Sterile saline (0.9 %, Dechra Veterinary products, UK), urethane (12.5 g in 50 mL of distilled water, (Sigma, UK)), atropine (Atrocare, Animal Care 600 μg/mL, UK), phenylephrine hydrochloride (Sigma, UK), homeothermic heating system (Harvard Apparatus, UK):

Tracheotomy—Polyethylene Tubing (Portex tubing, Smiths Medical, UK, 2.8 mm outer diameter, 30–40 mm long) , scalpel size 22 (Swann Morton, UK), 2* blunt tweezers (World Precision Instruments, UK), 4 hemostat clamps (World Precision Instruments, UK), suture (Size 3.0 Mersilk, Ethicon, UK), sprung micro-scissors (FST, Germany), Sterile swabs (Propax, 5 × 5 cm, 8 ply, BSN Medical, UK), sterile saline.

Femoral Vessel Cannulation—Polyethylene Tubing (Portex tubing, Smiths Medical, UK, 0.8 mm outer diameter, 900 mm long), scalpel size 22, 2* tweezers, 1* clamp, micro-clamp (FST, Germany), micro-scissors, heparin (25 units per mL), suture, 41 mL syringes, (Becton Dickinson, UK), cyanoacrylate, sterile swabs, sterile saline.

Rat stereotaxic frame (Kopf Inc.), scalpel, 4 hemostat clamps (World Precision Instruments), sutures, dental scraper tool, dental drill, sterile saline, sterile swabs.

Dental cement (two bottles, Duralay liquid and powder, Reliance Inc.), perspex ring, copper strip, cyanoacrylate, dental scraper tool.

3 Methods

3.1 Animal Preparation and Surgery

All aspects of these methods and their development were performed with UK Home Office approval under the Animals (Scientific Procedures) Act 1986. Female hooded Lister rats weighing between 250 and 400 g were used. Subjects were kept in a 12-h dark/light cycle environment at a temperature of 22 °C with food and water ad libitum. Prior to surgery, animals were anesthetized with urethane (1.25 g/kg intraperitoneal, i.p.). The hindpaw reflex response was conducted to assess whether depth of anesthesia was at an appropriate level to commence surgery. Additional doses (0.1 mL) were administered if necessary. Atropine was administered (0.4 mg/kg subcutaneously, s.c.) to reduce mucous secretions during surgery. Body temperature was maintained at 37 °C using a homeothermic blanket with rectal probe.

3.2 Tracheotomy

Animals were tracheotomized to allow artificial ventilation. The animal was placed on its back and forelimbs secured to the operating table for ease of access to the throat. A scalpel incision was made from approximately 10 mm below the mouth to the top of the rib cage. The skin and overlying muscle was blunt dissected and tied back with sutures to reveal the trachea. Clamps were used on the suture to open up the incision and allow easy access to the trachea for insertion of the cannula. Using the scalpel a small incision was made between two cartilage rings of the trachea. The diameter of the opening was increased using sprung micro-scissors and a prebeveled cannula inserted. The cannula is then sutured securely in place and wound closed.

3.3 Femoral Vessel Cannulation

The left and right femoral veins and arteries were cannulated. This allowed drug infusion and measurement of mean arterial blood pressure (MABP). With the animal still on its back and hindlimbs secured to the operating table a 30 mm incision was made from the top of the thigh at a 45° angle in a rostromedial direction. Blunt dissection of the skin, muscle, and fat revealed the femoral vein, artery, and nerve which were subsequently separated from one another using surgical tweezers. The procedure for cannulation of the vein and artery was identical. The vessel was sutured as distally as possible. The suture was clamped and pulled to extend and tension the vessel within the incized area. A sprung micro-clamp was attached on the proximal end of the vessel, ceasing blood flow into the operating area. A suture was then placed under the vessel to be tied off after insertion of the cannula. Using sprung micro-scissors a small V shaped cut was made into the top side of the vessel. A prebeveled cannula was then inserted into the vessel towards the heart. The length of the cannula is important to ensure access once the animal is placed in the center of the magnet bore. The cannula was attached to a 1 mL plastic syringe and prefilled with saline containing 50 units of heparin per mL. This prevents the blood from clotting within the small bore cannula. The micro-clamp was then removed and the cannula pushed through a further 20 mm before being securely tied using the previously placed suture. Once both vessels were cannulated both cannulae were secured to the hindlimb with another suture and the opening sutured together and cyanoacrylate and gauzing used over the wound to provide further hold for the cannula. The procedure was repeated on the opposite femoral area (*see* **Note 1**).

3.4 Thinned Cranial Window

A thinned cranial window allows direct optical measures of the cerebral cortex. Animals were placed in a stereotaxic frame (Kopf Instruments Ltd.). A midline incision of the head was made. Fascia was removed using the blunt end of a scalpel. The skin and remaining fascia was sutured back in four positions to reveal more of the skull. On the right-hand side the temporalis muscle covering the lateral aspect of the skull was blunt dissected with a dental tool and sutured back to expose more of the skeletal surface. The skull over a majority of the right somatosensory cortex was thinned to translucency (approximately 100–200 μm thick where vasculature can be clearly observed with the skull still intact) with a dental drill under constant cooling with saline. The thinned window always contained the middle cerebral artery (MCA) which bifurcates in the area of the somatosensory barrel cortex and can be used as a useful surface landmark to ensure the right area is being thinned. The thinned skull preparation preserves the integrity of the skull while allowing direct optical imaging of the underlying brain surface, without disrupting the central nervous system (*see* **Note 2**).

3.5 Attachment of Surface Coil/Well Assembly to Enable Concurrent 2D-OIS and fMRI

An RF surface coil, integrated into a 20 mm diameter Perspex well was fixed to the animals head using cyanoacrylate and then made watertight with dental cement, while ensuring that the thin window lay in the center of the well. The well also provided the means to hold the head firmly in the magnet bore (*see* **Note 3**). Upon completion of the coil attachment procedure the animal was secured within the magnet-compatible holding capsule (Fig. 1a). The surface coil was locked to a holding bridge, using a screw on a locking ring, thus suspending the head of the animal in the approximate center of the holding capsule, and therefore magnet center, following insertion. A non-magnetic endoscope (*see* below), inserted into a protective Perspex banjo, was subsequently positioned over the surface coil and held in place with locking screws (Fig. 1b). This formed a well which was filled with deuterium oxide (D_2O) having a similar refractive index to saline. This reduced optical specularities from the skull surface for 2D-OIS and air-tissue susceptibility artifacts (around the thinned cranial window) for high-field *f*MRI while not being excited by the 300 MHz RF pulses and consequently avoiding magnetic resonance. The RF feeder cables for the surface coil were attached to the tuning circuit.

3.6 Systemic Physiology Monitoring

Inside the capsule an electrically filtered and isolated heating blanket and rectal probe (Harvard Apparatus, UK) maintained the temperature of the animal. A pressure sensitive pad was placed under the animal to monitor breathing patterns while inside the magnet bore (SAII, USA—Model 1025 L monitoring and gating system). Animals were artificially ventilated at a rate of 1–1.3 Hz using a Zoovent animal lung ventilator (Triumph Technical Services Ltd—*see* **Note 4**) connected to either an air compressor

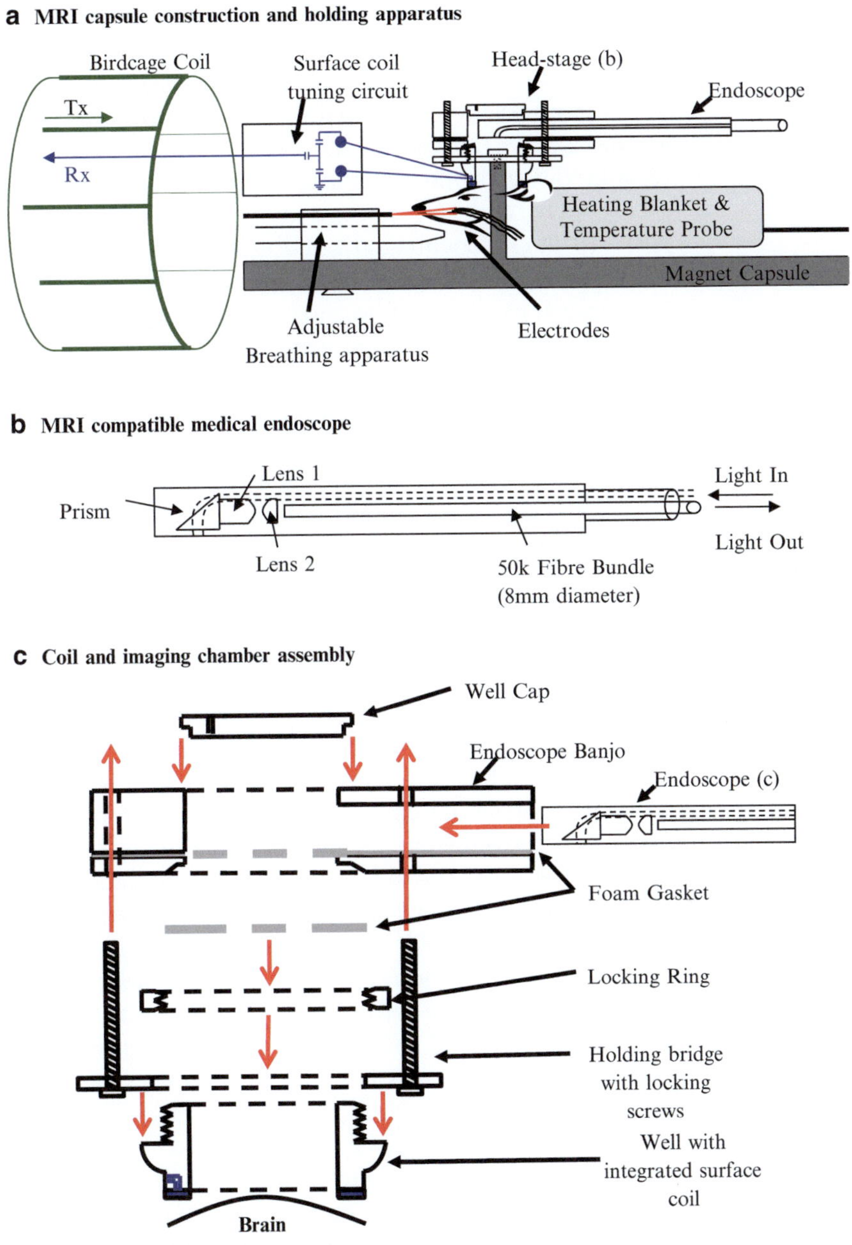

Fig. 1 Schematic diagrams of the experimental setup. (**a**) Magnet capsule containing all the physiological controls (heating blanket, breathing apparatus, blood pressure monitoring, and blood infusion). A large bird-cage coil is used for RF transmission with a circular copper receiver surface coil attached to the rat's head forming the lower aspect of the imaging chamber. (**b**) Non-ferrous adapted medical endoscope to transmit images of the surface cortex to the CCD camera placed outside the MRI cage. The endoscope consisted of a 50 K fiber optic bundle for and a series of prisms and lens to provide perpendicular imaging combined with fiber optics on either side that provided illumination of the cortex. (**c**) Perspex head stage. The head stage consisted of the well (with the surface coil at the bottom of this). This was securely attached to a holding bridge that eliminated movement artifact during the experiment. The endoscope assembly could then be lowered down and secured with screws. Foam washers ensured that the chamber remaining watertight for the addition of deuterium oxide that reduced air/tissue susceptibility artifacts

(Bambi, UK 150–500) or directly to a gas cylinder (BOC, medical air). The natural breathing rate for each animal was assessed and the ventilation parameters were adjusted accordingly to maintain blood gas measurements. Arterial blood from one of the femoral arteries was *let* (approximately 120 μL) into a glass capillary tube (3 mm outer diameter, Chiron, Diagnostics) and analyzed for pCO_2, pO_2, and sO_2 using a blood gas analyzer (Radiometer, ABL810-flex, Copenhagen). If abnormal, the experiment would be stopped (*see* **Note 4**) and ventilation parameters altered, allowing a return to a state of normoxia. Blood pressure was also monitored throughout the experiment. An isolated pressure transducer (Neurolog, NL108T1), filled with saline containing 50 units of heparin per mL, was attached to one of the femoral arterial cannulae. The signal was amplified (Neurolog, NL108) before being measured in real time using a high-performance data acquisition interface (CED LTD, plus 1401) and Spike2 software. A 0.13–0.26 mg/h dose of phenylephrine (Sigma, Aldrich) was infused using a syringe pump (WPI, sp2001) to maintain blood pressure between physiological limits (MABP, 100–110 mmHg) [37, 38] (*see* **Note 5**).

3.7 Four Wavelength 2D Optical Imaging Spectroscopy

2D-OIS was used to measure changes in the concentration of total, deoxy-, and oxy-hemoglobin (HbO_2, Hbr, and HbT, respectively). The cortex was illuminated with a white light source built into a switching galvanometer system (Lambda DG-4 Sutter Instruments Company) using 4 wavelength filters ($\lambda = 495 \pm 31$, 587 ± 9, 559 ± 16, and 575 ± 14 nm). Image data capture within the 310 mm bore of a 7 T magnet (Bruker BioSpecAvance, B/C 70/30 system) used a modified nonmagnetic endoscope (Endoscan Ltd, London. Fig. 1b). The 50 K fiber optic bundle of the endoscope had an active diameter of 1 mm and a 7 mm object at a distance of 9.12 mm to exactly fill the active window. The endoscope tip was a plastic square shaped sleeve ~100 mm long; so, it could be fixed rigidly to the Perspex banjo (Fig. 1c). The active circular window was 6 mm in diameter. The endoscope was attached to both the galvanometer (for light transmission) and a CCD camera (for light reception). The frame rate of the CCD camera was 32 Hz, giving an 8 Hz effective frame rate for each wavelength (*see* **Note 6**). The spectrographic data were recorded with a 2D spatial resolution of ~80×80 μm. Spectral analysis is based upon the path length scaling algorithm (PLSA) described previously [39]. This method uses a modified Beer-Lambert law incorporating either a homogeneous or heterogeneous tissue model. Detailed discussion of these two models is described in Kennerley et al. [27, 28] and only briefly summarized here. For the homogeneous tissue model baseline oxygen saturation (Υ_0) of 50 % was used and assumed to be homogeneous through the cortex. Baseline blood volume measurements were taken from vessel size index data at 3 T [26, 40]. For the homogeneous tissue model it was assumed that baseline blood volume fraction was 6 %, corresponding to 104 μM. In the heterogeneous model a 6 % baseline blood volume fraction in the superficial cortical layers (0–1 mm) dropping to 4 % in deeper layers and white matter (1–10 mm) was assumed. The effect of using separate tissue models will be summarized below.

3.8 Electrode Insertion for Stimulus Presentation

Once the endoscope allowing 2D-OIS data capture within the magnet bore was attached above the cranium, two nonmagnetic tungsten or stainless steel electrodes (Plastics One), insulated to within 2 mm of the tip, were inserted, in a posterior direction, between whisker rows A/B and C/D, ensuring the whole whisker pad was activated following electrical stimulation (*see* **Note 7**). The electrodes were then connected via an RF filter to the stimulator. The electrical pulse was provided by a constant current source stimulator (custom built). The output of the stimulator was sent through a CED1401 (CED, UK) input/output interface and recorded in real time (using spike2 software) ensuring that the timing and magnitude of the pulses remained constant.

3.9 Functional Magnetic Resonance Imaging

The magnet capsule was inserted into the bore of a 7 T magnet (Bruker BioSpecAvance, B/C 70/30 system) with pre-installed 12 channel RT-shim system (B-S30) and fitted with an actively shielded, 200 mm inner diameter, water cooled, three coil gradient system (Bruker BioSpin MRI GmbH B-GA20. 200 mT/m maximum strength per axis with 200 μs ramps). A ^{1}H quadrature volume resonator (Bruker 1P-T9561, 300 MHz, 1 kW max, outer diameter 200 mm/inner diameter 180 mm) was used for RF transmission and was actively decoupled from the custom built surface coil. A triplane fast gradient echo sequence running without phase encoding was used to ensure the surface coil (and thus subject) was positioned at the magnet isocenter for maximum signal gain. Both resonators were tuned and matched to 300 MHz using the Multlink ^{1}H preamplifier with a built-in tune/match display. Once the subject was localized, field shimming, off-resonance correction and RF gains were set and a triplane sequence (with phase encoding) was performed for subject localization. The resulting MR images were used to identify suitable coronal sections of the rat brain for high-resolution gradient echo (Fig. 2) scans (256×256 pixels, FOV = 30 mm, slice thickness = 2 mm, TR/TE = 1,000/15 ms, flip angle = 90°, 2 averages). Further high-resolution transverse scans covering the dorsal surface of the brain allowed accurate localization of the rodent whisker barrel cortex (2 mm back from the visible bregma line—Fig. 2a). A single oblique/topographic slice (covering the surface of the right cerebral cortex and in a similar plane to OIS data—Fig. 2a) was performed using the same scan parameters as above.

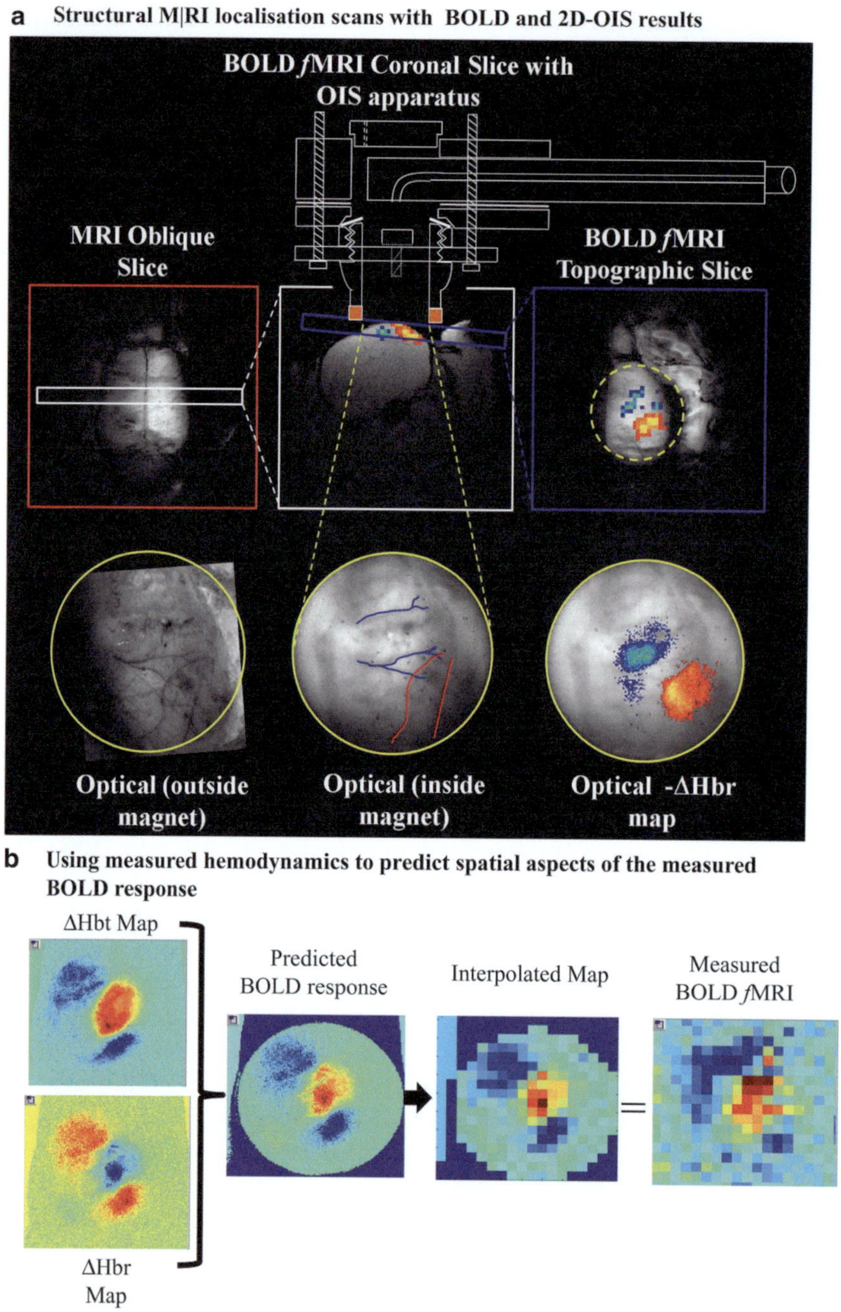

Fig. 2 Spatial results from left 16 s whisker pad stimulation. (**a**) Representative structural and functional imaging results. At the top of the diagram is a schematic of the endoscope, head stage assembly. The top row of images shows topographic structural scans (*left* and *right*) with a coronal scan in the middle. Positive and negative BOLD activation maps are superimposed over the structural scans. The bottom row of images shows a reference image of the cortex (*left*) taken outside the magnet, endoscope imaging (*center*) and the superimposed inverted Hbr response from the same experiment as the topographic results above. (**b**) Spatial predictions of BOLD from the underlying hemodynamics. Spatial maps of changes in Hbt and Hbr are put into a Monte Carlo simulation of MR signal attenuation to predict the BOLD signal. The resulting prediction can then by directly compared to the measured BOLD response

Functional data were acquired from suitable topographic slices using a single shot MBEST gradient echo–Echo planar imaging (GE-EPI) sequence during electrical stimulation of the whisker pad (raw data matrix = 64 × 64, data sampling interval = 5 μs, FOV = 30 mm, slice thickness = 2 mm, TR/TE = 1,000/12 ms, flip angle 90°, 10 dummy scans). Readout direction was left–right. Standard phase correction was used to minimize Nyquist ghosting [41]. The BOLD signal was calculated as a fractional change normalized by the mean of a 1 min preliminary baseline signal (*see* experimental design below).

3.10 General Experimental Design for Concurrent Experiments

Hemodynamic changes were induced by electrically stimulating the left whisker pad at an intensity of 1.2 mA and at a frequency of 5 Hz for 16 s. This stimulation frequency was chosen as it produces the largest hemodynamic response in the somatosensory cortex of anesthetized rodents while maintaining stable physiological parameters [42]. In all experiments an initial baseline of 60s was collected followed by 30 stimulation events, each with an inter-stimulus interval (ISI) of 70s. The BOLD signal was calculated as a fractional change normalized by the baseline mean, recorded during the initial 60 s control period. All stimulus and optical imaging timing control was performed using a CED1401 (using a Spike-CED, UK, interface) which was triggered to start using a TTL pulse indicating MR echo acquisition. Optical imaging data were set to be collected for all stimulus trials with a 10 s baseline period and for a duration of 70 s (covering the ISI). During spectral analysis 2D-OIS data averaged across trials.

For calibrated BOLD fMRI, or for a more robust whole brain response, hypercapnic challenges can be used. Animals were artificially ventilated with a normal gas mixture (20 % O_2, 80 % N_2). The animals were respired at a rate which matched their own spontaneous breathing frequency (~80 breaths per min). During hypercapnia challenge ($FICO_2$), a 5 or 10 % mixture of carbon dioxide was used. The oxygen and nitrogen ratio was kept constant. The concentrations of gases were maintained with purpose built flow meters. Following a 60 s baseline, respiratory challenges lasted for a maximum of 2 min. After each dose of $FICO_2$ physiological parameters (tested with blood gas measurements) were allowed to return to baseline before further stimuli/challenges were applied. Hypercapnia challenge is believed to be metabolically neutral [43–45]. Changes in blood flow and volume occur with no change in metabolic rate of oxygen consumption. It is therefore useful in understanding the underlying changes confounding the *BOLD fMRI* signal.

3.11 Data Analysis and Example Results

The analysis of both 2D-OIS and *f*MRI was conducted using Matlab (The MathWorks Inc., USA). Following functional (GE-EPI) to structural (GE) image registration all *f*MRI data were statistically analyzed using the general linear model (GLM) approach as in SPM [46]. The time series across each pixel was compared to a design matrix of a representative boxcar hemodynamic response function, a ramp and a DC offset. Activation *z*-scores were calculated on a voxel by voxel basis. These scores were then superimposed on detailed structural scans to show active regions. Regions were chosen by applying both positive and negative thresholds, which produced an "active" region having five or more adjacent voxels, with a z-score for each voxel greater than four (for positive BOLD) and less than minus two (for negative BOLD). Time series of responses were generated from voxels within each active region, with the mean across trials and subjects calculated for direct comparison with 2D-OIS data. Regions in 2D-OIS recordings were found using the same GLM and method as for the fMRI data and again were used to generate time series of positive and negative changes in $Hb0_2$, Hbr, and Hbt. Activity maps from the stimulation trials are shown for both fMRI and 2D-OIS. Results are shown for illustrative purposes (Fig. 2). By performing simultaneous acquisition the spatial hemodynamic response in the form of Hbt and Hbr activation maps (Figs. 2b and 3) can be fed into a mathematical model [47] to predict a BOLD response at 7 T. The spatial hemodynamic prediction can then be validated by comparison with the simultaneously measured BOLD response [28]. This can be taken a step further by comparing the time series of the model predicted and evoked BOLD response (Fig. 3) [27, 28]. It was found that when the data were analyzed with the homogeneous tissue model it resulted in an underestimation of the measured BOLD response (both positive and negative). Conversely, when the heterogeneous tissue model was used in the analysis of 2D-OIS and prediction of BOLD responses, the measured and predicted BOLD responses aligned. This is due to the model producing a larger change in Hbr with little difference in Hbt, when compared to the homogeneous model.

4 Notes

1. It was important to make sure that the tracheotomy cannula was not inserted past the carina of the trachea and the open end did not extend far beyond the mouth, so as to prevent damage to the bronchi and not encumber breathing due to respiratory "dead space." The tracheotomy cannula was sutured in place and following observation of breathing rhythm the incision was sutured shut. We routinely cannulate both femoral arteries to provide redundancy in case one becomes blocked. When tying off the distal portion of the vessel to be

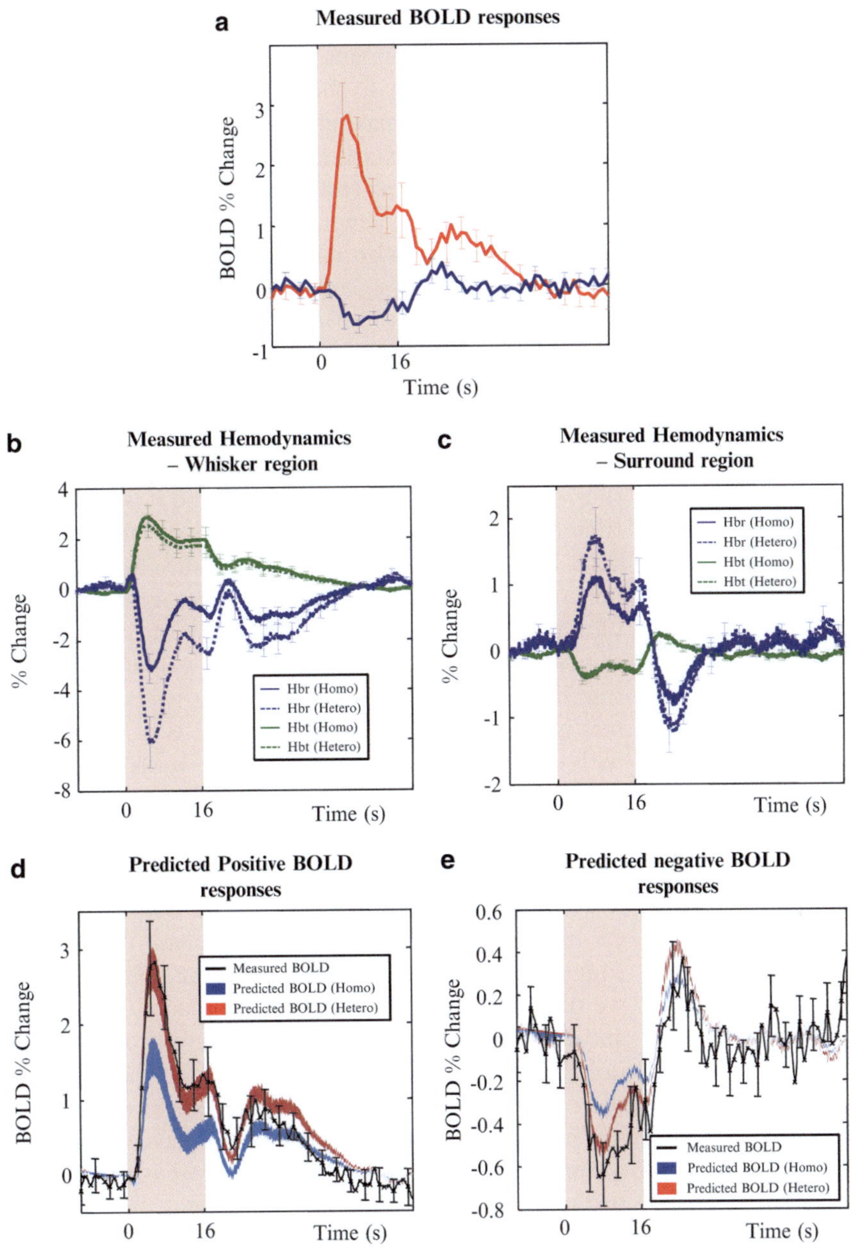

Fig. 3 Using different optical tissue models to predict temporal and magnitude aspects of the measure BOLD response. (**a**) Averaged measured positive and negative BOLD responses to 16 s whisker stimulation. (**b**) Whisker region Hbt and Hbr responses using the homogeneous or heterogeneous tissue models. (**c**) Surround region Hbt and Hbr responses using the homogeneous and heterogeneous tissue models. (**d**) Measured BOLD response in the whisker region compared to the predicted BOLD response using the different tissue models. (**e**) Measured BOLD response in the surround region compared to the predicted BOLD response using the different tissue models. It can be seen in both cases that the heterogeneous tissue model produces a far better fit to the measured data

cannulated it is important to make sure the knot does not twist the vessel. If this happens when the sprung scissors are used to cut a partial opening in the top of the vessel it can twist around to the underside making insertion of the cannula difficult. Initially tying an upward facing single knot and extending each end of the suture perpendicular to the vessel can minimize this possibility. Another general tip is to always make sure the vessels are kept moist with saline and not allowed to dry out. This not only avoids damage during tying but may also prevent the blood inside the vessels from coagulating with obvious deleterious effects.

2. If significant bruising occurs during the thinned cranial window surgery stop the experiment as this will significantly impair the ability of 2D-OIS to measure hemodynamic changes within the cortex. Once bruises have occurred it is difficult to remove them. At the stage of attaching the surface coil/well to the skull it is really important to make sure no areas of bleeding persist either from the skull or surrounding skin. Leaking blood into the well during the experiment will obscure the optical imaging. Always check the well is watertight before transfer to the MRI capsule as trying to fix this when the animal is in the magnet is difficult and is potentially hazardous with regards to using ferrous surgical tools.
3. The RF surface coil acts as an anchor to hold the end piece of the endoscope and deuterium-well assembly. It also serves another critical function: Most MRI capsules have the rat's head fixed in-line with the rest of the body often using plastic ear bars and a bite bar to secure the head. We assume this is to squeeze the rat into as small a volume coil as possible. When we tried this approach we routinely failed to measure BOLD and hemodynamic responses. By using the surface coil as the holding device we were able to replicate the standard position the rat is placed in when in a stereotaxic device. This sphinx like positions held on a flat plane and have subsequently led to a dramatic increase in the success rate of experiments following this change.
4. Normal blood gas values are essential for robust hemodynamic responses. As we operate our experiments with air ventilation (i.e., not an increased level of oxygen, which is the standard method with some gaseous anesthesia) the subject's pO2 should be between 80 and 100 mmHg and pCO_2 should measure between 30 and 40 mmHg. If the forced ventilation is too strong pO_2 increases and pCO_2 decreases. Lower the force of the ventilator and retest blood gasses. On occasion the animal may resist ventilation (as the Zoovent system used here is open to the atmosphere). If this happens, remove the

forced ventilation and monitor the normal breathing frequency of the animal and adjust the ventilator frequency as necessary. If this fails, the animal may have a blockage in the trachea (usually moisture) or the tracheal cannula has become displaced (i.e., the animal is breathing freely).

5. A side effect of using intraperitoneal urethane anesthesia is that it lowers systemic blood pressure. We use a constant infusion of phenylephrine to maintain blood pressure at normal physiological levels (100–110 mmHg). If the animal has low blood pressure a vascular oscillation often called vasomotion can dominate both 2D-OIS and BOLD measurements. This physiological effect, which was also termed the V-signal by our group [48], is an oscillation of blood flow, volume, and saturation that has a cycle of ~0.1 Hz (i.e., approximately once every 10 s). At high magnitudes it can completely mask stimulus-induced responses. Increasing baseline blood pressure is known to reduce the amplitude of vasomotion oscillations [49] and is essential, particularly when trying to monitor small evoked responses.
6. Although it seems obvious image capture must be synchronized with each wavelength exposure. The Lamda DG4 four wavelength light source used here cycles between each wavelength at 32 frames per second. We occasionally experience errors in data collection, missing a frame. This is normally due to data buffering when acquiring large images. This causes very serious problems for the spectroscopic analysis which relies on knowing the exact wavelength and sequence. All our camera code is developed in-house but one very useful boot strapping component is that a time series from each wavelength (which is an average intensity from all pixels collected in that frame) is recorded throughout the experiment. Any missed frames or skipped frames can be clearly identified by examining these four time series. Also, checking that the illumination starts and ends at the correct wavelength is also important.
7. It is important to check the quality and size of the stimulus before the animal is inserted into the MR scanner. Here a short electrical pulse is applied across the electrodes inserted into the rat whisker pad and the size of the whisker deflection is checked by the experimenter. If the deflection encompasses all of the whiskers and produces a sufficient whisker deflection the animal is then ready to enter the magnet bore. If the stimulus is too weak the electrodes are repositioned until this condition is achieved. Skipping this step may lead to no measurable BOLD/hemodynamic signal due to incorrect electrode placement, resulting in the animal being taken out of the scanner and all shimming and tuning redone, wasting experimental time.

References

1. Kwong KK, Belliveau JW, Chesler DA, Goldberg IE, Weisskoff RM, Poncelet BP, Kennedy DN, Hoppel BE, Cohen MS, Turner R et al (1992) Dynamic magnetic resonance imaging of human brain activity during primary sensory stimulation. Proc Natl Acad Sci U S A 89(12):5675–9
2. Ogawa S, Tank DW, Menon R, Ellermann JM, Kim SG, Merkle H, Ugurbil K (1992) Intrinsic signal changes accompanying sensory stimulation: functional brain mapping with magnetic resonance imaging. Proc Natl Acad Sci U S A 89(13):5951–5
3. Weisskoff RM, Kiihne S (1992) MRI susceptometry: image-based measurement of absolute susceptibility of MR contrast agents and human blood. Magn Reson Med 24(2): 375–83
4. Roy CS, Sherrington CS (1890) On the regulation of the blood-supply of the brain. J Physiol 11(1–2):85–158,17
5. Benar CG, Schon D, Grimault S, Nazarian B, Burle B, Roth M, Badier JM, Marquis P, Liegeois-Chauvel C, Anton JL (2007) Single-trial analysis of oddball event-related potentials in simultaneous EEG-fMRI. Hum Brain Mapp 28(7):602–13
6. Foucher JR, Otzenberger H, Gounot D (2004) Where arousal meets attention: a simultaneous fMRI and EEG recording study. Neuroimage 22(2):688–97
7. Garreffa G, Bianciardi M, Hagberg GE, Macaluso E, Marciani MG, Maraviglia B, Abbafati M, Carni M, Bruni I, Bianchi L (2004) Simultaneous EEG-fMRI acquisition: how far is it from being a standardized technique? Magn Reson Imaging 22(10):1445–55
8. Ritter P, Villringer A (2006) Simultaneous EEG-fMRI. Neurosci Biobehav Rev 30(6): 823–38
9. Rosenkranz K, Lemieux L (2010) Present and future of simultaneous EEG-fMRI. Magma 23(5–6):309–16
10. Berwick J, Johnston D, Jones M, Martindale J, Martin C, Kennerley AJ, Redgrave P, Mayhew JE (2008) Fine detail of neurovascular coupling revealed by spatiotemporal analysis of the hemodynamic response to single whisker stimulation in rat barrel cortex. J Neurophysiol 99(2):787–98
11. Boorman L, Kennerley AJ, Johnston D, Jones M, Zheng Y, Redgrave P, Berwick J (2010) Negative blood oxygen level dependence in the rat: a model for investigating the role of suppression in neurovascular coupling. J Neurosci 30(12):4285–94
12. Devor A, Hillman EM, Tian P, Waeber C, Teng IC, Ruvinskaya L, Shalinsky MH, Zhu H, Haslinger RH, Narayanan SN, Ulbert I, Dunn AK, Lo EH, Rosen BR, Dale AM, Kleinfeld D, Boas DA (2008) Stimulus-induced changes in blood flow and 2-deoxyglucose uptake dissociate in ipsilateral somatosensory cortex. J Neurosci 28(53):14347–57
13. Devor A, Tian P, Nishimura N, Teng IC, Hillman EM, Narayanan SN, Ulbert I, Boas DA, Kleinfeld D, Dale AM (2007) Suppressed neuronal activity and concurrent arteriolar vasoconstriction may explain negative blood oxygenation level-dependent signal. J Neurosci 27(16):4452–9
14. Devor A, Ulbert I, Dunn AK, Narayanan SN, Jones SR, Andermann ML, Boas DA, Dale AM (2005) Coupling of the cortical hemodynamic response to cortical and thalamic neuronal activity. Proc Natl Acad Sci U S A 102(10): 3822–7
15. Jones M, Devonshire IM, Berwick J, Martin C, Redgrave P, Mayhew J (2008) Altered neurovascular coupling during information-processing states. Eur J Neurosci 27(10):2758–72
16. Kennerley AJ, Harris S, Bruyns-Haylett M, Boorman L, Zheng Y, Jones M, Berwick J (2012) Early and late stimulus-evoked cortical hemodynamic responses provide insight into the neurogenic nature of neurovascular coupling. J Cereb Blood Flow Metab 32(3): 468–80
17. Masamoto K, Vazquez A, Wang P, Kim SG (2008) Trial-by-trial relationship between neural activity, oxygen consumption, and blood flow responses. Neuroimage 40(2):442–50
18. Vazquez AL, Masamoto K, Fukuda M, Kim SG (2010) Cerebral oxygen delivery and consumption during evoked neural activity. Front Neuroenergetics 2:11
19. Schridde U, Khubchandani M, Motelow JE, Sanganahalli BG, Hyder F, Blumenfeld H (2008) Negative BOLD with large increases in neuronal activity. Cereb Cortex 18(8): 1814–27
20. Goense J, Merkle H, Logothetis NK (2012) High-resolution fMRI reveals laminar differences in neurovascular coupling between positive and negative BOLD responses. Neuron 76(3):629–39

21. Goense JB, Logothetis NK (2008) Neurophysiology of the BOLD fMRI signal in awake monkeys. Curr Biol 18(9):631–40
22. Logothetis NK (2012) Intracortical recordings and fMRI: an attempt to study operational modules and networks simultaneously. Neuroimage 62(2):962–9
23. Logothetis NK, Wandell BA (2004) Interpreting the BOLD signal. Annu Rev Physiol 66:735–69
24. Scholvinck ML, Maier A, Ye FQ, Duyn JH, Leopold DA (2010) Neural basis of global resting-state fMRI activity. Proc Natl Acad Sci U S A 107(22):10238–43
25. Shmuel A, Augath M, Oeltermann A, Logothetis NK (2006) Negative functional MRI response correlates with decreases in neuronal activity in monkey visual area V1. Nat Neurosci 9(4):569–77
26. Kennerley AJ, Berwick J, Martindale J, Johnston D, Papadakis N, Mayhew JE (2005) Concurrent fMRI and optical measures for the investigation of the hemodynamic response function. Magn Reson Med 54(2):354–65
27. Kennerley AJ, Berwick J, Martindale J, Johnston D, Zheng Y, Mayhew JE (2009) Refinement of optical imaging spectroscopy algorithms using concurrent BOLD and CBV fMRI. Neuroimage 47(4):1608–19
28. Kennerley AJ, Mayhew JE, Boorman L, Zheng Y, Berwick J (2012) Is optical imaging spectroscopy a viable measurement technique for the investigation of the negative BOLD phenomenon? A concurrent optical imaging spectroscopy and fMRI study at high field (7T). Neuroimage 61(1):10–20
29. Davis TL, Kwong KK, Weisskoff RM, Rosen BR (1998) Calibrated functional MRI: mapping the dynamics of oxidative metabolism. Proc Natl Acad Sci U S A 95(4):1834–9
30. Hoge RD, Atkinson J, Gill B, Crelier GR, Marrett S, Pike GB (1999) Investigation of BOLD signal dependence on cerebral blood flow and oxygen consumption: the deoxyhemoglobin dilution model. Magn Reson Med 42(5):849–63
31. Hoge RD, Franceschini MA, Covolan RJ, Huppert T, Mandeville JB, Boas DA (2005) Simultaneous recording of task-induced changes in blood oxygenation, volume, and flow using diffuse optical imaging and arterial spin-labeling MRI. Neuroimage 25(3):701–7
32. Buxton RB, Wong EC, Frank LR (1998) Dynamics of blood flow and oxygenation changes during brain activation: the balloon model. Magn Reson Med 39(6):855–64
33. Cantin S, Villien M, Moreaud O, Tropres I, Keignart S, Chipon E, Le Bas JF, Warnking J, Krainik A (2011) Impaired cerebral vasoreactivity to CO2 in Alzheimer's disease using BOLD fMRI. Neuroimage 58(2):579–87
34. Fleisher AS, Sherzai A, Taylor C, Langbaum JB, Chen K, Buxton RB (2009) Resting-state BOLD networks versus task-associated functional MRI for distinguishing Alzheimer's disease risk groups. Neuroimage 47(4):1678–90
35. Greve JM (2011) BOLD MRI, applied to a murine model of peripheral artery disease. Methods Mol Biol 771:511–30
36. Paulsen JS, Zimbelman JL, Hinton SC, Langbehn DR, Leveroni CL, Benjamin ML, Reynolds NC, Rao SM (2004) fMRI biomarker of early neuronal dysfunction in presymptomatic Huntington's Disease. AJNR Am J Neuroradiol 25(10):1715–21
37. Golanov EV, Yamamoto S, Reis DJ (1994) Spontaneous waves of cerebral blood flow associated with a pattern of electrocortical activity. Am J Physiol 266(1 pt 2):R204–14
38. Nakai M, Maeda M (1999) Scopolamine-sensitive and resistant components of increase in cerebral cortical blood flow elicited by periaqueductal gray matter of rats. Neurosci Lett 270(3):173–6
39. Mayhew J, Zheng Y, Hou Y, Vuksanovic B, Berwick J, Askew S, Coffey P (1999) Spectroscopic analysis of changes in remitted illumination: the response to increased neural activity in brain. Neuroimage 10(3 pt 1):304–26
40. Tropres I, Grimault S, Vaeth A, Grillon E, Julien C, Payen JF, Lamalle L, Decorps M (2001) Vessel size imaging. Magn Reson Med 45(3):397–408
41. Bruder H, Fischer H, Reinfelder HE, Schmitt F (1992) Image reconstruction for echo planar imaging with nonequidistant k-space sampling. Magn Reson Med 23(2):311–23
42. Martin C, Martindale J, Berwick J, Mayhew J (2006) Investigating neural-hemodynamic coupling and the hemodynamic response function in the awake rat. Neuroimage 32(1): 33–48
43. Carlsson A, Holmin T, Lindqvist M, Siesjo BK (1977) Effect of hypercapnia and hypocapnia on tryptophan and tyrosine hydroxylation in rat brain. Acta Physiol Scand 99(4):503–9
44. Kety SS, Schmidt CF (1948) The effects of altered arterial tensions of carbon dioxide and oxygen on cerebral blood flow and cerebral oxygen consumption of normal young men. J Clin Invest 27(4):484–92

45. Sundt TM Jr, Anderson RE, Sharbrough FW (1976) Effect of hypocapnia, hypercapnia, and blood pressure on NADH fluorescence, electrical activity, and blood flow in normal and partially ischemic monkey cortex. J Neurochem 27(5):1125–33

46. Friston KJ, Frith CD, Liddle PF, Frackowiak RS (1991) Comparing functional (PET) images: the assessment of significant change. J Cereb Blood Flow Metab 11(4):690–9

47. Martindale J, Kennerley AJ, Johnston D, Zheng Y, Mayhew JE (2008) Theory and generalization of Monte Carlo models of the BOLD signal source. Magn Reson Med 59(3):607–18

48. Mayhew JE, Askew S, Zheng Y, Porrill J, Westby GW, Redgrave P, Rector DM, Harper RM (1996) Cerebral vasomotion: a 0.1-Hz oscillation in reflected light imaging of neural activity. Neuroimage 4(3 pt 1):183–93

49. Hudetz AG, Roman RJ, Harder DR (1992) Spontaneous flow oscillations in the cerebral cortex during acute changes in mean arterial pressure. J Cereb Blood Flow Metab 12(3): 491–9

Chapter 2

Neurovascular Coupling Investigated by Simultaneous Optical Coherence Tomography and Electrophysiology

Harsha Radhakrishnan, Maria Angela Franceschini, and Vivek J. Srinivasan

Abstract

A comprehensive understanding of the neurovascular coupling relationship requires the simultaneous measurement of neuronal and vascular responses and the capability to probe all layers of the cerebral cortex. Current macroscopic imaging techniques like laser Doppler imaging, diffuse optical tomography, fMRI, and PET lack spatial resolution. While two-photon microscopy is widely used in imaging the brain, it suffers from a lack of depth penetration and imaging speed. Optical coherence tomography (OCT) provides a platform for imaging the brain that potentially overcomes all of the above disadvantages, providing high-resolution cross-sectional images of light backscattered from cortical tissue. Here, we outline the experimental methods involved in simultaneous OCT (hemodynamic) and electrophysiological (neuronal) measurements to investigate neurovascular coupling in the rat somatosensory cortex. Using a spectral/Fourier domain OCT system, changes in cerebral blood flow and scattering were measured from multiple cortical layers. Simultaneous neuronal responses from layer IV using a tungsten microelectrode and surface potentials from a fire-polished ball electrode were also measured. This chapter provides details on animal preparation, instrumental setup, and data acquisition methods, and, finally, discusses potential limitations and pitfalls.

Key words Neurovascular coupling, Optical coherence tomography, Electrophysiology, Cerebral blood flow, anesthesia, somatosensory cortex

1 Introduction

The relationship between neuronal activity and the associated hemodynamic (blood flow, blood volume) changes is referred to as neurovascular coupling. The changes in neuronal activity due to an external stimulus are known to cause an increase in the cerebral blood flow (CBF) which supplies the activated regions of the brain with oxygenated hemoglobin [1], thereby maintaining homeostasis and normal brain function. The blood oxygen level-dependent signal obtained using functional magnetic resonance imaging (fMRI) is a measure of a deoxyhemoglobin change,

Mingrui Zhao et al. (eds.), *Neurovascular Coupling Methods*, Neuromethods, vol. 88,
DOI 10.1007/978-1-4939-0724-3_2, © Springer Science+Business Media New York 2014

resulting from hemodynamic or metabolic changes, and may reflect underlying changes in the neuronal activity [2]. Macroscopic optical imaging techniques such as laser Doppler flowmetry [3, 4], laser speckle imaging [5], diffuse optical imaging [6, 7], and 2D intrinsic optical imaging [8, 9] have been used to investigate neurovascular coupling. Because they use diffuse light to increase the depth penetration, these methods suffer from a lack of spatial resolution. Recently, two-photon microscopy has been used for cellular and subcellular imaging of neurovascular coupling [10], but this method still does not provide enough penetration depth and imaging speed to probe the entire cortical column at once. Optical coherence tomography (OCT), analogous to ultrasound, uses the principle of low-coherence interferometry to resolve the time delay of backscattered light [11]. Recent developments in spectral/Fourier domain OCT led to dramatic increases in sensitivity and imaging speeds [12–14]. With OCT, high-resolution measurements of RBC velocities [15] are possible by measuring the Doppler shifts in the scattered light arising from moving red blood cells (RBCs). Methods such as laser Doppler and laser speckle imaging do not provide absolute measures of blood flow, especially in tissues where photon paths and optical properties are unknown. By comparison, Doppler OCT is a path length-resolved technique that yields absolute measures of blood flow by detecting Doppler-shifted backscattered light in well-defined microscopic volumes. Recently, a promising method was developed that facilitates calculations of blood flow without requiring explicit knowledge of the vessel angle [16].

In this chapter, we present methods of simultaneously measuring hemodynamic changes, using OCT, and neuronal activity, using microelectrodes and ball electrodes, in the rat somatosensory cortex. Microelectrodes are inserted into the deep cortical layers while ball electrodes are used at the cortical surface. From microelectrode recordings, we measure local field potentials (LFPs), which is a weighted sum of transmembrane currents due to synaptic and dendritic activity [17], and multi-unit activity (MUA), which is the aggregate spiking activity of a population of neurons [18]. Considering the differences in anatomy and physiology between each layer in the cortex, simultaneous measurements provide more comprehensive understanding of neurovascular coupling across all layers of the cortical column. The Materials section of this chapter outlines the OCT and electrophysiological systems. The details of the animal surgery, region of interest localization, data acquisition techniques, and an overview of the data processing algorithms are provided in the Methods section. Finally the Notes section contains some useful hints on techniques to overcome issues that can arise from this type of multimodality imaging.

2 Materials

A schematic of the experimental setup is shown in Fig. 1.

2.1 OCT System

A 1,310 nm spectra/Fourier domain OCT system was used for imaging hemodynamics in the rat cortex [19]. The system used dual superluminescent diodes (SLD) to obtain a broadband spectral output. The central wavelength was 1,310 nm and the two SLDs were combined in a 50/50 coupler to yield a bandwidth of 170 nm. Due to lower scattering at longer wavelengths, the use of an infrared wavelength light source enabled the penetration of light to deeper tissue layers in a confocal imaging geometry. The axial (depth) resolution was 4.7 μm in air (3.5 μm in tissue, full-width-at-half-maximum) while the transverse resolution was 7.2 μm (full-width-at-half-maximum). A 5× objective (Mitsutoyu) with a working distance of 3 cm was used. Using an InGaAs line scan camera (Goodrich-Sensors Unlimited, Inc.), an imaging speed of 47,000 axial scans per second was achieved. For imaging through the thinned skull, the camera sensitivity was set to "medium" to achieve a higher dynamic range.

2.2 Electrophysiology Setup

For measurements of deep cortical layers, a 61–70 mm long, 2–4 MΩ impedance tungsten microelectrode with a standard fine taper (FHC, ME, USA) was used. This "needle" electrode was connected to one of the channels of a 32 channel custom-built amplifier system [20].

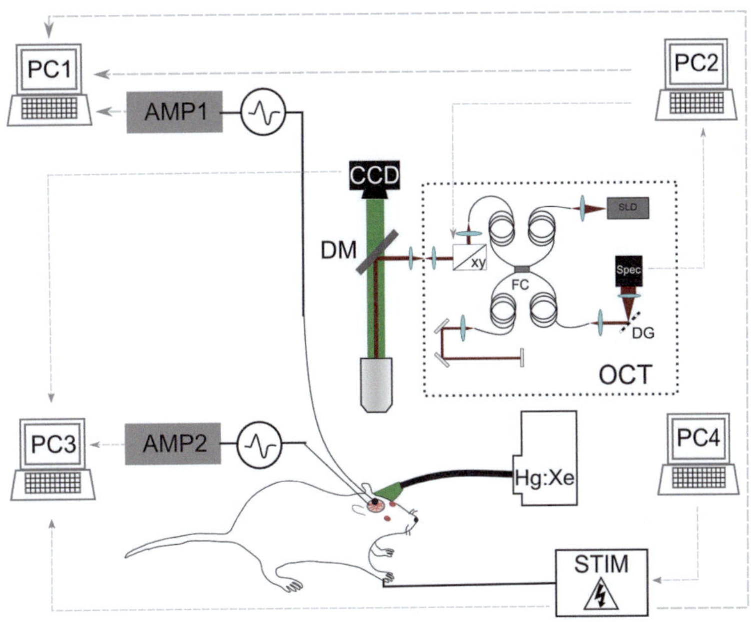

Fig. 1 Experimental setup and data acquisition diagram for the multimodal OCT-electrophysiology experiment. Four PCs were used to monitor and relay signals

The signals from the microelectrode were then band-pass filtered between 0.1 and 500 Hz to obtain LFPs and between 500 Hz and 5 kHz to obtain MUA. This filtered data were fed through two separate PCI-bus A/D cards—a 1.25 MS/s, 12-bit card for MUA (PCI-6071E, National Instruments, USA) and a 100 kS/s, 16-bit card for LFP data (PCI-6031E, National Instruments, USA). The data were recorded using a custom software written in LabVIEW™ and stored on a computer (PC3 on Fig. 1). An Ag/AgCl electrode (4 mm diameter, Warner Instruments, USA) was used as a reference electrode and placed subcutaneously in the neck. The circuit was grounded by connecting a wire with an alligator clip on one end to the stereotactic frame.

To obtain surface potentials, a ball electrode was made by fire-polishing a silver wire (Warner Instruments, USA). The ball electrode was attached, via a crimp connector, to the electrode holder of a battery powered bioamplifier (DAM80, WPI Inc., USA). Another Ag/AgCl electrode was used as a reference for the surface potential recordings and placed adjacent to the electrode that was used as a reference for the microelectrode recordings. The signal from the electrode was band-pass filtered between 0.1 Hz and 10 kHz, recorded, and stored on a laptop (PC1 in Fig. 1). Custom software written in Matlab™ was used to acquire these signals.

2.3 Intrinsic Optical Imaging Setup

Before OCT imaging and electrophysiology were performed, intrinsic optical imaging was used to determine the location of the maximal hemodynamic response. As shown in Fig. 1, a spectrally filtered Hg:Xe light source (570 ± 5 nm) illuminated the exposed cortical surface and 2-D images were obtained using a CCD camera (Infinity 2-1 M, Lumenera, CA). The CCD camera was connected by a USB cable to PC3 in Fig. 1, which acquired and saved the images. Due to the fact that oxygenated and deoxygenated hemoglobin have comparable molar extinction coefficients at 570 nm, reflectance changes around 570 nm indicated changes in blood volume. The images were later analyzed as a time series to determine the average response to a stimulus. For coarse localization of the maximal hemodynamic response, high temporal resolution was not required; hence, a lower frame rate of 1–2 Hz was used to minimize data processing time.

2.4 Data Acquisition Setup

The OCT data were written to a workstation (PC2 in Fig. 1) while the microelectrode electrophysiology data and the stimulus triggers were collected on another computer (PC3 in Fig. 1). The electrical pulses for the forepaw stimulus were triggered by custom software run on a laptop (PC4 in Fig. 1), and delivered by a current stimulus isolation unit (Model A360, WPI Inc., USA). The triggers for the OCT frame acquisition, the stimulus triggers, and the responses from the surface ball electrode were acquired by yet another laptop (PC1 in Fig. 1). Although the OCT data were acquired asynchronously, by carefully recording the locations of frames relative to the stimulus triggers, temporal information was incorporated into the linear regression procedure described below (Sect. 3.6).

3 Methods

3.1 Animal Surgery

The animal (rat) was initially anesthetized with isoflurane. Following this, it was placed on a homeothermic blanket (Harvard Apparatus, USA) and temperature was maintained at 37 °C. Under isoflurane, tracheostomy and cannulation of the femoral artery and vein were performed. Tracheostomy is a procedure to create an opening in the windpipe and insert a tracheal tube to control animal breathing (*see* **Note 1**). The arterial cannula was used to monitor blood pressure via a blood pressure transducer (SYS-BP1, WPI Inc., USA) and also perform blood glass analysis periodically to determine blood parameters such as pH, pO_2, and pCO_2. Cannulation of the vein was performed for administering the anesthetic during the imaging experiments. For hemodynamic measurements, it has been shown that alpha-chloralose is a suitable anesthetic [21] as it has minimal effects on respiratory and cardiovascular function [22]. Isoflurane shows minimal effects on neuronal activity but has the effect of vasodilation at higher doses [23] (*see* **Note 2**), while ketamine-xylazine has no impact on electrophysiology [24] and is widely used, but does impact hemodynamics [25, 26]. Anesthetics for neurovascular coupling experiments must therefore be chosen by carefully considering neuronal and vascular effects.

Following cannulation, rats were placed on a stereotaxic frame and the scalp was retracted. 2–2.5 % v/v isoflurane is administered during the craniotomy. A custom-built metal plate was then fixed on to the exposed skull using small bone screws and dental cement. A ~5 × 5 mm^2 area of skull over the left somatosensory cortex in the middle of the metal plate was thinned to translucency using a dental burr (*see* **Note 3**). A well of acrylic was built around the thinned skull area and filled with saline to keep the thinned skull region moist (*see* **Note 4**). Following the surgical procedure, the animal was switched from isoflurane to the anesthesia used for the experiment, e.g., alpha-chloralose. The animal should then be left to stabilize under the new anesthetic for 30 min to an hour. During this time, breathing of the animal should be monitored and blood gas analysis performed (*see* **Note 5**).

3.2 Localization of Neuronal and Vascular ROIs

Since microscopic methods are restricted to a small field of view, it is paramount that data are collected from the region of maximal neuronal and vascular responses. One way to ensure this is to localize the responses prior to collecting OCT data. For neuronal activity localization, responses from a fire-polished ball electrode (described in Sect. 2.2) during electrical forepaw stimulation pulses (200 μs pulses at 3 Hz for 4 s) delivered to the animal through hypodermic needles inserted in the contralateral forepaw were measured. The ball electrode was attached to the electrode holder of the stereotaxic frame and moved around to measure in a 3 × 3 or 4 × 4 grid within the somatosensory area where the skull has been thinned. The ball electrode was positioned in direct contact with the thinned skull surface, which was kept moist by topically adding aCSF. The goal here was to look for a response that resembles a standard mammalian-evoked potential response [27]—a large and narrow positive response component P1 followed within about 10 ms by a large and wide negative response N1 followed by two slower components P2 and N2 (*see* **Note 6**). For most anesthetics, a couple of measurement locations in the grid will clearly show a larger response. Once the neuronal activity has been localized, the skull at that location is further thinned and eventually removed along with the dura (*see* **Notes 3** and 7) to create space for inserting the needle electrode.

After localizing the neuronal response, the animal is moved under the OCT microscope (*see* **Note 8**), which also has the capability to perform optical intrinsic signal imaging as shown in Figure 1. To determine the region of maximal hemodynamic response, 2-D optical imaging using a CCD camera and a 570 nm Hg:Xe light source (described in Sect. 2.3) was used. Images were acquired during a 4 min run of electrical forepaw stimulation—equal to 12 trains of 200 μs pulses at 3 Hz for 4 s. Using custom software written in Matlab®, (Natick, MA) different regions of interest were selected. A region of decreased reflectance, resulting from increased blood volume, was then localized [28] (*see* **Note 9**). The region of maximal blood volume change was then used to determine the scan location for OCT imaging.

3.3 Setup for the Experiment

The skull should be kept moist at all times and should not be allowed to dry. For this purpose, aCSF or saline could be used. However, aCSF and saline evaporate over time, leading to a path length shift in the image that must later be corrected with post-processing. Conductive fluids that do not evaporate would thus be preferable for maintaining skull hydration. The CCD camera used to collect intrinsic images for vascular localization was used to orient the animal under the OCT objective to focus on the region of maximal blood volume change (obtained from the vascular localization). Generally speaking, the neuronal and vascular responses are well localized within a small region of approximately 1–2 mm in the somatosensory cortex for a forepaw stimulus. Thus care should be taken to ensure that the needle and the ball electrode, along with the OCT scan line, are within this small area. The working distance of the 5× objective used in this study is 3 cm. Therefore, it was imperative that extra attention be paid while inserting the needle electrode as perpendicular as possible to the cortical surface without nicking any of the vessels. It was also important to ensure that the needle did not obstruct the OCT image. The CCD camera image and the OCT display can be used as guides to accurately insert these probes.

The needle electrode was held in place with the help of an electrode holder connected to the custom-built animal stereotactic frame. The electrode holder was tilted at a certain angle that allows for the easy insertion of the electrode (*see* **Note 10**). Using the CCD image as a guide, the electrode was lowered to the region where the dura has been removed and carefully inserted down to about 600–700 μm from the brain surface until about layer IV of the cortical column. Layer IV receives the majority of the inputs to the cortex from the thalamus, which account for the positive peak P1 in the surface potential measurement. To determine if the needle is indeed in layer IV, a constant stimulus of 1 Hz was given to the animal and the responses are viewed on the screen on PC3 (in Fig. 1). It is well known that the LFP response from layer IV has an earlier response time and the largest negative amplitude than the other supragranular layers (above layer IV) [29] (*see* **Note 11**).

After the microelectrode was inserted, the OCT scan line position was defined. Doppler OCT images were obtained using a 5× objective lens, which has a working distance of about 3 cm. The power in the OCT reference arm was carefully adjusted such that the spectrum was below the saturation level of the spectrometer. The OCT scan location was chosen within the region of maximal hemodynamic response (described previously) but at a place with fewer large vessels and away from the inserted microelectrode. To check if the vessels in the OCT scan line are activated, Doppler responses to electrical forepaw stimulation (~2–3 Hz, the maximal response for alpha-chloralose anesthesia) were verified from the on-screen display. The vessels in the imaged region should show visible velocity increases and possibly diameter changes on the on-screen display. If the vessels are inactive, the OCT scan line needs to be moved to a location with a larger hemodynamic response (*see* **Note 12**). At all times during the experiment, the level of the liquid covering the skull must be monitored. There should not be a decrease in the level of the liquid due to evaporation. This was important for two reasons—the skull should not be allowed to dry out and the path length for the OCT images must be maintained. Other fluids can be used instead of aCSF and saline, such as agarose or a non-volatile conductive liquid. If there is a saturating reflection from the liquid surface on the OCT image, the head of the animal can be tilted a little or the power from the reference can be reduced.

Finally, the fire-polished ball electrode was placed on the other side of the OCT scan line, away from the microelectrode. The ball electrode must be placed such that it is in contact with the skull surface and sufficiently immersed in the liquid on the skull surface. Responses from the ball electrode must also be checked before beginning data collection (*see* **Note 13**).

3.4 Data Acquisition Procedure

Once the animal has been placed under the microscope and the electrodes and OCT scan line are positioned, a blood gas analysis must be performed to ensure that the animal is in good physiological condition. Baseline blood flow measurements are essential to establish the physiological condition of the animal. Therefore, a Doppler OCT image consisting of 4 consecutive volumes of 256 frames with 4,096 axial scans each was acquired. It required about 25 s for each volume to be recorded. The images were acquired with the focus approximately 100 μm below the pial surface. The CCD camera image, whose focus roughly coincided with the OCT focus, was used as a guide for the focal plane location.

Following the baseline measurement of flow, neuronal activity and OCT images were acquired during electrical forepaw stimulation. 2D OCT images ~1×1 mm^2 were obtained by acquiring images with 4,096 axial scans at 3.3 Hz. Varying frequency stimulation between 1 and 5 Hz was applied to the forepaw of the animal. An event-related stimulus paradigm was employed [30]. A standard 6 min run consisting of 23 stimulus trains with a mean interstimulus interval of 12 s was presented with a random order of stimulus frequencies. During the 6 min run, the animal physiology (blood pressure, heart rate) was continuously monitored and any changes in the reference plane of the OCT due to evaporation of saline on the skull were noted (these can be corrected during post-processing). Upon completion of the 6 min run, additional saline was applied to the skull surface. Also, after every few of runs, it is prudent to perform blood gas analysis to ensure that the animal physiology is stable. If at any point, there is a change in blood pressure or the animal breathing is not in synchrony with the ventilator, the measurement should be stopped immediately. Generally, the rest period between runs is about 5 min. A minimum of 5–6 runs were performed per experiment. This number can be extended provided that the animal vital parameters are optimum and both the neuronal and vascular responses are robust.

3.5 Flow Measurement Algorithm: Absolute Baseline Blood Flow and Relative Doppler Changes

A sample Doppler OCT image (averaged over all trials in a run) from the rat somatosensory cortex under alpha-chloralose anesthesia is shown in Fig. 2a. Large surface pial vessels and diving parenchymal vessels are marked on the image. Algorithms for determining Doppler OCT blood flow have been described previously [31]. A region of interest around a parenchymal vessel is shown in white. Figure 2b shows the blood flow responses from the marked parenchymal vessel due to stimulation. The time traces clearly show a habituation pattern with stimulus frequencies above 2 Hz.

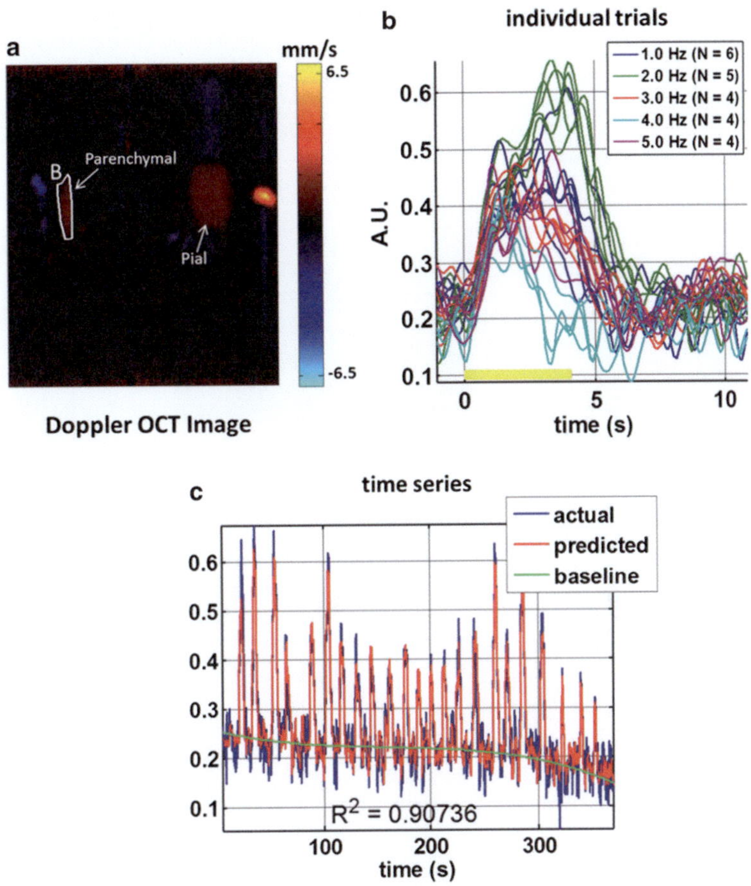

Fig. 2 Imaging of flow responses in individual vessels using Doppler OCT. (**a**) Doppler OCT image with labelled parenchymal and pial vessels. A region of interest around a parenchymal vessel is shown in *white*. (**b**) Individual trials for different stimulus frequencies in the parenchymal vessel are plotted. Clearly, heterogeneous but consistent flow responses are evoked by different stimulus frequencies. (**c**) The flow time series (*blue*) along with the time series prediction based on linear regression [32] is shown. A slow polynomial term to account for baseline fluctuations was included in the regression procedure. The results of the linear regression are time courses representing responses to stimulus frequencies between 1.0 and 5.0 Hz

Figure 2c shows the flow time series in blue along with the time series prediction based on linear regression [32]. A slow polynomial term to account for baseline fluctuations was included in the regression procedure. The results of the linear regression are time courses representing responses to stimulus frequencies between 1.0 and 5.0 Hz.

An image of the OCT amplitude response to function activation is shown in Fig. 3a. This logarithmic image corresponds to the Doppler OCT image shown in Fig. 2a. The changes seen in the

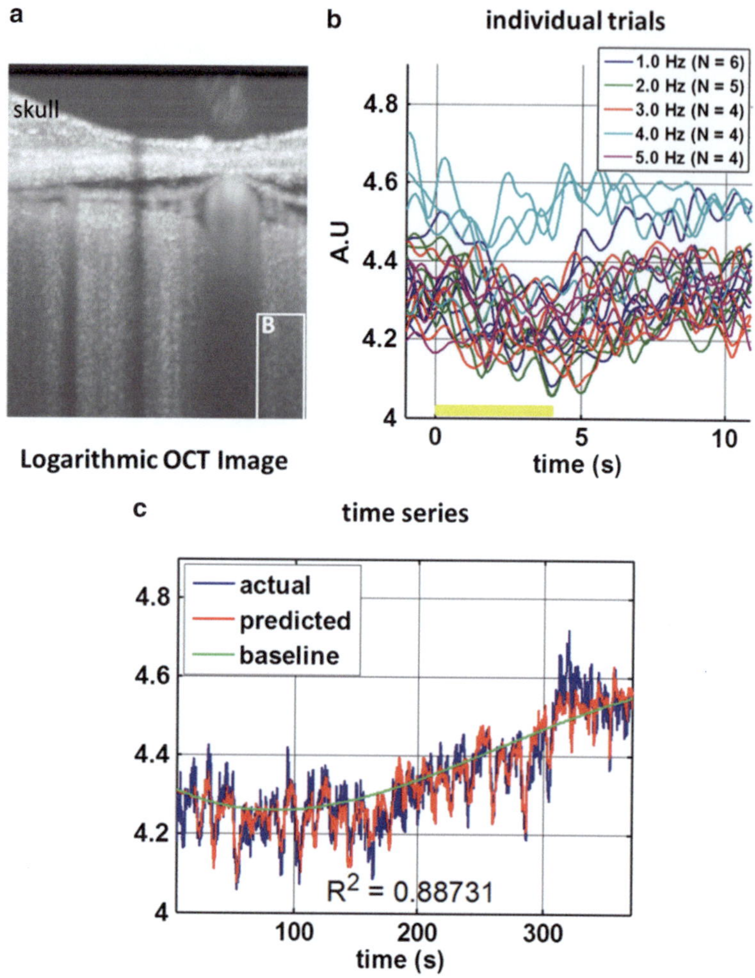

Fig. 3 Imaging of OCT amplitude response to functional activation, most likely caused by increases in RBC density or cell swelling. (**a**) OCT logarithmic image corresponding to the Doppler OCT image in Fig. 2a. A region of interest is shown in *white*. (**b**) Individual trials for different stimulus frequencies, showing the integrated amplitude within the region of interest, are plotted. (**c**) The time series (*blue*) along with the time series prediction based on linear regression is shown. A slow polynomial term to account for baseline fluctuations was included in the regression procedure. The results of the linear regression are time courses representing responses to stimulus frequencies between 1.0 and 5.0 Hz

amplitude image are probably due to increases in scattering from an increase in RBC density, or cell swelling. Figure 3b shows the responses from a region of interest marked in white in Fig. 3a. The traces are integrated amplitude responses from individual trials during parametric electrical stimulation. These responses are similar to the Doppler OCT responses in Fig. 2b where changes due to

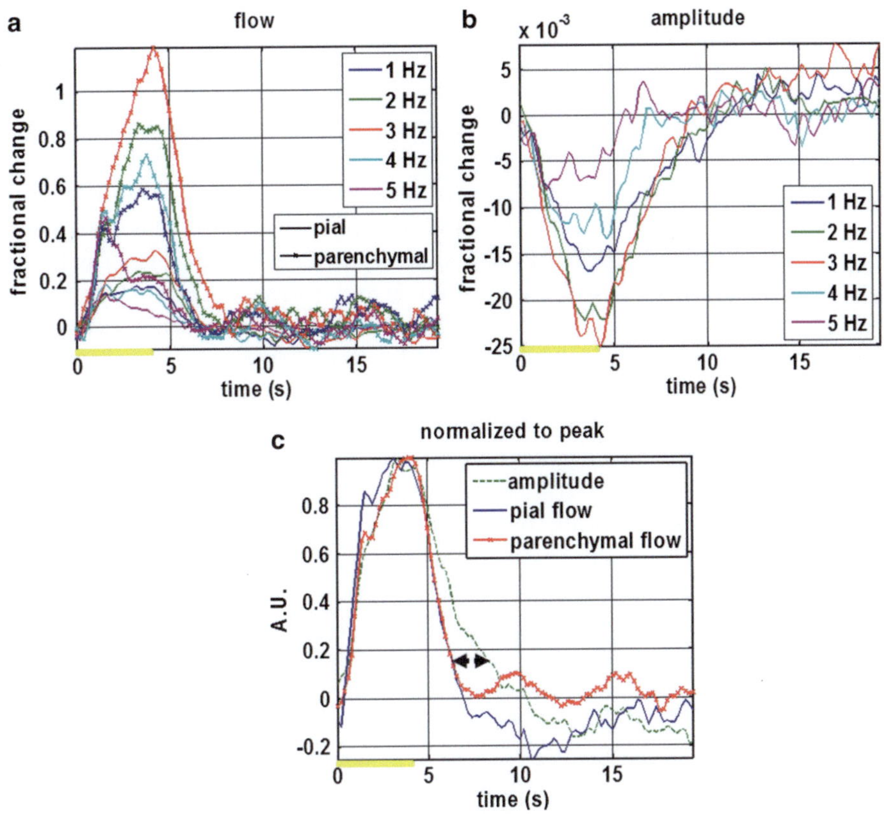

Fig. 4 Analysis of hemodynamic responses to functional activation in a single animal, averaged across six runs. (**a**) Parenchymal fractional flow responses are much larger than pial fractional flow responses, since pial vessels are less spatially specific. However, both vessel types show similar trends as the stimulus frequency is increased. (**b**) The OCT image amplitude, from regions of interest >200 μm below the surface, decreases in response to functional activation. This decrease in amplitude is likely due to an increase in scattering that accompanies an increase in red blood cell content. (**c**) Comparison of pial flow, parenchymal flow, and amplitude time courses, where all responses have been normalized to their peak. The black arrows show a delay between the flow responses and the amplitude response, which is consistent with a delayed washout of red blood cells after the stimulus ends

2–3 Hz stimulation were higher. In Fig. 3c, similar to Fig. 2c, the time series (blue) along with the time series prediction based on linear regression [32] is shown. A slow polynomial term to account for baseline fluctuations was included in the regression procedure. The results of the linear regression are time courses representing responses to stimulus frequencies between 1.0 and 5.0 Hz.

Figure 4a shows a sample group average in a different rat across 6 runs. From the figure, it is evident that the parenchymal fractional flow responses are much larger than pial fractional flow responses, since pial vessels are less spatially specific in the regions they supply or drain. Despite this difference, both vessels show a peak flow change at a 3 Hz stimulus, which is on par with changes seen in rats under alpha-chloralose anesthesia [6]. Figure 4b shows changes in OCT image amplitude in a region of interest below about 200 μm from the surface. The decrease in amplitude is likely due to an increase in scattering due to increased RBC content during activation [33]. Figure 4c displays a comparison between normalized pial flow, normalized parenchymal flow, and normalized relative amplitude time courses. Normalization is done with respect to the peak of the response (either positive or negative). The black dotted double headed arrow shows a delay between the flow responses and the amplitude response, which is consistent with a delayed washout of RBCs after the stimulus ends and flow has returned to baseline levels.

3.6 Neurovascular Coupling Analysis

To investigate neurovascular coupling, for each animal and all stimulus conditions, we assumed a linear convolution model between the measured electrical feature time series (MUA and LFP from the microelectrode and P1 and N1 surface potential components from ball electrode) and the measured hemodynamic time series from OCT. Either amplitude values, squares of peak amplitudes, or areas under the neuronal response curves can be used as inputs to the model [34]. Coefficients of determination (R^2) between the measured and predicted flows were determined. Statistical analysis can in principle be applied to determine the significance of differences between predictions of hemodynamics by different electrophysiological measurements and components [2].

Representative data from an animal anesthetized with alpha-chloralose are shown in Fig. 5. Using the electrophysiological measures as predictors of parenchymal flow responses as described above, we found that N1 yields higher R^2 values than MUA and P1. The results may be indicative of the fact that hemodynamic responses are predominantly driven by cortico-cortical transmission and not the initial thalamocortical activity in layer IV [35]. The impulse response in Fig. 4e, derived from the linear regression with N1 as a predictor of the fractional parenchymal flow response, shows a time to peak of approximately 1.2 s.

4 Notes

1. At times during the experiment, the animal may display some irregularities in breathing. One of the common problems is due to secretion into the tracheal tube, which needs to be cleaned. One way of reducing this secretion is with the injection of glycopyrrolate or atropine sulfate prior to starting the surgical procedures.
2. If isoflurane is the anesthetic used in the experiment, there is no need for femoral vein canulation. It is advised that the artery always be canulated to monitor blood pressure and draw

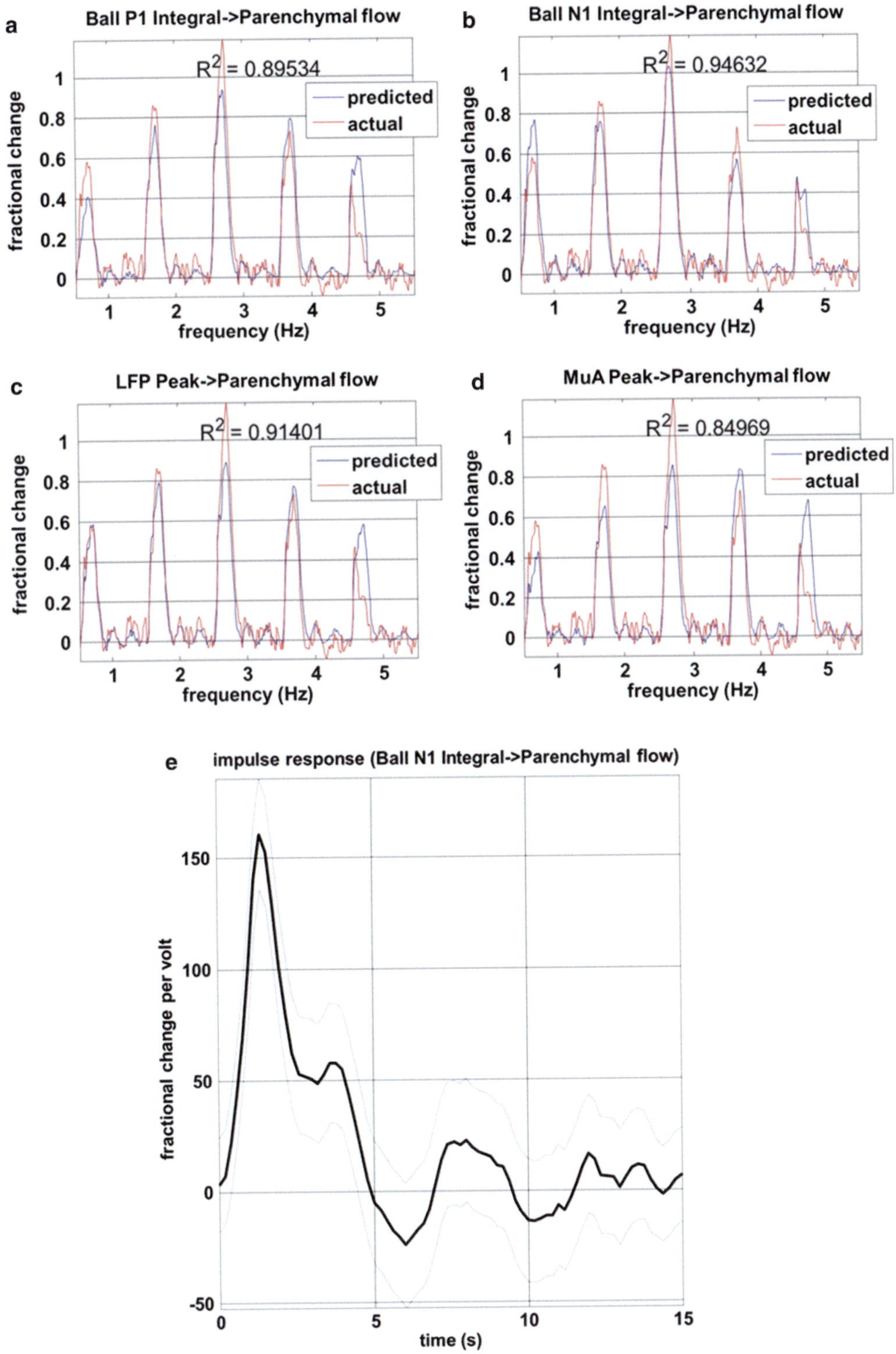

Fig. 5 Comparison of various electrophysiological measures as predictors of fractional parenchymal flow responses across stimulus conditions (**a–d**). The R^2 value is a measure of the goodness-of-fit. In this particular animal, anesthetized with alpha-chloralose, the integral of N1, obtained from a ball electrode on the surface of the skull, yields the highest R^2 (**b**). (**e**) The impulse response, derived from the integral of N1 as a predictor of the fractional parenchymal flow responses, shows a time to peak of approximately 1.2 s. The observation that the N1 yields better predictions than P1 is consistent with the theory that cortico-cortical transmission drives the hemodynamic response

blood to perform blood gas analysis. Increase in pCO_2 beyond physiological limits (40–45 mmHg) results in increased vasodilation which could hinder optimal hemodynamic responses.

3. Since the craniotomy could take up to an hour, a blood gas analysis to monitor the vital parameters should be performed during the surgery. Care should be taken to prevent drilling beyond 10 μm to prevent injuring the brain surface or causing dural bleeding.
4. During the experiment, the liquid in the well may evaporate and change the scanning plane of the OCT imaging. Care should be taken to ensure that the scanning plane is maintained by adding additional solution between stimulus runs. Preferably, a non-evaporating conductive fluid should be used.
5. The effects of isoflurane will diminish quickly after discontinuation as it is a volatile anesthetic. But it takes time for the animal to stabilize under any new anesthetic, such as alpha-chloralose. Data collected earlier than about 45 min after switching anesthetics may not be well correlated with the rest of the experiment.
6. Neuronal localization should be performed under the same anesthetic chosen for the rest of the experiment. While measuring responses during neuronal localization, responses from the ball electrode could either be viewed on an oscilloscope or recorded on a laptop. Also, with a fire-polished ball electrode, it is possible that the two slower components may not be seen in the responses due to the impedance of the electrode (in comparison to a surface disk electrode) or due to band-pass filter settings used in the experiment (*see* Sect. 2.2). One other important aspect is to ensure that there are no stimulus-induced motion artifacts. If the stimulus current is too high, the animal may twitch visibly. One way to avoid this is to start the current at zero and slowly increase it to a point where a neuronal response is seen with imperceptible twitching of the animal.
7. The dura mater under the region where the skull is removed also has to be removed with care so as to prevent bleeding. This is done to ensure that the microelectrode can be inserted without causing any bending or breakage of the tip of the electrode.
8. Standard stereotactic frames, such as those from Kopf Instruments, are thick and bulky and therefore cannot be used to hold the animal under the OCT microscope. A custom-built animal holder was used during imaging with our OCT system.
9. In our setup, the hemodynamic localization is done after the animal is moved from the surgical table to the microscope used for OCT imaging. To perform hemodynamic localization, the

brain surface is imaged onto a CCD camera through a dichroic mirror. Also, before the hemodynamic localization, a blood gas analysis is performed to ensure that the animal physiology is stable. High pCO_2 values (>40 mmHg) may cause baseline vasodilation, and thus poor vascular responses during stimulation. If optimal hemodynamic responses are not observed despite robust neuronal responses, make sure that all the vital parameters and blood gases are within range for the anesthesia used. The previous Notes describe possible problems that can arise during the experiment.

10. The microelectrode has a very fine tip with the sensor at the end. It is very easy to scratch the tip on the skull surface while moving the electrode to the location where it is to be inserted into the cortex. It is very important to gently maneuver the electrode to reach the location for recording and gradually lower the micromanipulator towards the brain surface by using the CCD image on PC3 as a guide. Another important guideline is to prevent the tip from nicking any blood vessels in its path. It is advisable to find a region with low vascular density within the localized area for microelectrode insertion.
11. While inserting the microelectrode, the cortical tissue is compressed. To prevent damage, insert the microelectrode slightly beyond the required depth and then bring it back up to the required position. Doing this would release some of the pressure applied during the insertion stage. Also, insertion of the microelectrode causes a spreading depolarization. Therefore, it is advisable to wait about 15–20 min (and preferably longer) after the insertion of the electrode before collecting data. One way to confirm recovery is to assess responses to forepaw stimulation a couple of minutes after the animal exhibits stable spontaneous activity.
12. Sometimes setting up of the experiment could take a while. The animal needs to be continuously monitored during this period. If there is no visible vessel dilation due to the stimulation, a blood gas analysis should be performed, and ventilation parameters should be changed accordingly. All the vital parameters should also be checked to ensure that they are within physiological ranges.
13. During the initial testing for responses from the various imaging modalities, it should be ensured that there is no coordinated increase in the animal blood pressure due to the stimulus. If this does occur, one possible cause may be insufficient anesthesia, which can be rectified by increasing the dose of the maintaining anesthetic. Another possible cause could be that the stimulus intensity is too high. Reducing the current should help prevent stimulus-induced blood pressure increases.

Acknowledgement

This research is supported by the US National Institutes of Health (NIH) grants R01-EB001954 and R00-NS067050.

References

1. Roy S, Sherrington S (1890) On the regulation of the blood-supply of the brain. J Physiol 11: 85–108
2. Logothetis NK et al (2001) Neurophysiological investigation of the basis of the fMRI signal. Nature 412:150–157
3. Fabricius M, Lauritzen M (1996) Laser-Doppler evaluation of rat brain microcirculation: comparison with the [^{14}C]-iodoantipyrine method suggests discordance during cerebral blood flow increases. J Cereb Blood Flow Metab 16:156–161
4. Dirnagl U et al (1989) Continuous measurement of cerebral cortical blood flow by laser-Doppler flowmetry in a rat stroke model. J Cereb Blood Flow Metab 9:589–596
5. Boas DA, Dunn AK (2010) Laser speckle contrast imaging in biomedical optics. J Biomed Opt 15:011109
6. Franceschini MA et al (2010) The effect of different anesthetics on neurovascular coupling. Neuroimage 51:1367–1377
7. Culver JP et al (2003) Diffuse optical tomography of cerebral blood flow, oxygenation, and metabolism in rat during focal ischemia. J Cereb Blood Flow Metab 23:911–924
8. Martin C et al (2013) Complex spatiotemporal haemodynamic response following sensory stimulation in the awake rat. Neuroimage 66: 1–8
9. Malonek D, Grinvald A (1996) Interactions between electrical activity and cortical microcirculation revealed by imaging spectroscopy: implications for functional brain mapping. Science 272:551–554
10. Shih AY et al (2012) Two-photon microscopy as a tool to study blood flow and neurovascular coupling in the rodent brain. J Cereb Blood Flow Metab 32:1277–1309
11. Huang D et al (1991) Optical coherence tomography. Science 254:1178–1181
12. Leitgeb R, Hitzenberger CK, Fercher AF (2003) Performance of Fourier domain vs. time domain optical coherence tomography. Opt Express 11:889–894
13. de Boer JF et al (2003) Improved signal-to-noise ratio in spectral-domain compared with time-domain optical coherence tomography. Opt Lett 28:2067–2069
14. Choma MA et al (2003) Sensitivity advantage of swept source and Fourier domain optical coherence tomography. Opt Express 11: 2183–2189
15. Chen Z et al (1997) Optical Doppler tomographic imaging of fluid flow velocity in highly scattering media. Opt Lett 22:64–66
16. Srinivasan VJ et al (2010) Quantitative cerebral blood flow with optical coherence tomography. Opt Express 18:2477–2494
17. Eccles JC (1951) Interpretation of action potentials evoked in the cerebral cortex. Electroencephalogr Clin Neurophysiol 3: 449–464
18. Mitzdorf U (1987) Properties of the evoked potential generators: current source-density analysis of visually evoked potentials in the cat cortex. Int J Neurosci 33:33–59
19. Srinivasan VJ et al (2010) Rapid volumetric angiography of cortical microvasculature with optical coherence tomography. Opt Lett 35: 43–45
20. Ulbert I et al (2001) Multiple microelectrode-recording system for human intracortical applications. J Neurosci Methods 106: 69–79
21. Austin VC et al (2005) Confounding effects of anesthesia on functional activation in rodent brain: a study of halothane and alpha-chloralose anesthesia. Neuroimage 24:92–100
22. Nakao Y et al (2001) Effects of anesthesia on functional activation of cerebral blood flow and metabolism. Proc Natl Acad Sci U S A 98: 7593–7598
23. Eger EI II (1984) The pharmacology of isoflurane. Br J Anaesth 56:71S–99S
24. Kochs E, Bischoff P (1994) Ketamine and evoked potentials. Anaesthesist 43:S8–S14
25. Crosby G, Crane AM, Sokoloff L (1982) Local changes in cerebral glucose utilization during ketamine anesthesia. Anesthesiology 56: 437–443
26. Lei H et al (2001) The effects of ketamine–xylazine anesthesia on cerebral blood flow and oxygenation observed using nuclear magnetic

resonance perfusion imaging and electron paramagnetic resonance oximetry. Brain Res 913:174–179

27. Di S, Barth DS (1991) Topographic analysis of field potentials in rat vibrissa/barrel cortex. Brain Res 546:106–112
28. Frostig RD et al (1990) Cortical functional architecture and local coupling between neuronal-activity and the microcirculation revealed by in vivo high-resolution optical imaging of intrinsic signals. Proc Natl Acad Sci U S A 87:6082–6086
29. Radhakrishnan H, Wu W, Franceschini MA (2011) Study of neurovascular coupling by modulating neuronal activity with GABA. Brain Res 1372:1–12
30. Dale AM (1999) Optimal experimental design for event related fMRI. Hum Brain Mapp 8: 109–114
31. Srinivasan VJ, Chan AC, Lam EY (2012) Doppler OCT and OCT angiography for in vivo imaging of vascular physiology. In: Liu G (ed) Selected topics in optical coherence tomography. InTech, Rijeka
32. Frykholm P et al (2005) Relationship between cerebral blood flow and oxygen metabolism, and extracellular glucose and lactate concentrations during middle cerebral artery occlusion and reperfusion: a microdialysis and positron emission tomography study in nonhuman primates. J Neurosurg 102:1076–1084
33. Srinivasan VJ et al (2009) Depth-resolved microscopy of cortical hemodynamics with optical coherence tomography. Opt Lett 34: 3086–3088
34. Franceschini MA et al (2008) Coupling between somatosensory evoked potentials and hemodynamic response in the rat. Neuroimage 41:189–2003
35. Iadecola C (2004) Neurovascular regulation in the normal brain and in Alzheimer's disease. Nat Rev Neurosci 5:347–360

Chapter 3

Studying Adenosine and Cerebral Blood Flow (CBF) Using Pharmacological Methods

H. Richard Winn and Kathryn Ko

Abstract

The authors review the use of pharmacological agents to investigate in rodents the role of adenosine (Ado) in the regulation of cerebral blood flow (CBF). Two techniques are highlighted: an in vitro preparation using perfused and pressurized penetrating brain arterioles (the PPP Prep) and in vivo cranial window preparation utilized to study CBF during neuronal activation evoked by contralateral sciatic nerve stimulation. Materials, Methods, and Notes are provided, as well as a data illustrating the utility and power of these techniques.

Key words Adenosine, CBF, Adenosine receptors, Arterioles, Cranial window and sciatic nerve stimulation

1 Introduction

This chapter will review the physiology of adenosine (Ado) and the use of pharmacological tool to investigate the role of Ado in the regulation of CBF. Ado is a purine nucleoside and a potent arterial dilator [1–4]. Consequently, it has been proposed as a regulator of blood flow in many organs including the brain [1, 5]. Its actions are mediated by specific cell surface receptors coupled to G proteins [6]. At least four adenosine receptors have been cloned and classified as A1, A2A, A2B, and A3 subtypes [6].

1.1 Physiology of Adenosine related to CBF

Ado and its breakdown products, inosine (Ino) and hypoxanthine (Hx), are ubiquitous in animals and plants [6]. Indicating Ado importance to functionality and survival, Ado is well preserved phylogenetically [7]. In animals, it is involved in a variety of functions in multiple organs and multiple metabolic pathways. For example, Ado plays a role in immunological regulation, cellular growth, and angiogenesis, to mention a few [7, 8].

Mingrui Zhao et al. (eds.), *Neurovascular Coupling Methods*, Neuromethods, vol. 88,
DOI 10.1007/978-1-4939-0724-3_3, © Springer Science+Business Media New York 2014

In the central nervous system, Ado affects two critical physiological functions [7]:

1. Neuronal activity by inhibiting excitability [9].
2. Arterial flow by decreasing arteriolar tone and thereby causing vasodilatation [5].

Taken together, adenosine's combined action in the CNS leads to a decreased metabolism (by depressing neural activity) while simultaneously causing an increase in oxygen availability (by vasodilatation) [5].

1.2 Adenosine Receptors

The physiological actions of Ado in the CNS (and elsewhere in the body) occur by means of a variety (Table 1) of surface receptors [6, 7]. The A1 receptor is the most sensitive receptor to Ado concentrations and is functionally involved in neural activity [6, 7]. Activation of A1 receptors, in general, attenuates neuronal excitability. In contrast the A2 receptor is less sensitive than A1 to Ado levels and is responsible functionally for Ado-induced vasodilatation [2–5, 10–24]. The A2 receptor has two subtypes (A2a and A2b) both of which affect vascular tone [6, 7]. The A1 receptor and A3 receptor do not participate directly in vasodilatation [3], but the A2 receptor has nonvascular action in brain [6, 7].

There are naturally occurring compounds which serve as Ado receptor antagonists, principally xanthine derivative such as caffeine (coffee, chocolates, and sodas) and theophylline (tea and coffee) [10]. In addition, over the past 50 years, a number of Ado receptor antagonists and agonists (Table 1) have been manufactured by pharmacology laboratories, but many of these compounds

Table 1
Adenosine receptors: agonists and antagonists

Receptor	Vascular effect: dilatation	Ado affinity (nM)	Agonist	Antagonist
A1	No	~70	CPA 2-Cl-CPA CVT-2759	DPCPX Theo Caffeine
A2a	Yes	~160	CGS-21680 DPMA NECA	ZM-241385 SCH-58261 Theo Caffeine
A2b	Yes	~5,100	NECA	IPDX Enprofylline Alloxazine
A3	No	~6,500	IB-MECA APNEA	MRS-1191 l-ABOPX

have difficulty gaining entry into the CNS, and their applicability to in vivo studies of CBF is thus limited [6, 11, 12].

Compounds that are useful in CBF studies are mainly within the A2 receptors category (Table 1). Unfortunately, there is a lack of specificity for A2 receptor and its subtype receptors (i.e., A2a and A2b) which limits the ability to definitively define the relative contribution of the A2a verses A2b to Ado-related CBF regulation. However, the use of pharmacological techniques (receptor antagonists and agonists) in genetically altered animals (i.e., knockout mice) provides a mechanism to address the relative contributions of A2 and A1 receptors.

2 Materials

A variety of initial steps are required before investigating the role of Ado in CBF regulation.

In Vitro Preparation of Perfused and Pressurized Penetrating Brain Arterioles [3, 4, 13–15] (PPP Prep)

The in vitro PPP preparation [3, 4, 16] requires creation of two solutions: one fluid is utilized in the periarteriolar bath and another solution is used for intraluminal perfusion of the arteriole.

1. Bath Solution: The composition of the buffered saline solution is composed of (mM): NaCl, 144; KCl, 3.0; $CaCl_2$, 2.5; $MgSO_4$, 1.5; glucose, 5.0; pyruvate, 2.0; EDTA, 0.02; MOPS, 2.0; and NaH2PO4, 1.21. This solution is temperature controlled as outlined in Sect. 3 and bubbled with gas with varying 5 % $CO_2/O_2/N_2$ balance mixture. The ratio of gas mixture is dependent on the experimental condition being investigated (i.e., ischemia, hypoxia, neuronal activation).
2. Intraluminal Solution: This solution consists of buffered saline (pH 7.3) containing 1 % albumin. As with the bath solution, the intraluminal perfusion fluid is temperature controlled as outlined in Sect. 3 and bubbled with gas with varying 5 % $CO_2/O_2/N_2$ balance mixture being analyzed.

2.1 *In Vivo: Cranial Window [2] and Sciatic Nerve Preparations [17, 18]*

Window Preparation [2] involves the preparation of an artificial CSF with the following composition: Na +156.5 mEq/L. K+2.95 mEq/L, Ca^{2+} 2.50 mEq/L, Mg^{2+} 1.33 mEq/L, C\ −138.7 mEq/L, HCO, −24.6 mEq/L, dextrose 66.5 mg/dl, and urea 40.2 mg/dl. The fluid is usually equilibrated with 10 % O_2/6 % CO_2/N_2 balance gas at 37 °C. When studying hypoxia, the oxygen is replaced by increasing N_2. The pH after equilibration is 7.34–7.35. We documented (2) in serial measurements that the PO_2, PCO_2, and osmolarity of five consecutive solutions are 77 ± 2 mmHg, 43 ± 1 mmHg, and 308.2 ± 2.8 mOsmo/L (mean ± sm, respectively). The artificial CSF is continuously bubbled with the mixtured gas and directly connected to a superfusion

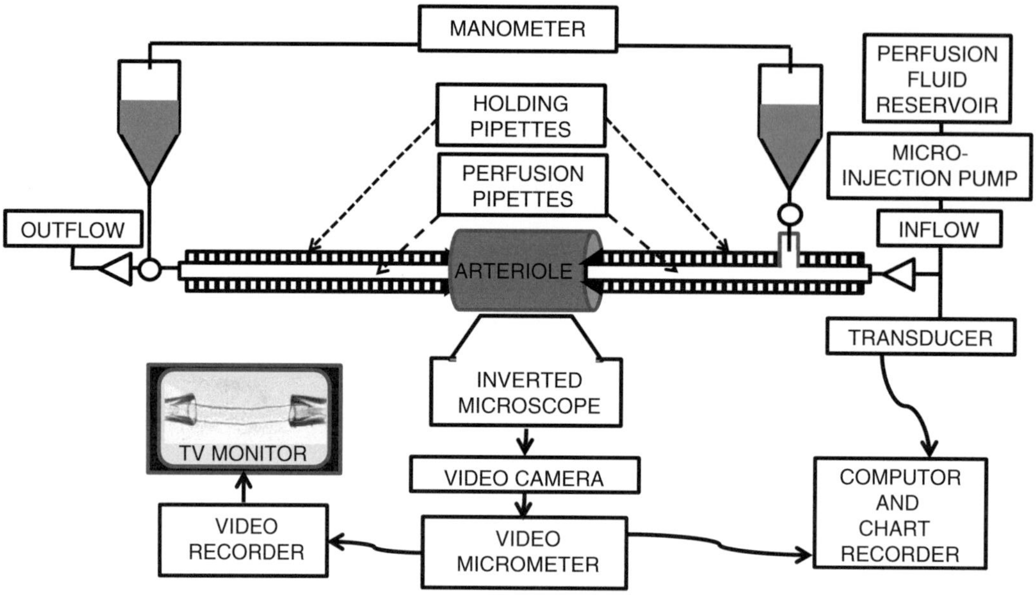

Fig. 1 Schematic of in vitro isolated penetrating perfused and pressurized (PPP) cerebral arteriolar preparation

system consisting of a pump and a warming device. To avoid gas diffusion into or out of the system, polyethylene, glass, or steel tubing is used. The speed of superfusion is 0.5 ml/min. During superfusion ICP is controlled by altering the level of the outflow tube (see below and Fig. 1).

2.2 Drug Preparation

1. In vitro (isolated arteriolar) preparation [3, 4, 15, 16]: Adenosine, ATP, and CGS-21680 (hydrochloride) are dissolved directly into the MOPS-buffered saline solution. CPX, CGS-15943, MRS-1191, and ZM-241385 are dissolved in DMSO to yield stock solutions of 1–10 mM. Subsequent dilutions are made with buffered saline, and pH is adjusted to 7.3. Because DMSO is used as a solvent as noted above, Ngai et al. [3] evaluated the effect of various concentrations of DMSO on intracerebral arteriole. Buffer containing as little as 0.5 % DMSO caused pronounced dilation of intracerebral arterioles, and consequently, DMSO concentration in all test solutions should not exceed 0.1 %.
2. In vivo preparation [16–18]: Adenosine, inosine, theophylline, and 8-SPT are dissolved directly into buffered solution. NECA, R-PIA, S-PIA, and PolyA ZM-241385 are initially dissolved in NaOH-polyethylene glycol 400 (50:50, vol/vol) and then diluted in saline before administration. Theo and 8SPT (40 mmol/l) are directly dissolved in isotonic saline. Stock solutions of 8SPT, CGS-21680, and adenosine are prepared in ultra-filtered water and then diluted in artificial cerebrospinal fluid (CSF) to reach final desired concentrations

for topical application. A stock solution of DPCPX (10 mM; 50 % ethanol-50% alkamus) is made in advance and subsequently diluted in artificial CSF to arrive at the desired final concentration. Vehicle concentration (ethanol–alkamus) is 0.005 % in all of the concentrations tested. Vehicle control studies for ZM-241385 or DPCPX revealed no effect of the vehicle on resting arteriolar diameter, hypercarbic vasodilatation, or SNS-induced vasodilatation [17]. All solutions are adjusted to a pH of 7.3.

3 Methods

There are two principal techniques, in vitro [3, 15] and in vivo [2, 17, 18], used to evaluate Ado role in CBF regulation.

3.1 In Vitro Preparation (PPP Prep) [3, 4, 13–15]

Briefly, male Sprague–Dawley rats weighing between 350 and 400 g are anesthetized with pentobarbital (50 mg/kg IP) and decapitated. The brain is rapidly removed from the skull, immersed in buffered saline solution containing 1 % dialyzed bovine serum albumin, and cooled to 4 °C. A piece of cerebral cortex approximately 2 mm thick containing the first portion of the middle cerebral artery is dissected from the brain. The pia mater and its attached penetrating intracerebral arterioles are then separated from the parenchyma and a segment of an intracerebral arteriole approximately 0.5 mm in length is severed from the pia and transferred to a temperature-controlled vessel chamber (volume, 1 mL) mounted on the stage of an inverted microscope.

As illustrated in Fig. 1, the isolated vessel is cannulated using a system of concentric glass pipettes mounted on micromanipulators attached to the microscope stage. The shape of these pipettes is designed to minimize flow resistance with tip diameters of 25–30 μm. By applying gentle suction, one end of the vessel is drawn into a holding pipette past a constriction at the latter's tip. A smaller inner pipette is inserted into the lumen of the vessel, wedging the vessel wall against the constriction in the holding pipette and sealing the lumen of the arteriole (Fig. 1). Both ends of the vessel are cannulated. The downstream end of the vessel is connected to a manometer used to set intraluminal pressure via an outflow reservoir. The upstream end is connected to a microinfusion pump for intraluminal fluid infusion. A fluid reservoir pressurized by the manometer could be inserted into the fluid path between the vessel and the pump by means of a three-way liquid switch. Upstream pressure is monitored with a pressure transducer. The cannulated vessel is observed on a video monitor, and lumen diameter is measured with a video analyzer. After cannulation, intraluminal pressure is set at 60 mmHg. The system is then checked for leaks: at zero flow, upstream pressure should be the same as downstream pressure. Leaky vessels are discarded. After measuring passive vessel diameter, the bath solution is changed to one without albumin, and bath temperature is raised to 37 °C. Over an equilibration period of 40 min, during which time the bath solution is changed once every 5–10 min, viable arterioles developed vasomotion and spontaneously contracted. Control diameter is measured after the development of steady tone. Reactivity of the vessels is assessed by changing the extraluminal pH from 7.3 to 6.8 and from 7.3 to 7.6. Vessels with weak pH response (<15 % dilation or constriction) are excluded from data collection.

3.2 In Vivo Preparations (Cranial Window and Sciatic Nerve Stimulation) [2, 16–21]

1. *Cranial Window*: Adult male Sprague–Dawley rats weighing 350–400 g are initially anesthetized with inspired gas anesthetic. The right femoral artery and vein are cannulated for monitoring MABP and intravenous administration of drugs. The rats are tracheostomized, immobilized with tubocurarine chloride (1 mg/kg i.v.), and mechanically ventilated. The concentration of halothane is decreased to at most 0.5 % and a mixture of air and oxygen inhaled. End-tidal CO_2 is continuously measured by a CO_2 analyzer, and a respirator is adjusted to maintain end-tidal CO_2 from 4.5 to 5.0 % and keep arterial blood gas tension within normal range. A small amount (0.2 ml) of arterial blood is anaerobically sampled for measurements of pH, $PaCO_2$, and PaO_2. Rectal temperature is monitored and maintained at 37 °C by a mat with thermostatically regulated circulating water. A femoral artery is cannulated for continuously monitoring blood pressure and sampling of blood for measurement of arterial blood gases and pH. In addition, a femoral vein is cannulated for administration of drugs.

Operative procedures: *Creation of Closed Cranial Window* (*Fig. 2a–c*):
One percent carbocaine is infiltrated subcutaneously around both external auditory meattus and the midline of the cranial vertex. After the head of the rat is fixed in a stereotaxic frame, the skull is exposed by a longitudinal skin incision from the occiput to the forehead. The periosteum is removed completely and the surface of the skull is cleaned and kept dry. Three polyethylene tubes (PE-50) are placed on the skull and embedded in dental acrylic. The tubes are subsequently used for measurement of intracranial pressure OCP) and superfusion of artificial CSF. The dental acrylic is mounted in a doughnut shape with thickness of I mm around the right parietal bone. Before the acrylic solidified, we pressed a glass slide onto the acrylic doughnut to obtain a flattened surface that would subsequently support a cover glass.

The craniectomy is then performed in the center of the acrylic doughnut, using power drill and micro rongeur. Under an operative microscope, the bone in the mediocaudal portion of parietal bone is thinned for an area 5 mm in diameter. This portion of the parietal bone is chosen because previous dissections had shown

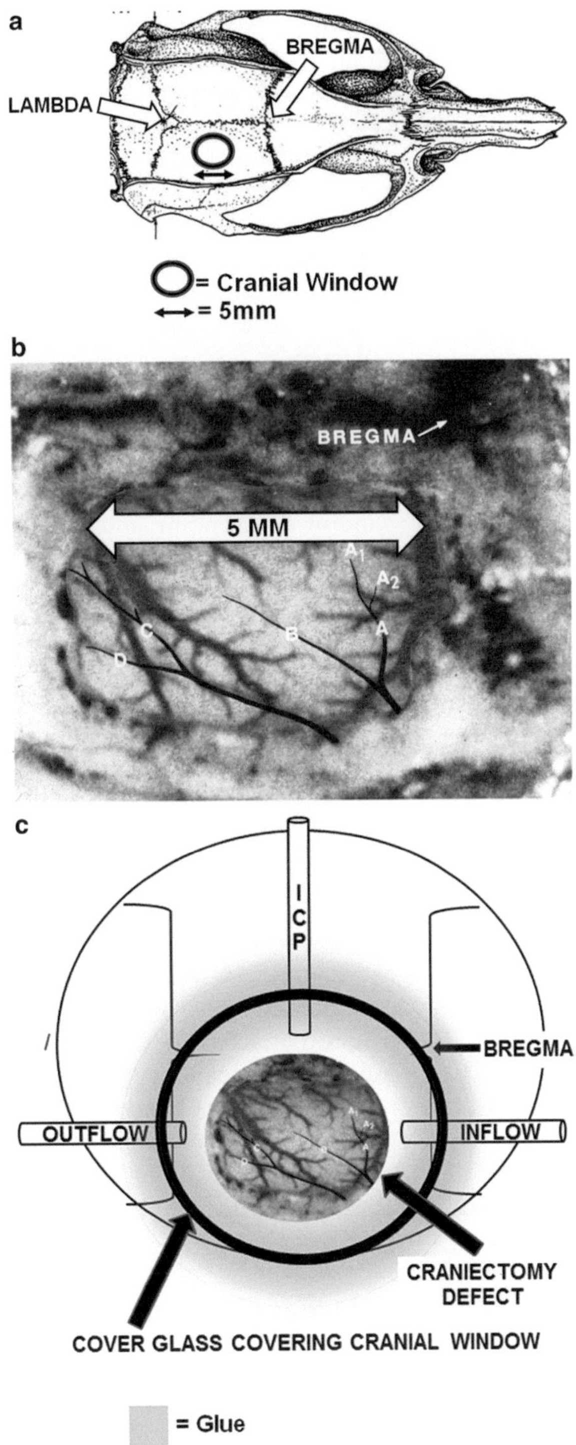

Fig. 2 (**a**) Rat Skull with location of cranial window. (**b**) Craniectomy: bone defect. (**c**) Details of cranial window with in port, out port, and ICP port

that major pial arterioles are not usually located in this area. The thinned bone is taken off by microforceps and the dura mater exposed. The parietal bone is removed piece by piece, care being taken not to depress the dura mater.

Complete hemostasis is achieved and then the dura mater incised with an ophthalmic scalpel using magnification. Since the dura mater and arachnoid membrane together contribute to the formation of a complex, continuous membrane layer, both tissues are incised together. The incision is extended to the edge of the bone window by microscissors, and the dura–arachnoid complex is reflected.

The average size of the cranial window is 5×6 mm. Artificial CSF is perfused over the exposed pial surface and a cover glass (12 mm in diameter and 0.13 mm thick) placed over the window, resting on the dentalacrylic. Air bubbles are removed from the window at this point, and the edge of the glass sealed to the acrylic with butyl-2-cyanoacrylate. On contact with water, this substance changes from a monomer to a polymer, forming a membrane instantaneously. The window is thus rapidly and completely sealed.

The level of the outflow tube is adjusted to create an ICP of 3–5 mmHg. At this constant pressure, brain herniation can be prevented. Through the window, branches of the middle cerebral arteries and the superior cerebral veins could be observed. These vessels are observed through a microscope. An automatic video dimension analyzer which tracks the width of red blood cell columns, allowed continuous online monitoring of pial vessel diameter.

2. *Sciatic Nerve Stimulation (SNS) Preparation: This is used to activate sensory hind limb cortex* [18]: The contralateral (to the cranial window) sciatic nerve is exposed, cut proximal to the bifurcation into tibial and peroneal nerves, and placed on stimulating electrodes that are immersed in a pool of warm mineral oil. The contralateral sciatic nerve is stimulated for 20 s (0.20 V, 5 Hz, 0.5 ms rectangular pulses) and vascular response to SNS is calculated as a percent increase over resting diameter.

4 Notes

A. In Vitro PPP Preparation

The investigator should be cognizant that time course of tone development in the isolated arterioles is an important indicator of vascular viability. Only vessels that displayed contractions within 5 min after the onset of bath warming have been found to generate satisfactory tone and responsiveness to vasoactive agents such as H' ion and acetylcholine [3, 4, 15].

In addition, we test the response of intracerebral arterioles to intraluminal application of acidotic buffer (pH 6.8) and also compare the response of these vessels to extraluminal and intraluminal application of 1 μM acetylcholine. After the development of spontaneous tone at 60 mmHg, the vessel is perfused at a rate of 2 μL/min with buffered saline (pH 7.3) containing 1 % albumin. Intraluminal flow at this rate did not affect control diameter. During intraluminal perfusion, downstream pressure is reduced so that upstream and downstream pressure changed by an equal and opposite magnitude, thus maintaining intravascular pressure constant at 60 mmHg.

Intraluminal albumin is necessary to preserve the integrity of the vascular wall and its responsiveness [22]. Our studies showed that it had no effect on vascular responses to adenosine.

At the end of each experiment, an Evans blue–albumin mixture (10 mg Evans blue and 10 mL of 1 % albumin buffer solution) is perfused into each vessel in order to confirm (1) successful intraluminal perfusion, (2) lack of leaks around the pipette tips, and (3) integrity of the cerebral endothelium.

B. In Vivo

1. *Cranial Window* Two aspects of the window creation are critical for success: non-traumatic craniectomy (bone removal) and a bloodless field. A traumatic craniectomy may result in injured cortex and/or nonresponsive pial arterioles. The exposure of the pial vessels to blood may alter the responsiveness and confound interpretation of data.

   During bone removal (i.e., craniectomy), the experimenter must exercise extreme caution to avoid damage to the underlying brain. Magnification with a microscope aids in visualization and decreases the frequency of cortical damage and alteration of pial physiological response. When thinning the bone with the high-speed drill, the drill bit should be continuously irrigated with cool saline. The alternative to a self-irrigating drill is an attentive assistant who constantly bathes the operative site with cool saline. A sharp bit will decrease the application of heavy manual pressure to the calvarium thereby decreasing the likelihood of injuring the underlying cortex. Injuries to the cortical surface can also occur upon opening of the dura. Consequently, this step must also be accomplished under the microscope.

   Hemostasis must be absolute since hemorrhage may alter resting pial arteriolar diameter, response to vasoactive compounds (e.g., Ado and its receptor agonists and antagonists), and reaction to physiological stimulation (e.g., CO_2), thereby confound data analysis. Bleeding from the bone is stopped by bone wax. Bleeding from the dural surface is controlled by temporary topical application of cotton pledgets. The dura mater is never cauterized with bipolar coagulation until the dura is reflected. Bleeding from the edge of the dura mater is cauterized over the bone only after reflection. A bipolar coagulator with the smallest electrical current is used. Meticulous attention is used to avoid injuries of pial vessels by electrical current.

2. *Sciatic Nerve Stimulation* Stimulation parameters can alter the vascular response, and consequently, these parameters must be precisely defined and maintained constant [18, 23–25]. For example, a study by Ngai et al. [18] revealed that specific stimulation parameters (0.20 V, 5 Hz, 0.5 ms rectangular pulses) optimize the vascular response without affecting systemic blood pressure. If pain fibers in the sciatic nerve are activated by increasing the intensity (V), blood pressure is increased and pail arterioles will constrict introducing variability and confounding interpretation of the data.

5 Demonstration of the Use of Pharmacological Agents Using the In Vitro and In Vivo Techniques to Investigate the Role of Ado in CBF Regulation

Using the closed cranial window, we [2, 26] visualized pial arterioles and venules and analyzed in vivo their response to topical application of Ado (Fig. 3).

With the PPP Prep, we [3, 4] demonstrated that Ado can cause potent dilatation of cerebral arterioles in vitro (Fig. 4a). Vasodilatation is discernible at ~10^{-9} M in isolated penetrating and reaches a maximum of ~10^{-5} to 10^{-4} M with an inflection point ~10^{-7} M. Resting concentrations of Ado in CSF and interstitial fluid are ~10^{-7} M. Thus small changes in ambient concentrations will have a significant effect on vessel diameter. Further studies [3] compared the concentration-dependent dilations of intracerebral arterioles of Ado to an A2 agonist, CGS-21680 (Fig. 4a).

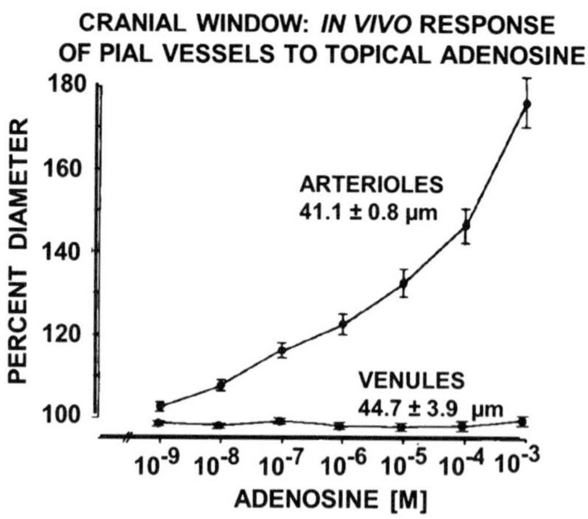

Fig. 3 In vivo response of pial arterioles and venules to topical adenosine (Ado)

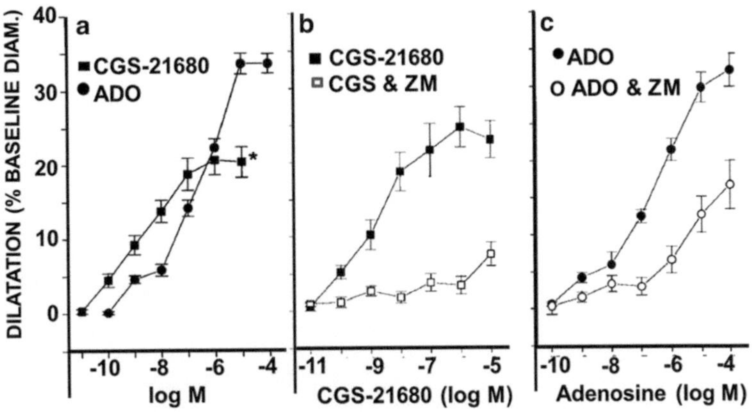

Fig. 4 In vitro response of isolated penetrating perfused and pressurized arterioles. (**a**) Ado vs. CGS-21680 (A2 agonist); (**b**) CGS-21680 vs. CGS and ZM (A2 antagonist); (**c**) Ado vs. Ado and ZM

The concentration–response curves showed significant ($P<0.05$) differences, with $-\log EC_{50}$ values of 6.5 ± 0.2 for adenosine and 8.6 ± 0.3 for CGS-21680. Although CGS-21680 has a lower dilator threshold (0.1 nM) than adenosine (1 nM), its dilator effect flattened out at 1 mM and is surpassed by adenosine at a concentration of 10 mM.

Adenosine concentration–response curves are reproducible when repeated in the same vessel and do not degrade with time (data not shown) [3]. Figure 4b, c compares the effects of 0.1 mM ZM-241385 (antagonist for A2a≫a2b) on dose–response curves for CGS-21680 and adenosine [3]. ZM-241385 almost completely inhibited CGS-21680-induced dilatation. On the other hand, it only partially attenuated dilation induced by adenosine. The almost complete inhibition of CGS-21680 suggests that ZM-241385 is an effective antagonist of the A2a receptor, but the residual vasodilatation with Ado in the presences of ZM-241385 suggests that A2b subtype receptors play a role in Ado-induced vasodilatation [3, 27].

In contrast, neither the A1 receptor-selective antagonist CPX (0.1 mM) nor the A3 receptor-selective antagonist MRS-1191 (0.1 mM) affected the dilator responses of intracerebral arterioles to adenosine (data not shown) [3].

Using the cranial window and sciatic nerve stimulation (SNS), we demonstrated a highly localized arteriolar response restricted to hind limb sensory cortex [18] (Fig. 5a, b). Arteriolar vasodilatation was measurable within 1.7 s after initiating SNS. A sharp peak occurs if stimulation exceeds 5 s with a lower but sustained "shoulder" of sustained vasodilatation related to the duration of stimulation (Fig. 6). With cessation of SNS, vasodilatation and CBF promptly returns to baseline.

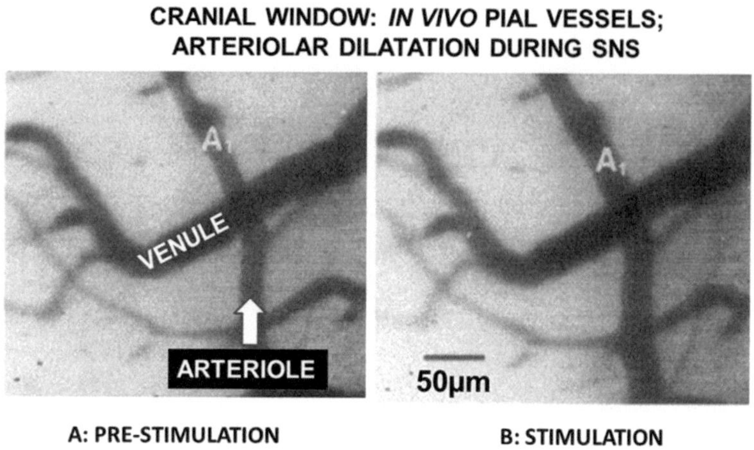

Fig. 5 Cranial window: in vivo pial arteriolar response to sensory cortical activation caused by contralateral sciatic nerve stimulation (SNS)

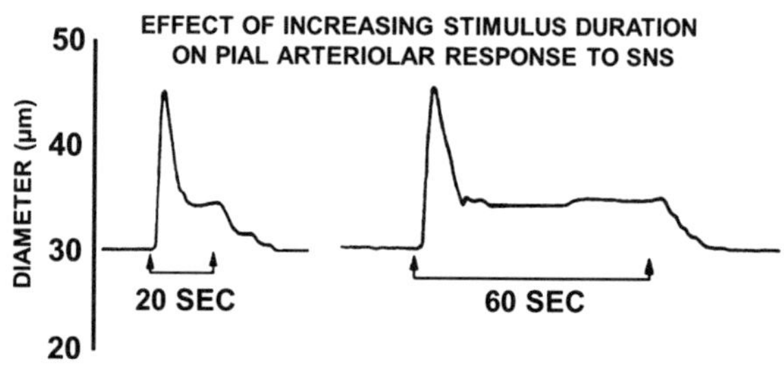

Fig. 6 Effects of increasing duration of sciatic nerve stimulation (SNS) on pial arteriolar response

Using the SNS preparation, we initially compared arteriolar response in the presence of mock CSF, topical theophylline, or topical dipyridamole. As expected, theophylline (Fig. 7) attenuated whereas dipyridamole (Fig. 8) exaggerated vasodilatation during neuronal activation, by monitoring SSEP, we noted that these pharmacological manipulations (i.e., theophylline and dipyridamole) [18] did not diminish (theophylline) or exaggerate (dipyridamole) neuronal excitability. Thus observed alteration of CBF during SNS is related to a direct vascular effect and not secondary to an effect on neuronal excitability.

Further studies [17] on the role of Ado in CBF during neuronal activation utilized systemic administration of 8SPT [8-(p-sulfophenyl)theophylline], a nonselective antagonist that does not penetrate the blood–brain barrier (BBB). 8STP had no effect on vasodilatation to SNS (Fig. 9). In contrast, systemic theophylline (intravenously administered in contrast to our earlier studies where we applied theophylline topically into the CSF [18]), which readily penetrates the BBB, nearly abolished the

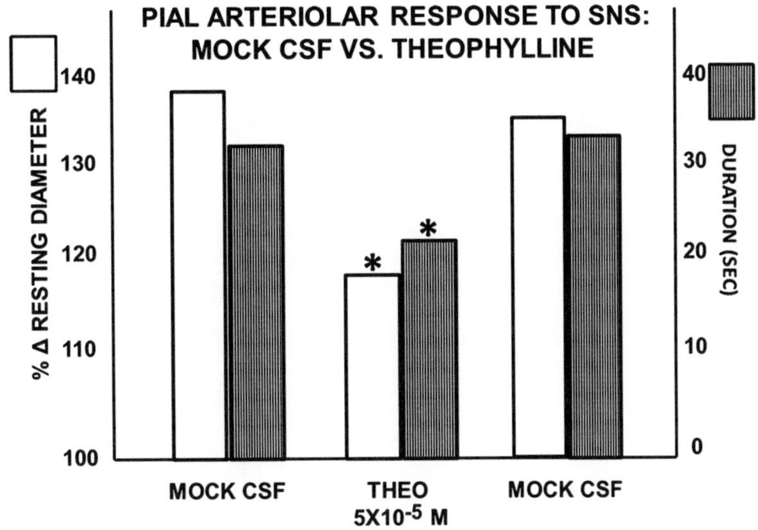

Fig. 7 Effect of Theophylline (A2 receptor antagonist) on pial arteriolar response to sciatic nerve stimulation (SNS)

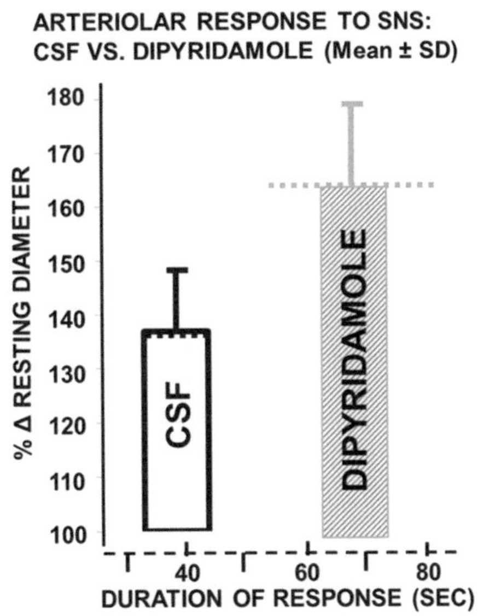

Fig. 8 Effect of Dipyridamole (DIP) on pial arteriolar response to sciatic nerve stimulation (SNS): DIP inhibits the uptake of Ado into cells and thereby increases extracellular concentrations of Ado

SNS-induced vasodilatation (Fig. 9). Topical superfusion of 8SPT significantly attenuated vasodilatation during SNS. Systemic administration of the relative select adenosine A2A receptor antagonist, ZM-241385, significantly ($P < 0.05$) attenuated the SNS-induced vasodilatation in a dose-related fashion (Fig. 9). Hypercarbic vasodilatation and somatosensory-evoked

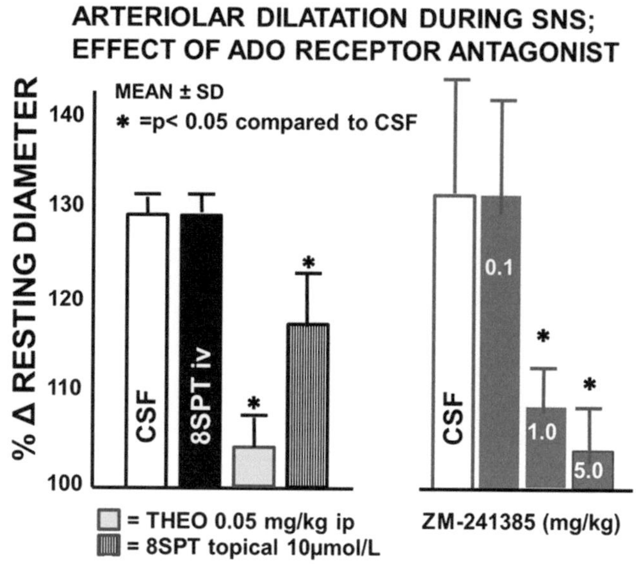

Fig. 9 Effects of 8SPT (iv), 8SPT (topically), theophylline (iv), and ZM (ip) (A2a > A2b antagonist) on pial arteriolar response to sciatic nerve stimulation (SNS): 8SPT does not cross the blood brain barrier (BBB); theophylline is a nonspecific A2 antagonist which crosses the BBB; ZM is relatively specific A2a > A2b receptor antagonist which crosses the BBB

potentials are not affected by any of the compounds tested. Lastly, topical superfusion of DPCPX [8-cyclopentyl-1,3-dipropylxanthine], a selective adenosine A1 receptor antagonist, significantly potentiated SNS-induced vasodilatation (data not shown).

Next, we [16] studied the effects of caffeine on cerebral arteries. Caffeine is the most widely used CNS stimulant in the world [28, 29]. We initially evaluated caffeine's actions in vitro on isolated arterioles and found that neither low (10 μmol/L) nor high (50 μmol/L) doses affected resting arteriolar diameter. The latter dose, but not the former, was capable of blocking Ado-induced vasodilatation when applied into the bath surrounding the isolated perfused and pressurized arterioles.

We then utilized our in vivo model to analyze caffeine's effect on resting arteriolar diameter and during SNS. Low doses (5 mg/kg) did not affect resting diameter and had a downward trend on arteriolar vasodilatation during SNS. Higher doses (10, 20, 30, and 40 mg/kg) significantly attenuated both resting diameter and CBF responses to somatosensory stimulation (Fig. 10). In a 70 kg human, a single cup of coffee (depending on brewing methodology) results in a serum level of 1.5–2.0 mg/kg of caffeine. Hypercarbic vasodilatation was unaffected by caffeine. The results indicate that caffeine significantly reduces cerebrovascular responses to somatosensory stimulation.

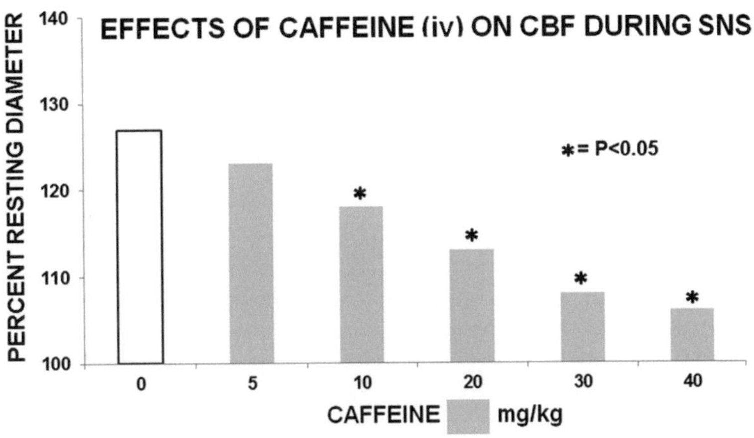

Fig. 10 Effect of caffeine on pial arteriolar response to sciatic nerve stimulation (SNS): caffeine is a nonspecific A2 receptor antagonist which crosses the BBB

Taken together, our findings strongly suggest that Ado is involved in regulating CBF during neuronal activation. In addition, we found that luminal endothelial adenosine receptors are not involved in the arteriolar response to SNS; that the adenosine A2A receptor subtype appears to be the major receptor responsible for vasodilation, but A2B receptor also is involved; and that the neuromodulatory action of adenosine, via the A1 receptor, significantly influences SNS-induced vasodilatation as demonstrated by DPCPX studies.

6 Conclusion

Using pharmacological agents, we review the proposed role of adenosine, a purine nucleoside, as a metabolic regulator of CBF. Highlighted are two techniques: the in vitro perfused and pressurized penetrating arteriolar preparation and the in vivo cranial window preparation used with contralateral SNS of sensory cortex. These techniques combined with pharmacological tools allow analysis of the role of adenosine in CBF regulation. The data support the hypothesis that adenosine plays a critical role of CBF under physiological conditions.

References

1. Berne RM, Rubio R (1974) Adenine nucleotide metabolism in the heart. Circ Res 35(suppl 3):109–120
2. Morii S, Ngai AC, Winn HR (1986) Reactivity of rat pial arterioles and venules to adenosine and carbon dioxide: with detailed description of the closed cranial window technique in rats. J Cereb Blood Flow Metab 6(1):34–41
3. Ngai AC, Coyne EF, Meno JR et al (2001) Receptor subtypes mediating adenosine-induced dilation of cerebral arterioles. Am J Physiol Heart Circ Physiol 280(5):H2329–H2335

4. Ngai AC, Winn HR (1993) Effects of adenosine and its analogues on isolated intracerebral arterioles. Extraluminal and intraluminal application. Circ Res 73(3):448–457
5. Winn HR, Rubio GR, Berne RM (1981) The role of adenosine in the regulation of cerebral blood flow. J Cereb Blood Flow Metab 1(3): 239–244
6. Fredholm BB, Abbracchio MP, Burnstock G et al (1994) Nomenclature and classification of purinoceptors. Pharmacol Rev 46(2):143–156
7. Fredholm BB (1995) Purinoceptors in the nervous system. Pharmacol Toxicol 76(4):228–239
8. Eltzschig HK, Sitkovsky MV, Robson SC (2012) Purinergic signaling during inflammation. N Engl J Med 367(24):2322–2333. doi:10.1056/NEJMra1205750
9. Dunwiddie TV, Masino SA (2001) The role and regulation of adenosine in the central nervous system. Annu Rev Neurosci 24:31–55, doi: 10.1146/annurev.neuro.24.1.31. pii: 24/1/31
10. Fredholm BB, Battig K, Holmen J et al (1999) Actions of caffeine in the brain with special reference to factors that contribute to its widespread use. Pharmacol Rev 51(1):83–133
11. Fredholm BB, Ijzerman AP, Jacobson KA et al (2001) International Union of Pharmacology. XXV. Nomenclature and classification of adenosine receptors. Pharmacol Rev 53(4):527–552
12. Fredholm BB, Arslan G, Halldner L et al (2000) Structure and function of adenosine receptors and their genes. Naunyn Schmiedebergs Arch Pharmacol 362(4–5):364–374
13. Dacey RG Jr, Duling BR (1984) Effect of norepinephrine on penetrating arterioles of rat cerebral cortex. Am J Physiol 246(3 pt 2):H380–H385
14. Duling BR, Gore RW, Dacey RG Jr et al (1981) Methods for isolation, cannulation, and in vitro study of single microvessels. Am J Physiol 241(1):H108–H116
15. Ngai AC, Winn HR (1995) Modulation of cerebral arteriolar diameter by intraluminal flow and pressure. Circ Res 77(4):832–840
16. Meno JR, Nguyen TS, Jensen EM et al (2005) Effect of caffeine on cerebral blood flow response to somatosensory stimulation. J Cereb Blood Flow Metab 25(6):775–784
17. Meno JR, Crum AV, Winn HR (2001) Effect of adenosine receptor blockade on pial arteriolar dilation during sciatic nerve stimulation. Am J Physiol Heart Circ Physiol 281(5): H2018–H2027
18. Ngai AC, Ko KR, Morii S et al (1988) Effect of sciatic nerve stimulation on pial arterioles in rats. Am J Physiol 254(1 pt 2):H133–H139
19. Meno JR, Ngai AC, Ibayashi S et al (1991) Adenosine release and changes in pial arteriolar diameter during transient cerebral ischemia and reperfusion. J Cereb Blood Flow Metab 11(6):986–993
20. Meno JR, Ngai AC, Winn HR (1993) Changes in pial arteriolar diameter and CSF adenosine concentrations during hypoxia. J Cereb Blood Flow Metab 13(2):214–220
21. Morii S, Ngai AC, Ko KR et al (1987) Role of adenosine in regulation of cerebral blood flow: effects of theophylline during normoxia and hypoxia. Am J Physiol 253(1 pt 2): H165–H175
22. Lew MJ, Rivers RJ, Duling BR (1989) Arteriolar smooth muscle responses are modulated by an intramural diffusion barrier. Am J Physiol 257(1 pt 2):H10–H16
23. Ngai AC, Jolley MA, D'Ambrosio R et al (1999) Frequency-dependent changes in cerebral blood flow and evoked potentials during somatosensory stimulation in the rat. Brain Res 837(1–2):221–228
24. Ngai AC, Meno JR, Winn HR (1995) Simultaneous measurements of pial arteriolar diameter and laser-Doppler flow during somatosensory stimulation. J Cereb Blood Flow Metab 15(1):124–127
25. Ngai AC, Meno JR, Ko KR et al (1998) Role of adenosine in cerebral vasodilator responses to sciatic nerve stimulation. J Cereb Blood Flow Metab 18(5):580–581
26. Ibayashi S, Ngai AC, Meno JR et al (1991) Effects of topical adenosine analogs and forskolin on rat pial arterioles in vivo. J Cereb Blood Flow Metab 11(1):72–76
27. Kusano Y, Echeverry G, Miekisiak G et al (2010) Role of adenosine A2 receptors in regulation of cerebral blood flow during induced hypotension. J Cereb Blood Flow Metab 30(4):808–815, doi: jcbfm2009244. pii: 10.1038/jcbfm.2009.244
28. Nehlig A (1999) Are we dependent upon coffee and caffeine? A review on human and animal data. Neurosci Biobehav Rev 23(4):563–576, doi: S0149-7634(98)00050-5
29. O'Brien TG, Kinnaird MF (2003) Conservation. Caffeine and conservation. Science 300(5619):587. doi:10.1126/science. 1082328 300/5619/587

Chapter 4

Imaging Vasodynamics in the Awake Mouse Brain with Two-Photon Microscopy

Andy Y. Shih, Patrick J. Drew, and David Kleinfeld

Abstract

In vivo two-photon laser-scanning microscopy is widely used to study brain structure and function in mice. Recent studies using awake mice have revealed a rich dynamism in cerebrovascular flow and neural activity that is significantly masked by anesthesia. Imaging of awake animals is necessary to fully understand neurovascular coupling during naturalistic sensorimotor activity. We provide detailed instructions to rigorously quantify blood flow at the level of single cortical vessels in awake mice. This includes a description of surgical techniques to obtain optical access to the cortex, improved head-restraint devices to reduce motion artifacts, and robust algorithms to quantify red blood cell flow and vessel caliber. Finally, we provide examples on how these techniques are used to measure sensory-evoked cortical hemodynamics.

Key words Arterioles, Blood flow, Capillary, Cortex, Cranial window, Mouse, Rat, Stroke, Transcranial window, Venules

1 Background

The moment-to-moment energy demands of neuronal computation in the mammalian brain are high, yet the brain has limited reserves of energy [1, 2]. A solution to this dilemma calls for a vascular system that functions as a reliable, pervasive supply chain. With regard to the cortical mantle, this system begins with the great cerebral arteries that emanate from the Circle of Willis and source a planar network of highly interconnected pial arterioles that span the surface of the mantle [3]. The inherent multitude of paths for flow to any given branch allows the surface network to distribute blood in a manner that is relatively insensitive to blockages in one or a few branches of the network [4, 5]. Further, the distribution can be dynamically shifted towards regions of heightened electrical activity and concurrent metabolic load [6, 7]. The surface network is connected to an underlying, three-dimensional network of microvessels by radially directed penetrating arterioles (Fig. 1).

Mingrui Zhao et al. (eds.), *Neurovascular Coupling Methods*, Neuromethods, vol. 88,
DOI 10.1007/978-1-4939-0724-3_4, © Springer Science+Business Media New York 2014

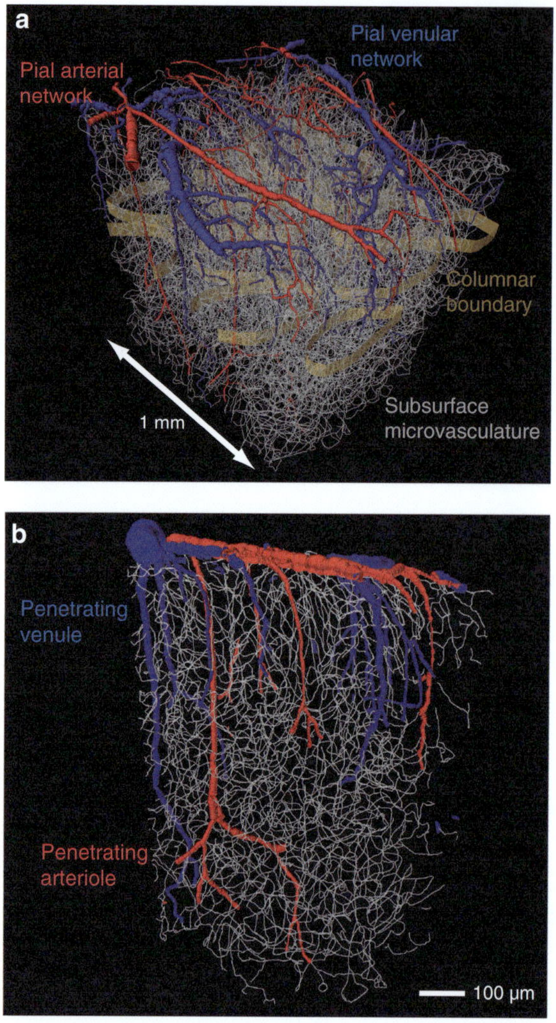

Fig. 1 The vasculature of the mouse cortex. (**a**) Example of vectorized data obtained throughout the full depth of cortex and extending into the white matter. Surface and penetrating arterioles are colored *red*, venules *blue*, and the borders of cortical columns denoted by a *golden band*. (**b**) A selected slice from the larger data set in panel **a** that illustrates the extent of penetrating vessels. Adapted from Blinder, Tsai, et al. [8]

The microvascular network forms a completely interconnected network such that flow may occur in any direction [8]. This network is drained by penetrating venules that return blood to the surface of cortex, where it empties into the central sinus to complete the supply chain (Fig. 1). While the microvascular network brokers the bulk of the exchange of metabolites and gases between brain cells and the blood stream, the mechanism for the control of flow in these vessels remains an open issue.

In vivo two-photon laser-scanning microscopy (TPLSM) is an established method to visualize cerebrovascular topology and quantify blood flow at the level of single vessels in the rodent cortex [9, 10]. In this respect, TPLSM has become a key tool to understand the basis of neurovascular coupling [11]. While a rich body of data has resulted from studies on anesthetized animals, there remains a striking paucity of studies on vasodynamics in the awake state. Anesthetics greatly dampen the activity of neurons [12] and astrocytes [13] and alter vascular dynamics in basal and stimulated states [14]. Anesthetics further preclude the study of vascular function during voluntary sensorimotor function. To image brain vasculature in the awake state, only minor enhancements upon routine in vivo imaging techniques are necessary. First, chronically implanted cranial windows are required to allow imaging well after recovery from initial surgeries. Second, a rigid head mount and gradual habituation to head restraint is needed to minimize movement during imaging. In this chapter, we describe methods to visualize and quantify blood flow at the level of individual vessels in the cerebral cortex of awake, head-restrained mice. We further provide an example of how this technique can be used to study the effects of sensory-evoked neuronal activity on vascular responses.

1.1 Visualizing the Cerebral Vasculature

A portion of the overlying skull must be thinned to translucency or removed entirely to gain optical access to the mouse cortex. Detailed methods have been described for chronic bone-removed cranial windows [15, 16] and semichronic thinned skull windows [17] and will not be discussed further. Thinned skull windows are preferred for experiments since exposure of the cortex to air, compression of the brain with an overlying coverslip, and mechanical disturbances can cause inflammation and gliosis that might affect the phenomenon under study [18]. However, thinned skull windows are limited in imaging depth, allowing optical access to the upper 250 μm of cortex with conventional TPLSM imaging systems. To study deeper cortical layers, one must use bone-removed cranial window, which can provide access to cortical structures as deep as 600–800 μm, or even the full 1 mm depth of cortex with special techniques [19]. Thus, the type of window used depends largely upon the goal of the experiment, and the factors to consider are imaging depth and potential affects of surgically induced inflammation.

Without structural support, the thinned skull is flexible and movement artifacts caused by heartbeat or breathing will affect imaging of small structures. Past studies have avoided this issue by generating very small windows in order to maintain rigidity, i.e., less than 200 μm in diameter [17]. However, larger windows are needed to study vasodynamics since the position of arterioles and venules cannot be readily predicted before window construction. We thus highlight a method to generate a relatively large, chronic thinned-skull cranial window in mice, i.e., a 2 mm in diameter

PoRTS window, where the skull is reinforced with a thin layer of cleat cement and cover glass for rigidity and to impede bone regrowth (Fig. 2) [20, 21]. The result is a cranial window that can be used to image pial arterioles, venules, and microvessels in the upper cortical layers (Fig. 2m) for over one month (Fig. 2n) [20].

2 Materials for PoRTS Window Preparation

2.1 Reagents

1. Isoflurane (029405; Butler Schein).
2. Betadine (6906950; Butler Schein).
3. Buprenorphine hydrochloride (Buprenex®) (031919; Butler Schein).
4. Fluorescein-dextran (FD2000S; Sigma) or Texas Red-dextran (D1830; Invitrogen).
5. Isopropyl alcohol, diluted to 70 % (v/v) in water (I9030; Sigma-Aldrich).
6. Lactated Ringer's solution (009846; Butler Schein).
7. Artificial cerebral spinal fluid (ACSF). Prepared from 125 mM NaCl, 5 mM KCl, 10 mM glucose, 10 mM HEPES, 3.1 mM $CaCl_2$, 1.3 mM $MgCl_2$, pH 7.4. Sterile filter and maintained as aliquots at 4 °C.

2.2 Disposables

1. Cotton-tipped applicators (23-400-100; Fisher Scientific).
2. Cover Glass, no. 0 thickness (6661B40; Thomas Scientific).
3. Cyanoacrylate cement (31428 H04308; ND Industries).
4. Bone wax (032005; Butler Schein).
5. Dental cement (S380; Metabond, Parkell).
6. Drill burrs (19007-05; Fine Science Tools).

Fig. 2 (continued) repeated imaging of the same location within the window. In this example, a connector was also implanted for repeated electrocorticogram recordings. (**k**) Schematic diagram showing dorsal view of the head mount and position of various components. Two #000 self-tapping screws can be added with the cross bar mount for added stability when imaging awake preparations. (**l**) Schematic diagram showing cross section of a PoRTS window. (**m**) Maximum z-axis projections along 65 μm of fluorescein-conjugated-dextran filled vasculature through a PoRTS window 90 days after surgery. Each image is the average of six frames, the z-step was 2 μm, the dwell time was 3 μs per pixel, and the average power was 25–120 mW. (**n**) Subsurface microvessels are stable under the PoRTS window for over a month after implantation. Images show maximum projections along 20–70 μm deep stacks using an average of five frames, a dwell time of 6 μs per pixel, and an average power of 30–45 mW. Panels a to l adapted from Shih et al. [10]; panels **m** and **n** adapted from Drew et al. [20]

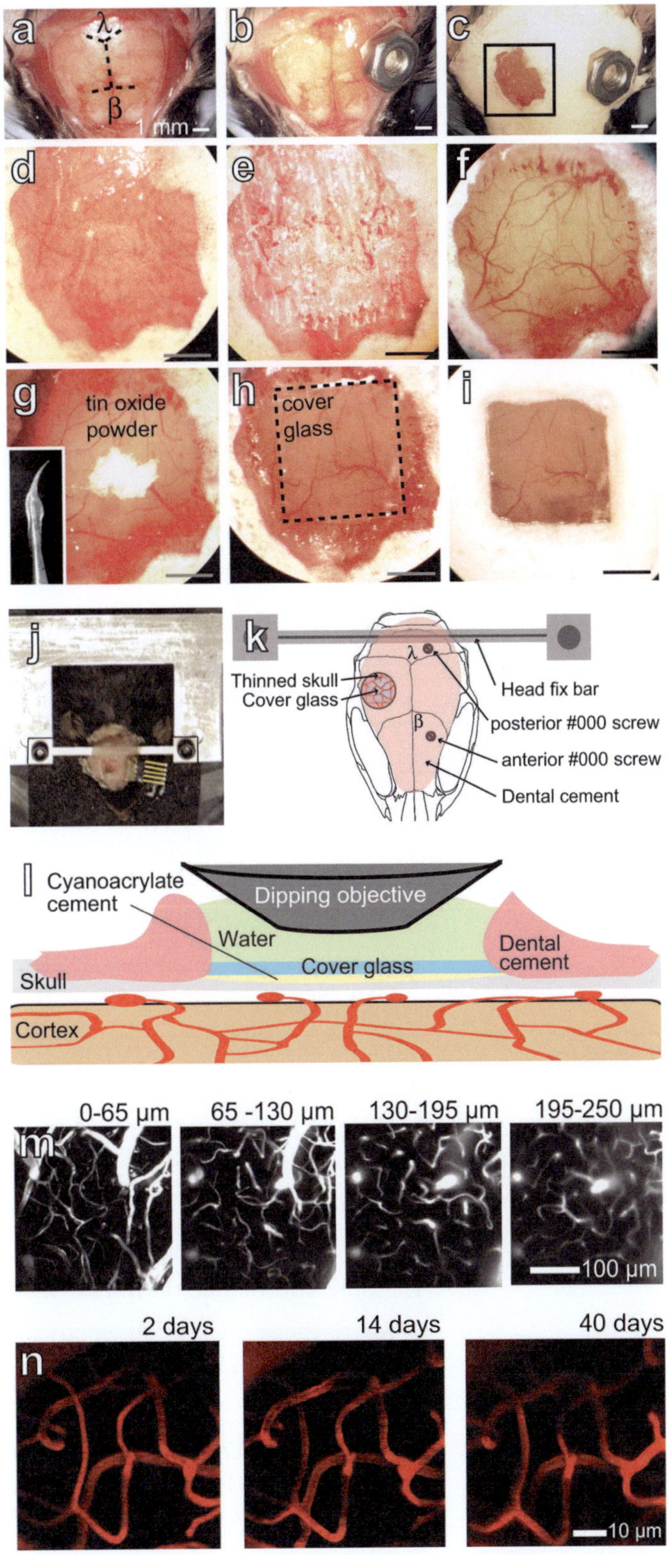

Fig. 2 Realization and benefits of a PoRTS transcranial window in mouse. (**a**–**i**) Images of the different stages in the procedure for generating a chronic PoRTS window. β = bregma and λ = lambda. (**j**) Custom-machined cross bar head mount for awake preparations. The cross bar greatly reduces degrees of freedom for

7. Insulin syringe, 0.3 mL volume with 29.5 gauge needle (309301; Fisher).
8. Kimwipes (06-666A; Fisher Scientific).
9. Opthalmic ointment (039886; Butler Schein).
10. Scalpel blades (12-460-448; Fisher Scientific).
11. Silicone aquarium sealant (31001; Perfecto Manufacturing).
12. Tin oxide powder (EQT-TINOX; Mama's Minerals).

2.3 Equipment

1. Electric powered dental drill (EXL-M40; Osada).
2. Stereo Microscope (SVX16; Olympus, or similar system).
3. Electrical razor (Series 8900; Wahl).
4. Forceps, Dumont no. 55 (11255-20; Fine Science Tools).
5. Forceps, serrated (11050-10; Fine Science Tools).
6. Glass scribe to cut cover glass (08-675; Fisher Scientific).
7. Heating pad with feedback regulation (Temperature control system, 40-90-8, rectal thermistor, 40-90-5D-02, and heat pad for mouse, 40-90-2-07; FHC Inc.).
8. Isoflurane vaporizer (IsoTec4; Datex-Ohmeda, GE Healthcare).
9. Stereotaxic frame (Model 900 stereotaxic frame, Model 923-B mouse gas anesthesia head holder, Model 922 non-rupture 60° tip ear bars; David Kopf).
10. Surgical scissors, blunt end (14078-10; Fine Science Tools).
11. Custom-made head-frame [20].

3 Procedure

3.1 Surgery for PoRTS Window Preparation

1. Induce anesthesia. Ensure a surgical plane of anesthesia by checking for lack of toe pinch reflex.
2. Secure animal in stereotaxic frame [22].
3. Maintain body temperature with feedback regulated heat pad.
4. Apply ophthalmic ointment to eyes to retain moisture.
5. Shave the scalp with an electrical razor.
6. Clean the scalp with betadine followed by 70 % (v/v) isopropyl alcohol.
7. Using sterilized tools, remove the scalp over the entire dorsal skull surface (Fig. 2a). Use a scalpel blade to remove the thin periosteum from the surface of the skull.
8. Clean and dry the skull surface. Gently score the skull surface with sharp forceps to aid glue adherence. Apply a very thin layer of cyanoacrylate glue to the surface and allow the glue to dry thoroughly (Fig. 2b).

9. Attach a metal head fix bar to the skull, away from the area of the desired window with a small dab of cyanoacrylate cement and allow the cement to dry thoroughly (Fig. 2k). Then secure the connector with a layer of dental cement.

   We adhere a custom-made bar, in this case with two attachment points (Fig. 2j, k). This greatly reduces the degrees of freedom and simplifies relocation of the same imaging field in longitudinal studies. A wide cross-bar gives ample room for electrode placement and stimulation of vibrissae, if desired. The exact design of the head mount has varied considerably between labs and is often custom designed to suit the needs of the experimenter. The head mount must be light weight and relatively compact so that it does not impact the animal's movement and ability to obtain food and water within the cage. It is typically made from aluminum, stainless steel, or titanium, precision machined or laser cut to a fit atop the animals head. A secondary piece (also custom designed) is necessary to affix the head to the restraint apparatus during imaging (Fig. 2j).
10. Cover the rest of the skull surface, excluding the location of the window, with a layer of dental cement (Fig. 2c, d). Ensure that all exposed edges of the skin are covered by cement.
11. Thin a 2 mm diameter region over the somatosensory cortex with a ½ mm burr. Alternate between wetting the skull with ACSF and then drying the skull surface with a gentle stream of air from a pressurized dusting can wet for cooling and dry for thinning. The skull may bleed from the vessels in the inner cancellous layer, but can be controlled by flushing with ACSF (Fig. 2e). The skull begins to flex without breaking under the slight pressure of the drill when it is ~50 μm thick, and the pial vessels should begin to be visible through the wet bone. Note that small white spots within the bone are visible immediately after the bone is moistened. The presence of these spots is used as a guide to estimate the thickness of the skull.
12. At this point, the bone must be thinned even further, and the speed and sharpness of the drill burr are critical. Change to a new drill burr and lightly shave the skull surface with small controlled movements while holding the drill like a pen. The shaving motions should move along the surface of the skull with the direction of bit rotation to minimize chatter; a drill speed of ~10,000 rpm is appropriate at this stage. The small white spots within the bone, normally visible when moistened bone is ~50 μm thick, will not be visible at the final skull thickness of ~10–15 μm (Fig. 2f).
13. Polish the window region with tin oxide powder (Fig. 2g). Attach a premade drill bit that has been dipped in silicone aquarium sealant and withdrawn, leaving a tapered whip (inset, Fig. 2g). Place a small pinch of powder on the window along

with a drop of ACSF. Agitate the slurry over the window for up to 10 min by gently touching the tip of the moving whip to the skull surface. Surface irregularities and adherent bone chips left by drilling in the previous steps should be removed after polishing. Flush away the tin oxide powder thoroughly from the window using ACSF and dry the bone thoroughly with a gentle stream of air. Note that remaining tin oxide particles can become luminescent and interfere while imaging with TPLSM.

14. Cut small square pieces of no. 0 cover glass, roughly 2 by 2 mm, by gently scoring separated horizontal and vertical lines in the cover glass with a scribe. Then place the cover glass in a petri dish and shake vigorously to separate the glass pieces.
15. Apply a small dab of cyanoacrylate cement over the dry window using the wooden tip of a broken cotton-tipped applicator; stirring the cement helps to prevent the formation of bubbles. Next, quickly place an appropriately sized piece of no. 0 cover glass atop the cement (Fig. 2h). Using forceps, gently push the glass downward so that it is in contact with the skull surface. Quickly wipe away any excess wet cement above the window with a Kimwipe twisted to a fine point. Allow the cement to dry thoroughly for 10 min before proceeding. Excess cyanoacrylate cement can be removed from the upper surface of the cover glass with a scalpel after it is dried.
16. Seal the edges of the cover glass with dental cement and form a slightly raised well around the window to hold water for the immersion objective (Fig. 2i).

    Metabond from Parkell bonds more strongly to the skull than other cements tested and has been effective without skull screws. However, if increased stability is necessary, two self-tapping #000 screws can be added to the contralateral hemisphere of the skull prior to application of the dental cement (Fig. 2k), as described in the past methods [10]
17. For some experiments, it may be desirable to inject dyes or insert electrodes into the tissue volume beneath the PoRTS window. A small hole can be added adjacent to the window, through which pipettes or electrodes can be introduced using a stereotaxic arm or Sutter manipulator [23]. This hole can be resealed with bone wax after the experiment if the animal is to be imaged again in future sessions.
18. Provide buprenorphine at a concentration of 0.03 mg per kg subcutaneously for analgesia and allow the animal to recover on a heat pad until fully awake.

3.2 Preparation for Imaging

1. 1–2 days following surgery, stabilize the animal in a restraint apparatus, using the frame as a head support (Fig. 2j). The apparatus typically consists of an optical breadboard with

miniature optomechanical components commercially available from Newport or ThorLabs.

2. Habituation to head fixation is important to reduced animal movement during imaging. A new animal can be gradually accustomed to head restraint over a period of 3–7 days, starting with 15 min sessions without imaging and working up to several hours [14]. Head-restrained mice tend to struggle less when their bodies are enclosed within a loose-fitting plastic or cardboard tube.

3. The first two-photon imaging session should be at least 3 days after the initiation of habituation when the animal has become more accustomed to the apparatus. For each imaging session, restraining the animal for 2 h at a time is reasonable, but times can differ for each animal. Discomfort should be gauged by the extent of struggling and vocalizations emitted, and the restraint time adjusted accordingly. For imaging sessions lasting hours, drinking water should be supplied with a pipette.

   In our experience, some animals, approximately 10 %, never become habituated to head fixation and cannot be used for awake imaging.

4. Before imaging, the mouse must be briefly anesthetized with isoflurane for an intravenous injection of fluorescent dextran dye. Inject 25–50 μL of 5 % (w/v) fluorescent-dextran dye dissolved in saline either through the tail vein or infraorbital vein to label the blood serum. Use an ultrafine 0.3 mL insulin syringe with 29.5 gauge needle. The dye will remain in circulation for approximately 3–4 h, and supplements can be given as necessary if the animal is reanesthetized. Intravenous injections are painful and must be done under general anesthesia.

5. Physical movement of the animal can be a source of artifacts in the data. In practice, a properly habituated animal will provide movement-free data for several minutes at a time even during sensory stimulation. Typically, several trials are collected for each stimulus paradigm and trials with excessive movement will be discarded. As a quality control measure, data acquisition can be tested on capillaries, which are only 3–5 μm in diameter and thus most sensitive to motion. Similarly, a surface arteriole, which may move due to dilations/constriction, can be scanned simultaneously with a neighboring venule that should exhibit little or no change in lumen diameter. A movement in both vessels would indicate movement of the animal. Low-cost accelerometers can be used to monitor movement of the animal and be recorded alongside the vasodynamic measurements to identify data sets with excessive movement [14].

3.3 Measurement of Blood Flow Dynamics in Single Cortical Vessels

When the vasculature is labelled with an intravenous bolus of fluorescent-conjugated dextran, red blood cells (RBCs) exclude the high molecular weight dextran dye and will appear as dark shadows moving against a bright fluorescent background. This differential staining is the basis for measuring RBC velocity using laser-scanning microscopy [9, 24]. Conventionally, single line scans are used to sample the RBC velocity and diameter of a vessel separately [4, 25]. However, blood flow is dynamic and could change within the time between each scan. In order to collect more samples simultaneously and with varying trajectories within the imaging plane, we use custom software to direct the imaging laser beam in a user-defined path (Fig. 3a) [26, 27], following earlier work [28–30]. Linear segments of constant scan speed traverse along the length of the center of the vessel and across the width of the vessel to measure RBC speed and lumen diameter, respectively (right panel in Fig. 3a). These linear scan segments are connected by cubic splines, where connecting portions of the scan are accelerated to allow for rapid data collection across multiple vessels (right panel in Fig. 3a) [27]. The resulting line scan is a space-time image, typically displayed with the individual scan lines stacked on top of each other (Fig. 3b).

3.3.1 Calculating Red Blood Cell Velocity

Portions of the scan path along the centerline of the vessel lumen reveal angled streaks within the cascade image (right panel in Fig. 3b). Moving RBCs in flowing vessels sampled at a sufficient rate will appear as diagonal streaks. A stationary or stalled flow will result in vertical streaks. In the limiting case of extremely fast flowing vessels, the streaks will become horizontal and velocity data cannot be extracted. The centerline velocity is inversely proportional to the slope of the RBC streaks. This slope can be

Fig. 3 (continued) The animal is anesthetized with alpha-chloralose. (**a**) Image of fluorescently stained vessels. The forelimb and hindlimb representations across cortex were mapped using intrinsic optical imaging. The expanded view shows the scan path, with the colored segments corresponding to scan regions when the mirrors moved at constant velocity and interconnecting white lines as regions when the mirrors were accelerated to reduce total scan time. (**b**) The scanning procedure. The top trace is the mirror speed as a function of time. Portions used to acquire diameter and velocity data are constant speed. The line-scans generated from the path are stacked sequentially as a function of time to produce a raw cascade image. The diameter is calculated as the full-width half-max of a time average of several scans across the width of a vessel. Red blood cell velocity is calculated from the angle of the RBC streaks. (**c**) Data traces of diameter, velocity, and flux for the arteriole and venule, before, during, and after simulation of the forelimb, processed to remove heart rate and smoothed with a running window. Both vessels show an increase in flux in response to stimulation. Adapted from Driscoll et al. [27]

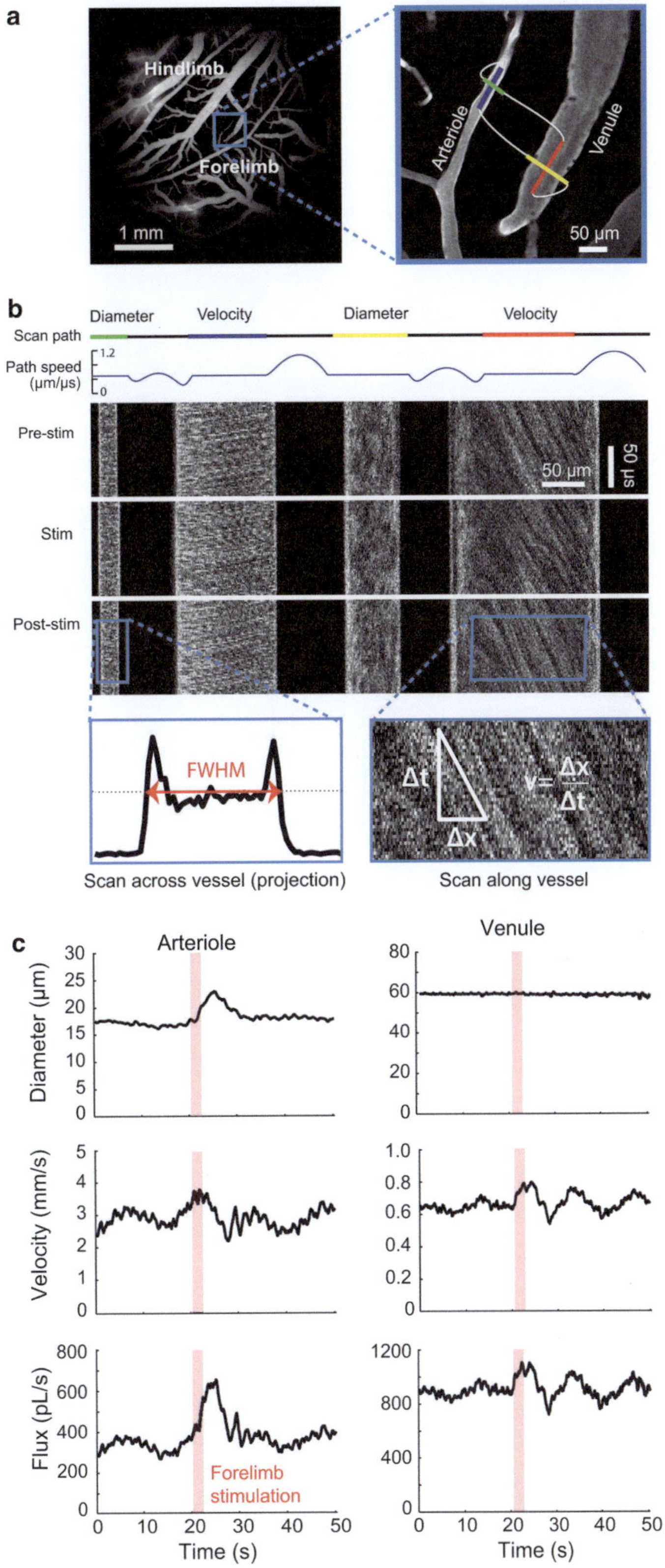

Fig. 3 Simultaneous measurement of diameter and velocity in two vessels in somatosensory cortex of a Sprague Dawley rat using a spatially optimized line scan.

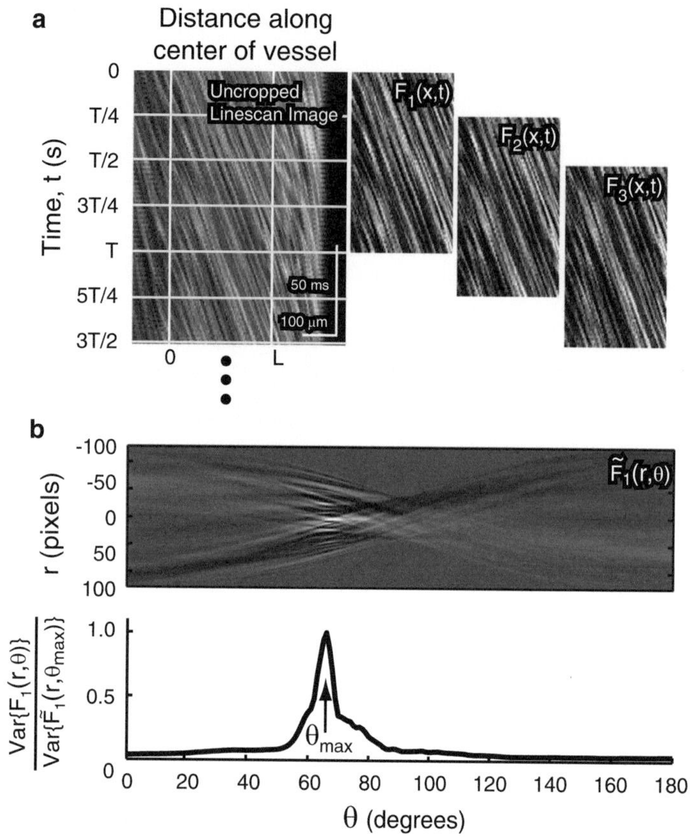

Fig. 4 Illustration of the Radon method to determine RBC velocity. (**a**) Raw line-scan image data from a subsurface vessel. *White, vertical lines* show the spatial range selected for analysis. The *yellow lines* show the temporal windowing. The data blocks to the right are overlapping windows of data that are ready for processing, denoted $F_i(x, t)$ where i is an index in units of $T/4$, where $T = 25$ ms is the duration of the data block and $L = 300$ μm is the length of the usable line-scan. (**b**) The top image shows the Radon transform of the first data block. The plot below shows the normalized variance of the Radon transform. The peak is at $\theta_{max} = 65°$, which corresponds to an RBC velocity of $v(0) \equiv (\Delta x/\Delta t)\cdot\cot(\theta_{max})$, where $\Delta x = 1.0$ μm is the pixel width and $\Delta t = 0.5$ ms is time per line, so that $v(0) = 0.93$ mm/s is calculated. Adapted from Drew et al. [31]

efficiently determined with a Radon transform of windowed portions of the data [31], which is available in the MATLAB™ Image Processing Toolbox (Mathworks) (Fig. 4). The direction of flow can be extracted from the sign of the slope and the direction of the line scan sweep.

A velocity time series is calculated by taking successive time-windowed portions of the line scan (Fig. 4a). The size of the window must be sufficiently short to resolve the highest velocity modulation frequency, the heart rate, which is 8–10 Hz for awake mice [14]. In addition, the window size must be large enough to capture enough streak lines so that the Radon transform has sufficient data to calculate an accurate velocity value, but small enough to prevent smoothing of the higher frequencies of the velocity data (Fig. 4b). We find that a window size of 40 ms is a good compromise, which yields a Nyquist frequency of 12.5 Hz. We further use a window spacing of 10 ms.

Oscillating physiological rhythms within the RBC velocity are a first indicator of good data quality. In addition to heart rate, other physiological signals detected in the RBC velocity may include breathing at ~1–2 Hz and vasomotion at ~0.1–1 Hz [32, 33]. Breathing rate, however, is not always detectable in the flow data collected from awake mice (Fig. 5) [14].

Measurement of RBC velocity is the limiting factor for the measurement of flow in larger arterioles on the cortical surface. In practice, conventional galvanometric mirrors are sufficiently fast to simultaneously capture RBC velocity and diameter from 1 to 2 penetrating arterioles at a time in the anesthetized state. However, in the awake state RBC velocity tends to increase, and thus atypical fast scanning and/or special analysis techniques [34] may be necessary or smaller vessels should be sampled. Venules and deep microvessels exhibit slower RBC velocity and thus 3–4 vessels can be measured simultaneously with user-defined line scans. Greater distances traversed by the laser will reduce the sampling frequency and a 1–2 kHz line scan rate is typically required for accurate sampling of the movement of RBCs in pial arterioles [27].

3.3.2 Calculating Lumen Diameter

As with the velocity calculation, the diameter calculation is taken from a time-windowed portion of the data. The same time-window and spacing used for velocity is typically used for diameter so that both parameters can be calculated on the same time scale. Vessel diameter is calculated as full-width at half-max of the vessel profile for each window (left panel in Fig. 3b). Note the intensity profile tends to increase near the edges due to the repulsion of RBCs from the glycocalyx, which generates two peaks. The two outermost half-max points of these peaks are used to calculate the vessel boundary. Linear interpolation is used to add subpixel accuracy to the diameter measurement.

3.3.3 Estimating the Flux of Red Blood Cells

There are two limiting cases for the flux. When the diameter of the vessel allows the passage of RBCs only in single file, as for the case of capillaries, the flux of RBCs is just the number of RBCs that pass per second [9, 35]. In this limit, the blood plasma has an essentially constant velocity profile as a function of distance from the center of the vessel, then the velocity falls rapidly to zero near the walls. The thin layer of plasma near the walls acts as a lubricating layer that, together with the glycocalyx, minimizes friction [36].

When the diameter of the vessel is much greater than that of the RBC, the flow is laminar and nearly parabolic as RBCs flow in

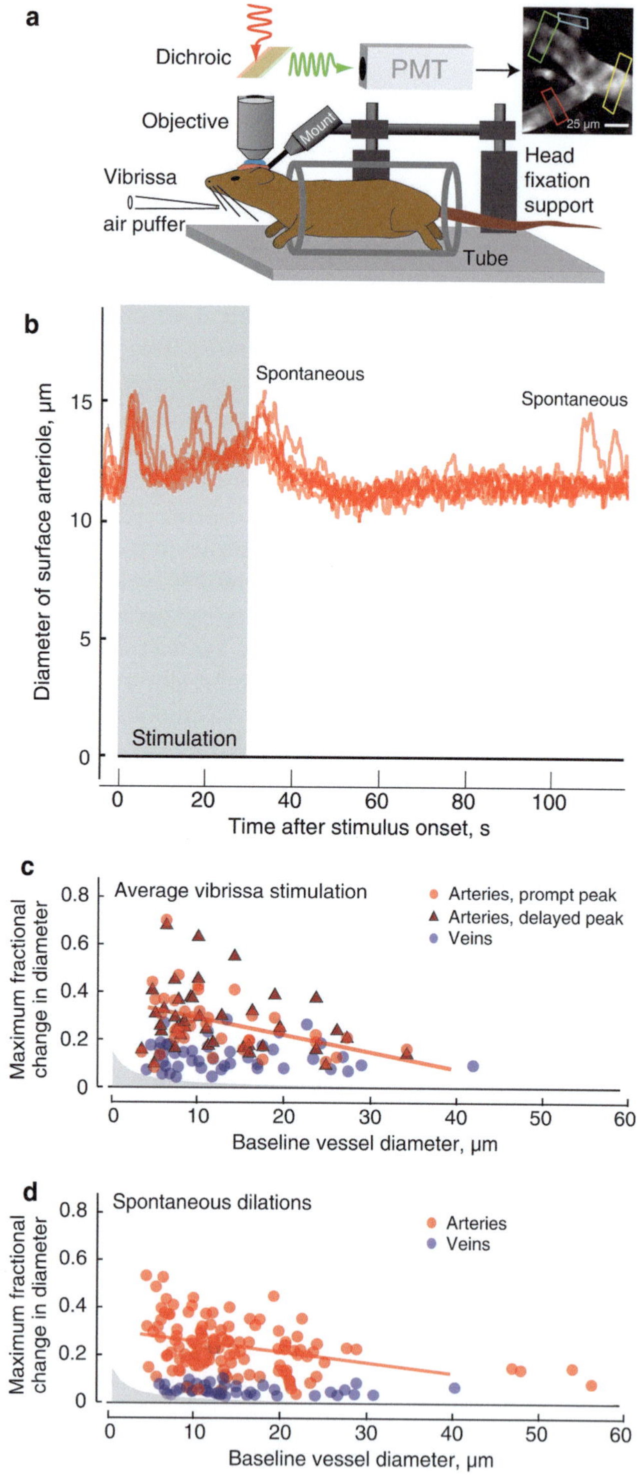

Fig. 5 Spontaneous and stimulus-induced vascular dynamics in the cortex of awake mouse. (**a**) Schematic of the experimental setup. The awake mouse is head fixed by means of a bolt and sits passively in an acrylic cylinder beneath the

parallel streams [4, 37]. The two vascular parameters, RBC velocity and lumen diameter, are combined to calculate the volume flux, i.e., RBCs and plasma, for each vessel. Flux is a more complete description of blood flow in single vessels, as RBC speed and lumen diameter can change independently of each other [25, 38]. The volume flux through the vessel is given by

$$\vec{F} = \vec{v}\ A = \frac{\pi}{8}\vec{v}(0)\ d^2$$

where $\vec{v}$ is the average RBC velocity over A, the cross-sectional area of the vessel lumen, $\vec{v}(0)$ is the time-averaged RBC velocity at the center line of the vessel, and d is the lumen diameter. Note that this formula underestimates the flux as the nonzero spatial extent of the RBC flattens the parabola of Poiseuille flow; empirical corrections have been discussed [39] and a first-order correction was estimated [10].

4 Typical/Anticipated Results

4.1 Stimulus-Induced Changes in Flux in the Anesthetized Rat

We measure the change in flux at the level of single pial vessels in response to somatosensory stimulation in rat anesthetized with alpha-chloralose (Fig. 3c). A line scan frequency of 733 Hz was used to capture the lumen diameter and RBC velocity of an arteriole and a venule nearly simultaneously. Both diameter and RBC velocity in the arteriole respond to stimulation (left column in Fig. 3c). The flux through the arteriole increases to a peak of 86 % over baseline and contains contributions from both a change in diameter of 29 % and a change in velocity of 24 % taken (near) simultaneously from the same region of the vasculature. In contrast to the arteriole, a neighboring venule exhibits no change in lumen diameter, but a 23 % change in RBC velocity. Note that in

two-photon microscope. An air puffer for sensory stimulation is aimed at the vibrissae. The inset shows an image of surface vessels. (**b**) Single trial time series of the diameter of an arteriole in response to a 30 s period of stimulation. Note the spontaneous dilations. (**c**) Plot of peak-averaged dilation responses to 30 s vibrissae stimulation. Early arterial peaks, in the 0–10 s interval after stimulation, are denoted by *red circles*; regression slope = 0.007 μm^{-1} (r^2 = 0.15, p < 0.02). Late arterial peaks, greater than 10 s after onset, are denoted by *red triangles*; the linear regression is not significant. Venules are denoted by *blue dots*; the linear regression is not significant. (**d**) Plot of peak spontaneous dilations for arteries, in *red*, and veins, in *blue*. *Grey* area shows the 0.2 μm resolution limit of detectable changes. *Lines* show linear regressions; slope of −0.004 μm^{-1} for arterioles is significant (r^2 = 0.13, p < 0.001), while that for veins is not significantly different from zero. Adapted from Drew et al. [14]

both cases the increase in RBC velocity is partially masked by a peak in the underlying vasomotor fluctuation, but remains a significant increase over an average 1-min period of basal activity.

4.2 Vasodynamics in Awake Mice

Mice were habituated to head fixation (Fig. 5a) and blood flow was measured in single vessels through a chronically implanted PoRTS window (Fig. 2). The speed of RBCs within arterioles is high during the awake state and thus a line-scan frequency greater than 1 kHz was required to capture data. In this example, we focus on the changes in diameter. Robust arteriole dilations could be evoked by prolonged contralateral whisker stimulation with puffs of air (Fig. 5b). Dilatory events consisted of an early peak followed by a secondary slowly rising plateau, in averaged data, during sustained stimulation (Fig. 5c). The dilatory affect returned to baseline over 10 s following cessation of stimulation. Notably, spontaneous dilations observed during slow basal vasomotor activity were similar in magnitude to dilations evoked by stimulation, possibly indicating a common driving source (Fig. 5d). In both cases, the magnitude of dilation was larger with smaller diameter arterioles, i.e., penetrating arterioles or vessels immediately feeding penetrating arterioles, rather than larger arterial branches. Pial venules, typically thought to be static in terms of diameter, show a weak dilation in the awake state (Fig. 5c). Overall, these data suggest that functional hyperemia changes detected by blood oxygen level-dependent functional magnetic resonance imaging (BOLD fMRI) may be dominated by large changes in arteriole volume, in agreement with recent studies [40], rather than in venules as predicted by the "balloon" model [41].

Both basal and stimulated responses of arterioles and venules were strongly suppressed by the use of urethane, a common anesthetic [14], that can attenuate the magnitude of the overall dilatory response by up to 75 % in the anesthetized state. Other commonly used anesthetics, including isoflurane [42], α-chloralose [43], and propofol [44], similarly suppress the neurovascular response. In addition to the direct neural effects, anesthetics exert strong depressing effects on cardiovascular function [45]. Consistent with the depressive effects of anesthetics on neural and cardiovascular system, optical imaging studies of neurovascular coupling have found differences between the awake and anesthetized animal [46–48].

5 Future Directions

The methodology described here fills an important gap in the existing range of tools for blood flow quantification and enables the study of vasodynamics in awake mice. While intrinsic optical imaging, laser speckle, and BOLD fMRI enable measurement of flow over broader areas of brain, two-photon microscopy complements these techniques and provides a view of dynamics at the resolution of individual small cortical vessels. Further, TPLSM can be concurrently used to examine activity of surrounding cells that could be driving vascular activity [49, 50]. This is important as a means to dissect the basis of neurovascular coupling and bears direct relevance on clinical tools such as BOLD fMRI, as well as to examine vascular responses to disease states such as ischemic stroke.

It is becoming increasingly important to study cortical blood flow in the awake state, since anesthetics dampen the rich interplay between components of the neurovascular unit. To fully exploit vascular imaging in awake mice, some further advances will be necessary. First, faster laser-scanning is necessary to capture RBC velocity within the largest cortical arterioles. Acousto-optic modulators [51] can in principle scan vessels with faster flow as well as sample from larger populations of vessels simultaneously. Recent advances in processing of line-scan data using particle image velocimetry will also be useful to assess RBC velocity in larger vessels [40]. Second, physiological parameters most relevant to blood flow, including blood pressure and blood gas remain difficult to measure in awake mice. In particular, even when these parameters are measured invasively from catheters in peripheral vessels, they may not reflect what is happening in the cerebral vessel being examined. Novel imaging probes need to be developed to assess real-time changes in blood pressure, gas, and pH at the vessel being examined. Finally, new tools to manipulate cell-specific vasoactive signaling cascades will be an important step in dissecting the chemical basis of neurovascular coupling [52].

Acknowledgement

We thank the National Institutes of Health for generous support through grants NS078168 and NS079737 to PJD and EB003832, MH085499, MH072570, and OD006831 to DK.

References

1. Magistretti PJ, Pellerin L (1999) Cellular mechanisms of brain energy metabolism and their relevance to functional brain imaging. Philos Trans R Soc Lond B Biol Sci 354:1155–1163
2. Attwell D, Laughlin SB (2001) An energy budget for signaling in the grey matter of the brain. J Cereb Blood Flow Metab 21:1133–1145
3. Mchedlishvili G (1963) Arterial behavior and blood circulation in the brain. Consultants Bureau, New York
4. Schaffer CB, Friedman B, Nishimura N, Schroeder LF, Tsai PS, Ebner FF, Lyden PD, Kleinfeld D (2006) Two-photon imaging of cortical surface microvessels reveals a robust redistribution in blood flow after vascular occlusion. PLoS Biol 4:258–270
5. Blinder P, Shih AY, Rafie CA, Kleinfeld D (2010) Topological basis for the robust distribution of blood to rodent neocortex. Proc Natl Acad Sci U S A 107:12670–12675

6. Devor A, Tian P, Nishimura N, Teng IC, Hillman EM, Narayanan SN, Ulbert I, Boas DA, Kleinfeld D, Dale AM (2007) Suppressed neuronal activity and concurrent arteriolar vasoconstriction may explain negative blood oxygenation level-dependent signaling. J Neurosci 27:4452–4459
7. Derdikman D, Hildesheim R, Ahissar E, Arieli A, Grinvald A (2003) Imaging spatiotemporal dynamics of surround inhibition in the barrels somatosensory cortex. J Neurosci 23: 3100–3105
8. Blinder P, Tsai PS, Kaufhold JP, Knutsen PM, Suhl H, Kleinfeld D (2013) The murine cortical angiome: a fully interconnected network of microvasculature that supports restricted blood flow. Nat Neurosci. 16:889–897
9. Kleinfeld D, Mitra PP, Helmchen F, Denk W (1998) Fluctuations and stimulus-induced changes in blood flow observed in individual capillaries in layers 2 through 4 of rat neocortex. Proc Natl Acad Sci U S A 95: 15741–15746
10. Shih AY, Driscoll JD, Drew PJ, Nishimura N, Schaffer CB, Kleinfeld D (2012) Two-photon microscopy as a tool to study blood flow and neurovascular coupling in the rodent brain. J Cereb Blood Flow Metab 32:1277–1309
11. Helmchen F, Kleinfeld D (2008) In vivo measurements of blood flow and glial cell function with two-photon laser scanning microscopy. Methods Enzymol 444:231–254
12. Franks NP (2008) General anaesthesia: from molecular targets to neuronal pathways of sleep and arousal. Nat Rev Neurosci 9:370–386
13. Thrane AS, Rangroo-Thrane V, Zeppenfeld D, Loua N, Xua Q, Nagelhus EA, Nedergaard M (2012) General anesthesia selectively disrupts astrocyte calcium signaling in the awake mouse cortex. Proc Natl Acad Sci U S A 109: 18974–18979
14. Drew PJ, Shih AY, Kleinfeld D (2011) Fluctuating and sensory-induced vasodynamics in rodent cortex extends arteriole capacity. Proc Natl Acad Sci U S A 108:8473–8478
15. Holtmaat A, Bonhoeffer T, Chow DK, Chuckowree J, De Paola V, Hofer SB, Hübener M, Keck T, Knott G, Lee WC, Mostany R, Mrsic-Flogel TD, Nedivi E, Portera-Cailliau C, Svoboda K, Trachtenberg JT, Wilbrecht L (2009) Long-term, high-resolution imaging in the mouse neocortex through a chronic cranial window. Nat Protoc 4:1128–1144
16. Mostany R, Portera-Cailliau C (2008) A method for 2-photon imaging of blood flow in the neocortex through a cranial window. J Vis Exp 12. http://www.jove.com/video/678
17. Yang G, Pan F, Parkhurst CN, Grutzendler J, Gan WB (2010) Thinned-skull cranial window technique for long-term imaging of the cortex in live mice. Nat Protoc 5:201–208
18. Xu HT, Pan F, Yang G, Gan WB (2007) Choice of cranial window type for in vivo imaging affects dendritic spine turnover in the cortex. Nat Neurosci 10:549–551
19. Kobat D, Durst ME, Nishimura N, Wong AW, Schaffer CB, Xu C (2009) Deep tissue multiphoton microscopy using longer wavelength excitation. Opt Express 17:13354–13364
20. Drew PJ, Shih AY, Driscoll JD, Knutsen PM, Davalos D, Blinder P, Akassoglou K, Tsai PS, Kleinfeld D (2010) Chronic optical access through a polished and reinforced thinned skull. Nat Methods 7:981–984
21. Shih AY, Drew PJ, Mateo C, Tsai PS, Kleinfeld D (2012) A polished and reinforced thinned skull window for long-term imaging and optical manipulation of the mouse cortex. J Vis Exp. http://www.jove.com/video/3742
22. Cetin A, Komai S, Eliava M, Seeburg PH, Osten P (2006) Stereotaxic gene delivery in the rodent brain. Nat Protoc 1:3166–3173
23. Stosiek C, Garaschuk O, Holthoff K, Konnerth A (2003) In vivo two-photon calcium imaging of neuronal networks. Proc Natl Acad Sci U S A 100:7319–7324
24. Villringer A, Haberl RL, Dirnagl U, Anneser F, Verst M, Einhaupl KM (1989) Confocal laser microscopy to study microcirculation on the rat brain surface in vivo. Brain Res 504:159–160
25. Shih AY, Friedman B, Drew PJ, Tsai PS, Lyden PD, Kleinfeld D (2009) Active dilation of penetrating arterioles restores red blood cell flux to penumbral neocortex after focal stroke. J Cereb Blood Flow Metab 29:738–751
26. Valmianski I, Shih AY, Driscoll J, Matthews DM, Freund Y, Kleinfeld D (2010) Automatic identification of fluorescently labeled brain cells for rapid functional imaging. J Neurophysiol 104:1803–1811
27. Driscoll JD, Shih AY, Drew PJ, Cauwenberghs G, Kleinfeld D (2011) Two-photon imaging of blood flow in cortex. In: Helmchen F, Konnerth A, Yuste R (eds) Imaging in neuroscience: a laboratory manual. Cold Spring Harbor Laboratory Press, New York, pp 927–938
28. Göbel W, Kampa BM, Helmchen F (2007) Imaging cellular network dynamics in three dimensions using fast 3D laser scanning. Nat Methods 4:73–79
29. Göbel W, Helmchen F (2007) New angles on neuronal dendrites in vivo. J Neurophysiol 98: 3770–3779

30. Lillis KP, Eng A, White JA, Mertz J (2008) Two-photon imaging of spatially extended neuronal network dynamics with high temporal resolution. J Neurosci Methods 172:178–184
31. Drew PJ, Blinder P, Cauwenberghs G, Shih AY, Kleinfeld D (2010) Rapid determination of particle velocity from space-time images using the Radon transform. J Comput Neurosci 29: 5–11
32. Kleinfeld D, Mitra PP (2011) Applications of spectral methods in functional brain imaging. In: Yuste R (ed) Imaging: a laboratory manual. Cold Spring Harbor Laboratory Press, New York, pp 12.1–12.7
33. Mayhew JEW, Askew S, Zeng Y, Porrill J, Westby GWM, Redgrave P, Rector DM, Harper RM (1996) Cerebral vasomotion: 0.1 Hz oscillation in reflectance imaging of neural activity. Neuroimage 4:183–193
34. Kim TN, Goodwill PW, Chen Y, Conolly SM, Schaffer CB, Liepmann D, Wang RA (2012) Line-scanning particle image velocimetry: an optical approach for quantifying a wide range of blood flow speeds in live animals. PLoS One 7:e38590
35. Chuquet J, Hollender L, Nimchinsky EA (2007) High-resolution in vivo imaging of the neurovascular unit during spreading depression. J Neurosci 27:4036–4044
36. Secomb TW, Hsu R, Pries AR (1998) A model for red blood cell motion in glycocalyx-lined capillaries. Am J Physiol 274:H1016–H1022
37. Rovainen CM, Woolsey TA, Blocher NC, Wang D-B, Robinson OF (1993) Blood flow in single surface arterioles and venules on the mouse somatosensory cortex measured with videomicroscopy, fluorescent dextrans, nonoccluding fluorescent beads, and computer-assisted image analysis. J Cereb Blood Flow Metab 13:359–371
38. Kontos HA (1989) Validity of cerebral arterial blood flow calculations from velocity measurements. Stroke 20:1–3
39. Nishimura N, Rosidi NL, Iadecola C, Schaffer CB (2010) Limitations of collateral flow after occlusion of a single cortical penetrating arteriole. J Cereb Blood Flow Metab 30: 1914–1927
40. Kim T, Kim SG (2011) Temporal dynamics and spatial specificity of aterial and venous blood volume changes during visual stimulation: implication for BOLD quantification. J Cereb Blood Flow Metab 31:1211–1222
41. Buxton RB, Wong EC, Frank LR (1998) Dynamics of blood flow and oxygenation changes during brain activation: the balloon model. Magn Reson Med 39:855–864
42. Takuwaa H, Matsuura T, Obataa T, Kawaguchia H, Kannoa I, Ito H (2012) Hemodynamic changes during somatosensory stimulation in awake and isoflurane-anesthetized mice measured by laser-Doppler flowmetry. Brain Res 1472:107–112
43. Dudley RE, Nelson SR, Samson F (1982) Influence of chloralose on brain regional glucose utilization. Brain Res 233:173–180
44. Alkire MT, Haier RJ, Barker SJ, Shah NK, Wu JC, Kao YJ (1995) Cerebral metabolism during propofol anesthesia in humans studied with positron emission tomography. Anesthesiology 82:393–403
45. Janssen BJ, De Celle T, Debets JJ, Brouns AE, Callahan MF, Smith TL (2004) Effects of anesthetics on systemic hemodynamics in mice. Am J Physiol Heart Circ Physiol 287:H1618–H1624
46. Nakao Y, Itoh Y, Kuang TY, Cook M, Jehle J, Sokoloff L (2001) Effects of anesthesia on functional activation of cerebral blood flow and metabolism. Proc Natl Acad Sci U S A 98: 7593–7598
47. Jones M, Devonshire IM, Berwick J, Martin C, Redgrave P, Mayhew J (2008) Altered neurovascular coupling during information-processing states. Eur J Neurosci 27: 2758–2772
48. Martin C, Jones M, Martindale J (2006) Haemodynamic and neural responses to hypercapnia in the awake rat. Eur J Neurosci 24: 2601–2610
49. Takano T, Tian GF, Peng W, Lou N, Libionka W, Han X, Nedergaard M (2006) Astrocyte-mediated control of cerebral blood flow. Nat Neurosci 9:260–267
50. Nizar K, Uhlirova H, Tian P, Saisan PA, Cheng Q, Reznichenko L, Weldy KL, Steed TC, Sridhar VB, MacDonald CL, Cui J, Gratiy SL, Sakadzić S, Boas DA, Beka TI, Einevoll GT, Chen J, Masliah E, Dale AM, Silva GA, Devor A (2013) Stimulus-induced vasodilation occurs without IP3 receptor activation and may precede astrocytic calcium increase. J Neurosci 33:8411–8422
51. Lechleiter JD, Lin DT, Sieneart I (2002) Multi-photon laser scanning microscopy using an acoustic optical deflector. Biophys J 83: 2292–2299
52. Kleinfeld D, Blinder P, Drew PJ, Driscoll JD, Muller A, Tsai PS, Shih AY (2011) A guide to delineate the logic of neurovascular signaling in the brain. Front Neuroenergetics 1:1–9

Part II

Neurovascular Coupling Techniques in Normal Visual Processing

Chapter 5

Neurovascular Coupling in the Deep Brain Using Confocal Fiber-Optic Endomicroscopy

Samuel Bélanger, Bruno Oliveira Ferreira de Souza, Philippe Pouliot, Christian Casanova, and Frédéric Lesage

Abstract

Developing fast functional imaging approaches of subcortical structures is essential to make progress in our understanding of brain function and diseases. Positron emission tomography and functional magnetic resonance imaging have been used to improve our understanding of brain function and integration of neuronal activity between deeper structures and the superficial cortex but limitations remain associated with signal interpretation. This work describes the design and utilization of confocal microendoscopy techniques to image brain structures involved in visual processing, either deep or on the surface of the cortex. Also, multiple examples using different experimental approaches are described.

Key words Confocal microendoscopy, Calcium imaging

1 Background and Historical Overview

The investigation of neurovascular coupling remains at the forefront of current brain research due to its role in fundamental brain mechanisms and pathologies but also to link these mechanisms with macroscopic imaging observations. Multiple studies have explored the role of cell types and their topology in different cortical layers [1–3]. Separately, advances in our knowledge of brain vasculature [4, 5] and its function have also been achieved [6–8]. The aim of research on neurovascular coupling is to link these two areas with the goal of characterizing the system as a whole [9–13]. A better understanding of this coupling and how it is reflected in clinical imaging may help understand and diagnose diseases such as strokes (CVA) [14, 15], Alzheimer (AD) [16, 17], and attention deficits [18], to name a few. However, in these pathologies or even with aging, some of the key degenerative mechanisms are not only affecting the cortex but also subcortical areas. Therefore, an understanding of brain function in subcortical

Mingrui Zhao et al. (eds.), *Neurovascular Coupling Methods*, Neuromethods, vol. 88,
DOI 10.1007/978-1-4939-0724-3_5, © Springer Science+Business Media New York 2014

areas and their connection to the cortex is of interest. Imaging remains the tool of choice for these investigations.

Imaging tools used to study brain function can be divided into two distinct categories related to the spatial scale of the acquired information: macroscopic and microscopic techniques. Macroscopic devices have been used to measure the relation between different functional areas of the brain [19] or even to map functional organization of a particular cortical area [20, 21]. Hemodynamic methods, such as functional magnetic resonance imaging (fMRI) and diffuse optical and intrinsic signal imaging (ISI), use macroscopic scale hemodynamic fluctuations to infer neuronal activity. While their spatial resolution is not high enough to measure single cell activity, they have proven very useful to quantify local activity on large spatial scales and to map brain function indirectly [22]. However limitations remain and questions persist about the origin of the measured signals. Recent work suggests a role for astrocytes in blood flow control [23, 24], as well as in neuronal communication [25–28]. Macroscopic imaging is inadequate to uncover this potential role since distinguishing neurons from glia in the generated data from hemodynamic modalities is not possible.

At the microscopic scale, some of these questions can be addressed. Depending on the method used, the recording spatial extent can encompass from a single to a few hundred cells. By imaging cell ensembles, neuronal networks were investigated including the relationship between different cell types [29] or between layers of the cortex [1, 30–32]. Such high-resolution imaging using two-photon (2P) [33] or confocal microscopy [34] is mostly performed with the use of fluorescent dyes. Both techniques offer subcellular resolution and the possibility to acquire on a multidimensional area containing a few neurons. Different fluorescent indicators can be used depending on the application. Vascular architecture can be imaged by injecting a dye in the circulatory system [35–37]; when combined with specialized scanning techniques it can further be used to measure blood flow [38–40]. Oxygen consumption can also be measured with phosphorescent quenching probes [5]. To increase specificity, compounds targeting specific types of cells have been used to identify a single cell or multiple neurons in an area (with multicellular bolus loading (MCBL) injections) [41, 42]. In this case, fluorescence signal yields either anatomical data (e.g., in the case of SR-101 and astrocytes labelling [43]) or cell activity through its intensity variations (e.g., Oregon Green Bapta-1 (OGB-1) and calcium imaging [44, 45]).

Kerr et al. [46] demonstrated the ability of 2P microscopy with a calcium indicator (OGB-1) to map the spatial organization of the rat barrel cortex and to identify specific neural networks. By analyzing the occurrence of spiking activity for each neuron related to stimulation of whiskers, they were able to establish a spatial map of the deflection sensitivity tuning in this cortical area. Using a pair-wise correlation of neuronal temporal response, networks were identified and analyzed in order to establish the columnar organization in layers 2/3 of the rat barrel cortex. Calcium imaging of astrocytes and their function represent another area of intense activity to investigate their role in neurovascular coupling [11, 13, 23, 24]. Specific astrocyte stains have been developed [43], making it possible to outline their role in the neurovascular coupling and in selected clinical applications like AD [47–49].

Further potential of microscopic imaging was demonstrated in another study by Wilson et al. [50] who added an optogenetic [51, 52] component to the 2P microscope in order to activate directly the targeted cells during imaging sessions and to measure the effect of these activations on local networks in the visual cortex. There, it was shown that two different classes of inhibitory cells, parvalbumin (PV) and somatostatin (SOM) expressing neurons, have distinct functions and also belong to different network types. The responses of target cells to directional stimulation were modulated by PV neurons as their influence was proportional to the cell response. On the other hand, SOM inhibitory cells reduced the response equally for all directions (subtraction). This finding led the authors to conclude that SOMs might be useful to sharpen the selectivity of target cells while PVs control the gain on excitatory neurons.

The possibility to use 2P or confocal microscopy to measure oxygen delivery spatially in blood vessels and brain tissue was demonstrated in a recent work by Sakadzic et al. and Yaseen et al. [5, 53]. Using a gated illumination source and a photon counting detector, they were able to measure the spatially varying phosphorescence decay of an oxygen quenching probe to evaluate the local oxygen pressure in vessels (pO_2). With the 2P microscope, Sakadzic et al. were also able to follow this measurement both in vessels and tissue at different depths.

All studies above used microscopy to image the surface of the cortex and reached a maximum depth of a few hundred microns. These limitations are due to the optical properties of tissue (scattering and absorption) combined with the intrinsic design of the optics used in these instruments. Yet, many questions remain unexplored in subcortical areas: functional organization [54, 55], relationship with afferent and efferent connections [56, 57], and the central role these regions may have in brain diseases. While it is possible to work with tissue slices to answer some of these questions, one of the major concerns with ex vivo slices is the absence of cell–vasculature interaction. Transferring slice results in vivo is not trivial. The importance of this interaction is also underlined by new evidence leading to the idea that some of these subcortical regions may have a major influences on blood flow control [58]. Hamel et al. have identified different nuclei that are involved in a vasoactive pathway [58, 59]. This forms the rationale for this chapter: the development of an imaging system that enables to record single cell activity deep in the brain and over a large area.

1.1 Confocal Microendoscopy

Confocal microendoscopy is based on the classic confocal technique. The main idea behind the confocal microscope is to eliminate most of the light originating from structures placed over and under the focal plane under the microscope objective. This is achieved by placing a pinhole (a hole with a diameter of a few tenths of microns) between an imaging lens and the detector (i.e., a photomultiplier tube (PMT)) [60–62] (*see* Fig. 1a). The thickness of the optical section (depth of view) in the sample being imaged is then directly proportional to the diameter of the pinhole used.

Two-dimensional images are created by moving the laser spot on the sample and by sampling the output of the detector at a given frequency. The position of the laser spot is determined by the angle of the laser beam into the objective (Fig. 1c), controlled by galvanometric mirrors (or another scanning device) and an optical telescope between the objective and the laser source. The telescope has two main functions: first, it is used to realign the laser beam with the desired angle towards the objective; second, it enables adjusting the size of the beam in order to set the focal spot

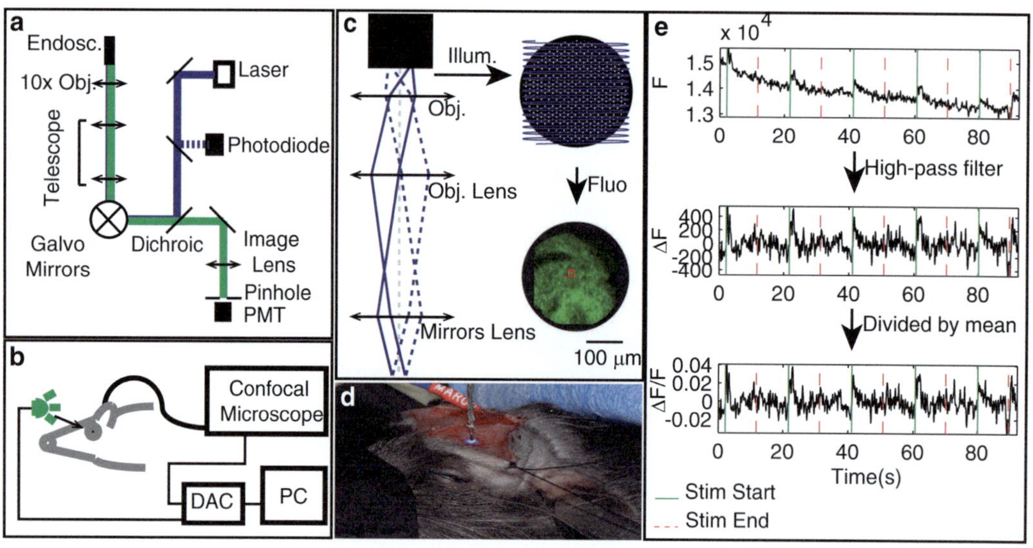

Fig. 1 (**a**) Schematic of the confocal microscope. Light comes from the laser source at 473 nm. A PMT is used to read fluorescent light coming back from the endoscope. Galvo mirrors are used to scan through each fiber of the bundle. (**b**) Example of a typical experimental setup. (**c**) Telescope part of the optical pathway presented in (**a**). The scanning pattern on the optic fiber bundle is shown with the fluorescence return. (**d**) Installation of the fiber on the head of the animal for validation (without cannulation). (**e**) Filtering process of the signal red from pixels (*F*) to recover fluorescence variations ($\Delta F/F$)

diameter. Fluorescence coming back from the sample that will reach the pinhole will come out of the objective lens with the same angle as the excitation beam. A dichroic mirror is placed in the pathway to split the two wavelengths (excitation at 488 nm and emission at 535 nm for OGB-1). From there, the emission light will go through the pinhole and the detector.

To image deep structures of the brain, simple modifications can be done to the confocal microscope: Essentially, by translating the optical plane using a miniature optic fiber bundle (e.g., Sumitomo IGN-035/6 or Mitsubishi Cable Industries Silica Image Guide), the latter can be used as a microendoscope with imaging performed at the distal end. With a classic confocal design, this can be done by placing the tip of the image guide right at the focal plane of the objective [62, 63]. Since the image guide is flexible, its tip can be moved anywhere within its length. By positioning a cannula from the top of the skull down to the desired recording area, the microendoscope can be inserted in the cannula to form images. Injecting the desired fluorescent dye on site and then sliding the endoscope into the cannula, neuronal activity can be recorded (Fig. 1d, e).

2 Equipment, Materials, and Setup

Table 1 provides the list of reagents, materials, and equipment required to realize a typical confocal microendoscopy experiment. Other fluorescent dyes can be used than OGB-1; nevertheless, it is important to make sure that excitation and emission wavelengths are compatible with the filters installed on the confocal microscope. Anesthetic choice may have a strong influence on neuronal responses [64]. According to the region being imaged and the type of stimulation used here, isoflurane was chosen for the experiments presented below. It may not be the best choice in all circumstances.

3 Procedures

All experiments below were performed on rats (adult Long Evans, 250–350 g). All procedures were performed in accordance with the directives of the Canadian Council for the Protection of Animals and the Ethics review board of the Université de Montréal.

Prior to the surgery, the exact position of the targeted area in the brain and the best path to reach the brain structure had to be determined. In planning the cannula implantation surgery, it was important to avoid brain structures that are functionally related to the target to ensure minimum damage to the function that was being investigated. In this regard, it might be better in some cases

Table 1
List of reagents, materials, and equipment required to realize a typical confocal microendoscopy experiment

Material (reagents)
Anesthetic (isoflurane)
Anti-inflammatory (dexamethasone sodium phosphate and carprofen)
Local anesthetic (xylocaine and epinephrine)
Skin disinfectant (iodine solution)
Ethanol (70 %)
Dental cement
Oregon Green Bapta-1 AM (Invitrogen)
Pluronic (Sigma Aldrich)
DMSO (Sigma Aldrich)
Calcium-free standard pipette solution (150 mM NaCl, 2.5 mM KCl, and 10 mM Hepes)
ACSF (124 mM NaCl, 36 mM $NaHCO_3$, 3 mM KCl, 2.5 mM $MgSO_4$, 1.25 mM KH_2PO_4, 1.8 g/L $C_6H_{12}O_6$, 3.4 mM $CaCl_2$)
Atropine 1 % (eyedrops)
Equipment
Standard surgery tools
Micro drill
Cannula (PlasticOne)
Cannula holder for stereotaxic device (Kopf)
Precision Pump (WPI, Ultra-Micro pump 3 and Micro 4 controler)
Syringe for small volume (Hamilton, 10 μL, Model 1701 RN SYR and Neuros Adapter Kit)
Priming kit to fill syringe (Hamilton, Priming Kit)
Image guide (Sumitomo, IGN-035/06 or Mitsubishi, Silicia Image Guide)
Setup
Confocal microscope (custom built to be used with an endoscope)
Stereotaxic system with stereotaxic arm and electrode holder (Kopf)
Visual stimulation system (controllable LED and video projector used with Vpixx software)

to insert the cannula towards the desired area with an angle to avoid other relevant structures. In the results below, the target was the superior colliculus (SC), situated at −5 mm from bregma in the sagittal axis, ±1 mm on the lateral axis (depending on whether stimulation was ipsilateral or contralateral), and 3 mm deep. For implantation, a straight 90° angle was chosen since visual areas are more lateral than the SC and projections from them to the SC are minimal.

Finally, all surgery tools and cannula, as any other component put in direct contact with the animal, were sterilized with alcohol before use.

3.1 Animal Preparation

One hour before anesthesia, the rat was administered dexamethasone (4 mg/kg, IM) and carprofen (Rimadyl, 5 mg/kg, SC) in order to reduce inflammation and to facilitate an efficient recovery. Anesthesia induction was done in a small chamber, ventilated with a mixture of isoflurane (5 %) and oxygen (500 mL/min), for a period of 10 min. After testing reflexes on one hindpaw, the animal was installed on a stereotaxic frame as quickly as possible and equipped with a respiratory mask supplied with the same isoflurane mixture (set to 3 % during surgery). With the animal installed and head fixed on the stereotaxic frame, vital signs monitoring was initiated and maintained according to normal rat physiology: heart rate (320–420 beat per minute), respiratory rate (40–60 breaths/min), and oxygen saturation (over 95 %). Adjustments to anesthesia mixture were done accordingly. Before starting the surgery protocol, a xylocaine and epinephrine mix was injected locally in the surgical area (xylocaine 0.2 % with epinephrine at 0.001 mg/mL, 0.3 mL SC).

As a first step, the head of the rat was shaved and disinfected to avoid any contamination. Then, skin and muscles over the area of interest were removed in order to have an access for cannula implantation. Tissue cutting or removal was minimized in order to limit bleeding and inflammation. With the position well identified on the cranium, a small hole was drilled using the micro-drill (less than 1 mm of diameter) to expose the dura mater.

3.2 Installation of a Cannula

The cannula was cut to the desired length (3 mm for SC) and sterilized before the surgery. It was then slowly put in place using stereotaxic guidance, making sure that it was lowered according to the selected parameters (position of the target and angle of the path towards it). The cannula was then fixed in place by the application of dental cement. Muscles and tissues were put back in place and skin was sutured to close the wound. After this procedure, the animal was transferred to its cage and given 1–2 weeks of rest. During this period, daily monitoring of the recovery was done to make sure that the animal did not suffer from any infections. Also, a topical ointment (polysporin) was applied over the scar and a daily dose of anti-inflammatory was given (carprofen, 5 mg/kg, over a 3 days period)

3.3 Dye Injection

On the day of the experiment, animals were anesthetized again and put back into the stereotaxic device, following the same procedure as in Sect. 3.1. Once positioned, the cap from the cannula was removed. If necessary, the area around the cannula was carefully washed with artificial cerebrospinal fluid (ACSF) in order to remove anything that could eventually obstruct it. If any liquid was seen inside the cannula, ACSF was also used to delicately wash it out.

Dye was prepared ahead of time. The detailed method for the dye preparation (Oregon Green Bapta-1 AM in this case) can be found in the literature [65–67]. A solution of OGB-1, diluted in a 20 % pluronic acid in DMSO, was mixed with a calcium-free HEPES preparation in order to get a 1 mM final concentration. For every injection site, 1 μL of this OGB-1 preparation was needed, according to the protocol used here. Since the dose of dye had to be injected slowly (0.1 μL/min, over a 10 min period), a pump (WPI, Ultra-Micro pump 3 and Micro 4 controller) was used with a precision syringe. This setup was used with either a pulled glass micropipette (50 μm opening) or with a small gauge needle (Hamilton, Neuro-syringe kit with needle gauge 33) with no significant difference in results. The solution was loaded in the injecting needle using a priming kit.

The injection system was then mounted on a stereotaxic arm over the cannula and the needle was aligned with it. It was very important to set the angle on the stereotaxic arm to fit that of the cannula in order to avoid any pressure on the needle during the lowering since it could cause damage to the injection tip or displace the cannula within tissue. Then, the tip of the needle (or micropipette) was lowered a few microns (100–300 μm) past the end cannula in the target area. During this step, any bleeding was monitored. Blood clots at the bottom of the cannula would compromise the ability to get a good image. After the whole 1 μL had been delivered, a rest period of 10 min was necessary to let the pressure in the injector come back at its normal level and avoid any loss of dye during retraction of the needle. Even after this period, removing the needle was done in small steps of a hundred microns, with pauses of 1 min in-between each step. At the end of this process, the animal was put in complete darkness for a whole hour to let OGB-1 diffuse inside the cells.

3.4 Installation of the Image Guide

The optic fiber bundle was installed on a stereotaxic holder (WPI). All manipulations of the image guide had to be done with great care to avoid any scratching or chipping of the tip. In order to have the most stable signal possible, the laser had to be turned on at least 5 min before imaging. The bundle was then lowered down the cannula, until it reached the region to be imaged. During this step, the position of the image guide tip was monitored through the fluorescence images. To avoid any bleaching, the power of the laser was adjusted as required during this step.

3.5 Visual Stimulation and Experimental Protocol

A final step prior to starting the experiment was to put one drop of atropine in the contralateral eye of the rat. This was used to dilate the pupil of the animal to maximize the response of the retina to light. From this moment to the end of the experiment, it was important to keep the room in the dark to insure activity measurements originated from stimulation only.

After a 10–15 min rest period, the visual stimulation protocol was initiated. With a controllable LED system, trains of light pulses (10 ms pulses at 5 Hz during 5 s) were directed towards the eye, interlaced with rest periods of 25 s. Stimulation wavelengths were either green (535 nm) or amber (590 nm). All signals from the confocal microscope and from the stimulation system were synchronized and recorded on a computer using a custom Labview interface (National Instruments; code available upon request). Post-processing analysis was done using homemade Matlab software (Mathworks).

4 Analysis

The confocal microendoscope software was designed to save all the signals needed to reconstruct frame by frame the acquisition and also to be able to synchronize the data with stimulation. Both reconstruction and analysis were done in post-processing using the following signals:

- The position of the two galvanometric mirrors
- The measured laser output power
- Stimulation status (active or rest)
- The photomultiplier tube (PMT) measured output

The acquisition rate of the system had to be high enough in order to have a fluorescence reading for each individual fiber of the endoscope on every frame. For this, the system used resonant galvanometric scanners with a fixed frame rate of 15 fps. Frame dimensions varied from 100 × 100 to 400 × 400 pixels, depending on the number of individual fiber in the fiber guide used as a microendoscope. However, the resolution of each pixel was fixed since individual fibers were always 4 μm of diameter and no additional GRIN lens was used in these experiments. Using a 10,000 elements bundle, and in order to get the desired 15 Hz frame rate, the input sampling rate of the different signals used to rebuild images was set to 1.2 MHz.

As a first step to reconstruct the fluorescence signal, the PMT output was corrected for laser power variations that might occur. These fluctuations could be misinterpreted as neuronal activity if not rectified and the PMT signal was divided by the time-varying averaged laser output signal. The PMT signal was then reshaped into a three-dimensional matrix (X, Y, and time) using mirror positions recorded synchronously with the data. Each single frame was then assembled to form a 3D matrix representing the evolution in time of fluorescence within a 2D slice. Then, a high-pass filter was applied along the time direction on each pixel to remove photobleaching artifacts making the signal decay exponentially over time. Since higher laser powers lead to higher photobleaching, it is best to adjust illumination to minimize this effect. Finally, to get the measure of interest, i.e., calcium-induced fluorescence variations caused by neuronal activity ($\Delta F/F$), the intensity measured on each frame is divided by its average over the full acquisition, pixel per pixel.

5 Typical Results

5.1 Validation of Neuronal Activity Measures

When used with OGB-1, the confocal microendoscope will measure fluorescence signals indicating fluctuating calcium levels in cells under the tip of the endoscope. Comparing stimulation and rest periods, it is possible to quantify the activity induced by a specific stimulus.

In a first experiment, the system was used to detect neuronal activity in the visual cortex of an adult rat. The stimulation protocol used was a train of green flashes going on for a period of 5 s at a frequency of 5 Hz, every 30 s (Fig. 2b). The experiment was divided into three different trials (Fig. 2c): (1) stimulation on contralateral eye, (2) stimulation on ipsilateral eye, (3) stimulation on contralateral eye.

As shown in Fig. 2a, the scanned area presented significant fluorescence indicating large neuronal density which diminished contrast between somas and neuropils. Still, individual neurons can be observed as brighter spots on the 2D image. A region of interest was chosen encompassing one such neuron (red box in Fig. 2a). When computing the mean signal of this ROI averaged over stimulation and resting periods for the three trials (Fig. 2d), the ROI displayed a 5 % increase of fluorescence during stimulations on the contralateral eye as expected.

5.2 Neuronal Activity in the Superior Colliculus of the Rat

As described in the methodology section, the installation of a cannula was mandatory in order to image deep in the brain. Figure 3a, b shows the stereotaxic coordinates for this session determined with the use of a brain atlas. In this example, the position was chosen to avoid the visual areas of the rat (V2 and V1). Once installed, the cannula provided direct access to the brain structure under study (Fig. 3b). Validation of positioning could be performed ex vivo with histology following the experiment (Fig. 3c). The area where the interior part of the cannula was installed into the brain was marked with the red arrow. The yellow arrow showed the SC

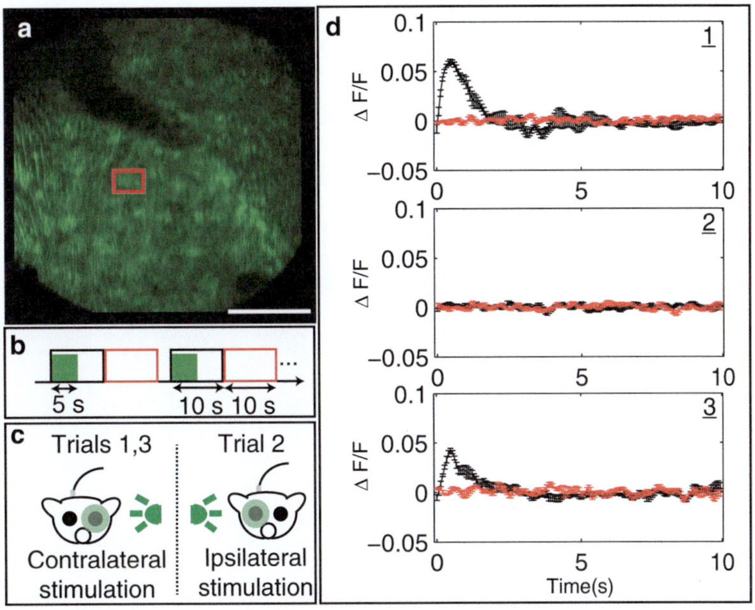

Fig. 2 (**a**) Image from the optic fiber bundle. Scale bar: 100 μm. *Red box* is the region from where the signal in D came from. (**b**) Experimental protocol: 5 s stimulation (light flashes of 10 ms, at 5 Hz), followed by a 25 s pause period. Black and red signals in D represent the mean of $\Delta F/F$ signal taken during the period showed by the *boxes* in **b**. (**c**) First and third trials were made with contralateral stimulation of the eye, second trial was on ipsilateral side. (**d**) Presentation of $\Delta F/F$ signals for the different trials for the stimulation period (*black*) and the poststimulation period (*red*)

(stained with Chicago Blue). Dangling tissue was seen in the path of the cannula due to aspiration by the cannula removal prior to histology.

A typical image of fluorescence coming back from the SC is shown in Fig. 3d. The brighter areas on the image correspond to cells (neurons or glia) and neuronal activity was recovered as described in Sect. 4. A typical example of neuronal activity is provided in Fig. 3e displaying $\Delta F/F$ for the red box area of Fig. 3d.

5.3 Hemodynamic Signals

The hemodynamic response could also be recorded with the confocal microendoscope in this configuration. Since the filter placed just before the detector was a bandpass between 515 and 560 nm, hemodynamic fluctuations through slow signal changes were observed as a modulator of measured fluorescence. Furthermore, based on the spectrum of both oxygenated hemoglobin (HbO) and reduced hemoglobin (HbR) (Fig. 4a), it could be expected that signal changes modulated by hemodynamics should reflect mostly cerebral blood volume with a smaller dependence on HbO/HbR concentration variations.

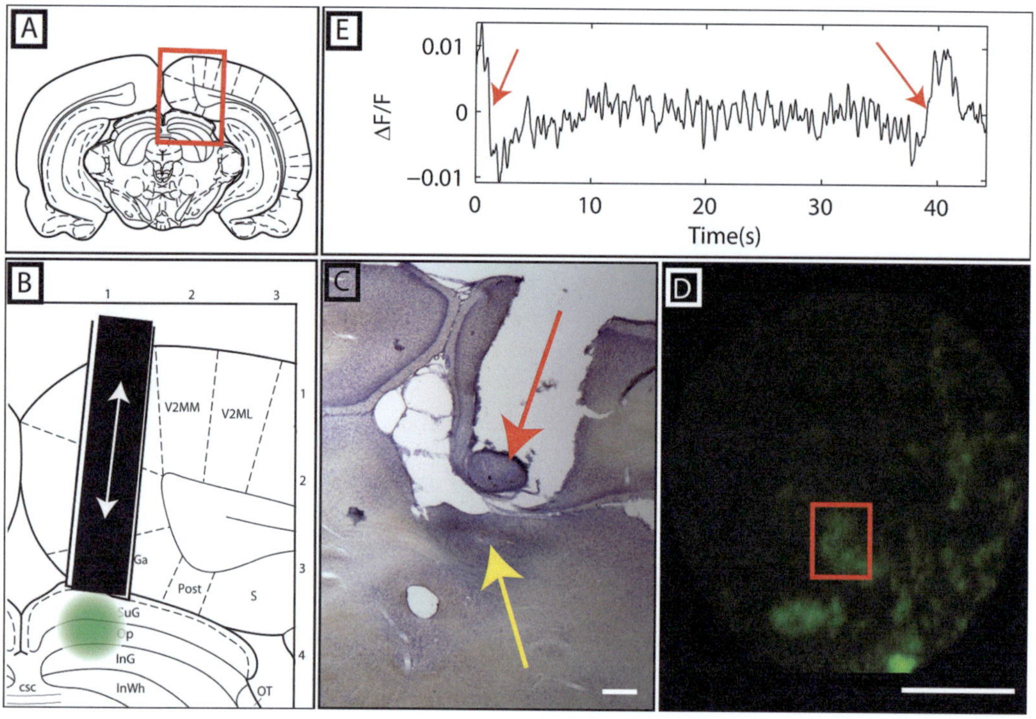

Fig. 3 (**a**) Stereotaxic coordinates from a rat brain atlas [68] (close-up in **b**, with the injection site in *green* and the endoscope represented by the *black rectangle*). (**c**) Histological slice showing the position of the cannula in the brain (*red arrow*). The *yellow arrow* shows the position of the superior colliculus that was stained with Chicago Blue at the end of the experiment. (**d**) Image of the fluorescence from the endoscope. In red, neuron from which the signal showed in (**e**) was taken. Scale bar: 100 μm. (**e**) $\Delta F/F$ signal. *Red arrows* target neuronal activity (10 % increase in fluorescence signal)

The hemodynamic response is much slower than neuronal activity and the peak response is usually found 4–10 s after the stimulus. Using the same stimulation protocol as Sect. 5.1, changes in fluorescence using ROIs located in a cortical area and in blood vessels were investigated. Figure 4b showed that when taking an average over multiple stimulation and rest cycles (30 s cycles over a 10 min experiment for a total of 20 blocks), a small variation in the signal over the blood vessel could be observed (1.5 %, 4–6 s after the end of the stimulation).

In Fig. 4c, a frame by frame sequence showed the difference in the signal for each pixel of the endoscope, showing a stronger decrease for the pixels corresponding to the blood vessel area. It was noted however that signal-to-noise ratio (SNR) over vessels would always be lower because a large amount of the emitted fluorescence was absorbed by the blood.

5.4 Continuous Stimulation

Continuous stimulation paradigms, using a periodic stimulation at a constant frequency, could also be used to investigate neural cell selectivity with calcium imaging. Fourier analysis could then be

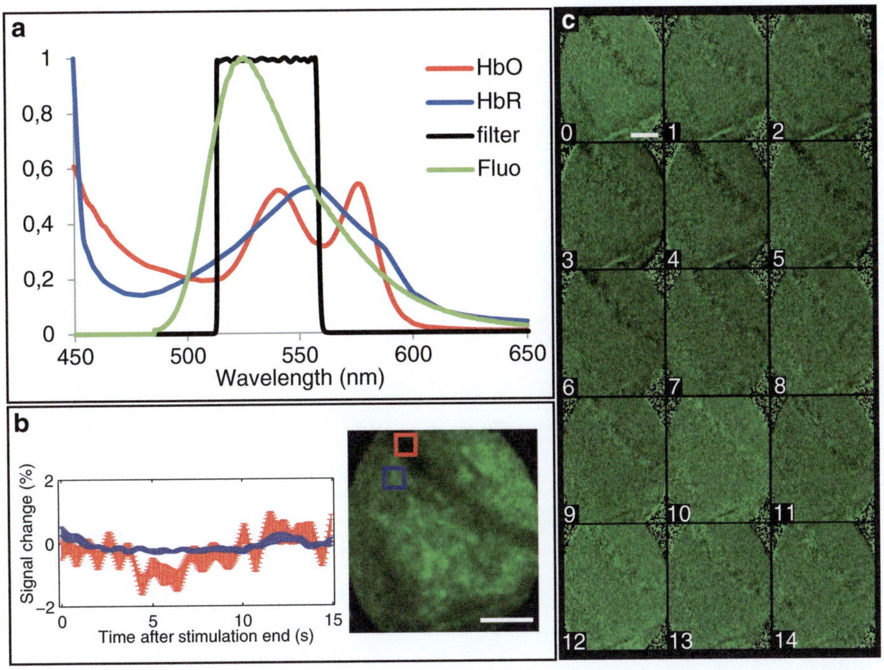

Fig. 4 (**a**) Superposition of the HbO/HbR absorption curves, fluorescence emission for OGB-1 and filter transmission curve. Most of the fluorescence light corresponds to an isobestic area of HbO/HbR. To be noted that the excitation wavelength used here is at 473 nm, which is not an isobestic point. (**b**) Fluorescence signals change over a cortical area (*blue*) and a blood vessel (*red*), in the 15 s after stimulation. Vessel signal shows hemodynamic response. (**c**) Frame by frame representation of the fluorescence variations (between −2 and 2 %). All scale bars are 100 μm

performed to gather organizational maps by looking at the phase of the calcium signal at the stimulation frequency. In Fig. 5, a retinotopic mapping experiment was presented while imaging with the confocal microendoscope. The rat was placed in front of the screen with a bar moving at a frequency of 0.2 Hz (5 s period to move through the screen). For each position of the bar on the screen, a phase was computed (between $-\pi$ and π) (Fig. 5a). Repeating this along two opposite directions of the screen, it was then possible to find the position on the screen associated with the neurons being imaged (called receptor field) (Fig. 5b–d).

6 Troubleshooting

6.1 Fluorescence Intensity Fades Quickly

As explained in Sect. 4, photobleaching is a phenomenon that has to be minimized as much as possible. This phenomenon is caused by the photochemical destruction of the dye molecules and is not reversible once the molecule has changed conformation. When an area is bleached, the only practical way to get signal from this area again is to inject the dye once again. The two principal causes for

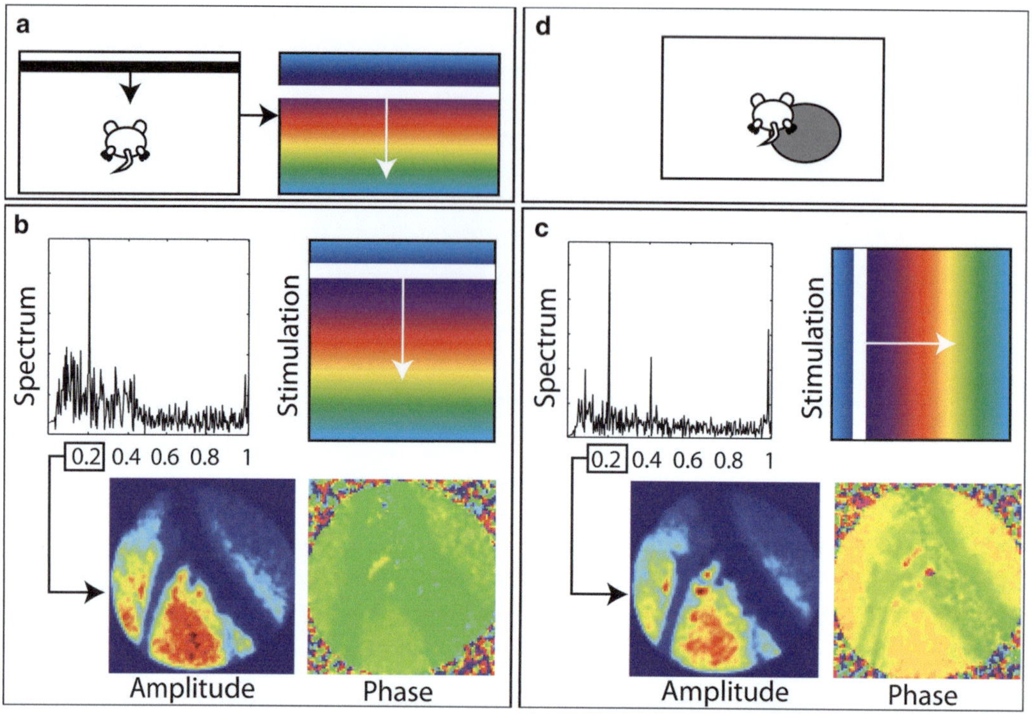

Fig. 5 (**a**) Representation of the experimental setup. The rat is placed on a stereotaxic device, with a screen in front. The bar (usually white on a black background) moves in one direction at a constant frequency. In Fourier analysis, the phase at the stimulation frequency corresponds to a specific position on the screen, for which a color is associated. (**b**) Results for a stimulation going from the top of the screen to the bottom (0.2 Hz). The spectrum is calculated from the mean of an ROI taken over cortical area. A strong response is observed at the 0.2 Hz stimulation frequency. Both amplitude and phase map are presented. From the phase map, it can be seen that most of the neurons imaged are associated with an area in the $-\pi/2$ (bottom quarter of the screen). (**c**) Same presentation as in (**b**), but for a left to right stimulation. The phase map shows that the area related to this cortical region is situated in the 2/3 of the screen ($-2\pi/3$). (**d**) From these two experiments, the receptor field linked to this cortical area is represented by a *gray circle* on the screen

photobleaching are high laser power and poor injections. Laser power: Since the destruction process of the fluorescent dye is proportional to the light level received, it is important to monitor and minimize laser power. Injection: If the amount of dye absorbed by the tissue is too low, it has been observed that the exponential decay of fluorescence level is more rapid.

Possible explanations for bad injections are:

- Error in the preparation of the dye
- Error in the injection pump's settings
- Clotted needle or micropipette
- Needle or micropipette tip not deep enough past the end of the cannula

6.2 Noise Level

Fluctuations of fluorescence during neuronal activation vary from 2 % to around 10 %, depending on stimulation parameters and on the region of the brain being imaged. According to this range, a SNR limit to efficiently measure activity can be estimated to be less than 1 %. Noise sources in a confocal microscope setup are many, but here is a list of the most common points to verify when noise level is too high.

6.2.1 Laser Power

Laser power has to be as low as possible to avoid photobleaching (which will reduce fluorescence level and, consequently, lower the SNR). However, if the power is set too low, the low levels of fluorescence coming back to the PMT will require higher PMT gains. Associated Poisson noise (shot noise) will then limit SNR, typical of low light level application.

6.2.2 Light Interference

Structured noise (always on the same pixels of each frame) that seems to increase (or even saturate) PMT readings is typical of light interference coming from outside the optical pathway. A practical way to identify this is to measure black noise (run the microscope with laser off) and to find the unwanted source of light (most of the time, from a joint around the microscope enclosure). Once found, a way to block this source has to be put in place before doing an experiment.

6.2.3 Cleaning Optics

Noise may also originate from optics that needs to be cleaned. Dry air can be used to remove dust from some optical component. Otherwise, one should follow usual procedures to clean lenses (soft cloth and methanol) and check with lens manufacturers to make sure that the procedure is coating safe.

6.3 Blurry Image

After the installation of the endoscope inside the cannula, the image may appear blurry or out of focus. In this situation, the ability to distinguish neurons from each other is greatly reduced. Also, the resulting $\Delta F/F$ signal might be of lower intensity, making it closer to the noise level.

6.3.1 Position of Tip from Objective

The first thing to check is the position of the tip of the endoscope on the microscope side. The face of the image guide has to be as close as possible to the focal plane of the objective in order to be inside the optical slice of the confocal microscope. A good way to adjust this position is a micrometric stage that moves the endoscope tip towards the optical axis direction.

6.3.2 Cleaning Fiber Bundle

After the few days recovery period following the cannula installation, clots or thick liquid might be located at the bottom. The cannula itself may be cleaned by gently pumping in and out some ACSF. However, some substances stick to the end of the endoscope over time. In this case, a gentle wash with a soft cloth and some ACSF is recommended to remove any particles that could block light. After the experiment, methanol can also be used to clean efficiently. Finally, in the case where the tip of the endoscope is damaged, a fine repolishing should be performed.

6.4 No Neuronal Signal

During the recording session, it may happen that no fluctuations of fluorescence are observable from the brain. Variations in the response quality from experiment to experiment are normal. But in the case where this problem is recurrent, here is a list of what should be investigated.

6.4.1 PMT Saturated

PMT gain can be adjusted to optimize SNR during acquisition. Nevertheless, too high a gain can lead to saturation of the fluorescence signal in which case it will not be possible to measure signal increases. Gain should be set in order to make sure that the highest level of fluorescence at rest is at least 20 % under the maximum of the dynamic range of the PMT. Otherwise, any strong neuronal activation in this area would saturate the output of the device and lead to wrong analysis.

6.4.2 Position in the Brain

Subcortical structures are not always easy to target due to their small size and hidden location. Any error in the installation of the cannula could lead to testing the wrong brain region. A possible method to make sure that the installation procedure is not the cause of the absence of neuronal recording activity is to do a histological study postmortem. Some dyes will not resist to the histology protocol and it may thus be necessary to inject a small amount of a secondary dye to stain the area under the cannula before the end of the experiment.

6.4.3 Damaged Cortex

The resting period after cannula installation surgery is necessary to help reduce inflammation and to provide time for the cortex to heal. However, if the cannula is not fixed properly on the head of the animal, it might move (even slightly) during this period, causing swelling and generating inflammation. The animal may also have a tendency to scratch itself in the area of the surgery since the healing process of the skin can be irritating. There are two ways to control these issues: First, the dental cement base can be solidified by adding some anchor points on the skull (e.g., small screws). Second, a head cone can be installed to protect the head from any scratching.

7 Conclusion

Exploring the neurovascular coupling of deep-seated brain areas remains challenging, but new techniques that will enable unprecedented explorations of these brain structures are emerging. Confocal microendoscopy is a relatively new technique that gives the ability to image neuronal activity in deep areas of the brain. This tool provides new opportunities to better investigate brain mechanisms and the role of these structures in information integration. Further improvements may even lead to the possibility to distinguish between different cell types and to activate (or to inhibit) targeted cells with the use of optogenetics.

References

1. Olsen SR, Bortone DS, Adesnik H, Scanziani M (2012) Gain control by layer six in cortical circuits of vision. Nature 483(7387):47–52
2. Ghose GM, Ohzawa I, Freeman RD, DeAngelis GC (1999) Functional micro-organization of primary visual cortex: receptive field analysis of nearby neurons. J Neurosci 19(10):4046–4064
3. Nelson SB, Le Vay S (1985) Topographic organization of the optic radiation of the cat. J Comp Neurol 240(3):322–330
4. Hirsch S, Reichold J, Schneider M, Székely G, Weber B (2012) Topology and hemodynamics of the cortical cerebrovascular system. J Cereb Blood Flow Metab 32(6):952–67. http://www.ncbi.nlm.nih.gov/pubmed/22472613. Accessed 1 Jun 2012
5. Sakadžić S, Roussakis E, Yaseen MA, Mandeville ET, Srinivasan VJ, Arai K et al (2010) Two-photon high-resolution measurement of partial pressure of oxygen in cerebral vasculature and tissue. Nat Methods 7(9):755–759
6. Chen BR, Bouchard MB, McCaslin AFH, Burgess SA, Hillman EMC (2011) High-speed vascular dynamics of the hemodynamic response. Neuroimage 54(2):1021–1030
7. Srinivasan VJ, Sakadzić S, Gorczynska I, Ruvinskaya S, Wu W, Fujimoto JG et al (2010) Quantitative cerebral blood flow with optical coherence tomography. Opt Express 18(3): 2477–2494
8. Devor A, Hillman EMC, Tian P, Waeber C, Teng IC, Ruvinskaya L et al (2008) Stimulus-induced changes in blood flow and 2-deoxyglucose uptake dissociate in ipsilateral somatosensory cortex. J Neurosci 28(53): 14347–14357
9. Koehler RC, Roman RJ, Harder DR (2009) Astrocytes and the regulation of cerebral blood flow. Trends Neurosci 32(3):160–169
10. Cauli B, Hamel E (2010) Revisiting the role of neurons in neurovascular coupling. Front Neuroenergetics 2:9. http://www.ncbi.nlm.nih.gov/pmc/articles/PMC2899521/. Accessed 10 Sep 2012
11. Carmignoto G, Gómez-Gonzalo M (2010) The contribution of astrocyte signalling to neurovascular coupling. Brain Res Rev 63(1–2):138–148
12. Devonshire IM, Papadakis NG, Port M, Berwick J, Kennerley AJ, Mayhew JEW et al (2012) Neurovascular coupling is brain region-dependent. Neuroimage 59(3):1997–2006
13. Petzold GC, Murthy VN (2011) Role of astrocytes in neurovascular coupling. Neuron 71(5):782–797
14. Blicher JU, Stagg CJ, O'Shea J, Østergaard L, MacIntosh BJ, Johansen-Berg H, et al (2012) Visualization of altered neurovascular coupling in chronic stroke patients using multimodal functional MRI. J Cereb Blood Flow Metab 32(11):2044–54. http://www.nature.com/jcbfm/journal/vaop/ncurrent/full/jcbfm2012105a.html. Accessed 21 Sep 2012
15. Lin WH, Hao Q, Rosengarten B, Leung WH, Wong KS (2011) Impaired neurovascular coupling in ischaemic stroke patients with large or small vessel disease. Eur J Neurol 18(5): 731–736
16. Girouard H, Iadecola C (2006) Neurovascular coupling in the normal brain and in hypertension, stroke, and Alzheimer disease. J Appl Physiol 100(1):328–335
17. Lin AJ, Konecky SD, Rice TB, Green KN, Choi B, Durkin AJ, et al (2012) Towards spatial frequency domain optical imaging of neurovascular coupling in a mouse model of Alzheimer's disease. Proc SPIE 8207:82074U
18. Da Silva N, Szobot CM, Anselmi CE, Jackowski AP, Chi SM, Hoexter MQ et al (2011) Attention deficit/hyperactivity disorder. Clin Nucl Med 36(8):656–660
19. White BR, Bauer AQ, Snyder AZ, Schlaggar BL, Lee J-M, Culver JP (2011) Imaging of functional connectivity in the mouse brain. PLoS One 6(1):e16322
20. Ayling OGS, Harrison TC, Boyd JD, Goroshkov A, Murphy TH (2009) Automated light-based mapping of motor cortex by photoactivation of channelrhodopsin-2 transgenic mice. Nat Methods 6(3):219–224
21. Vanni MP, Provost J, Casanova C, Lesage F (2010) Bimodal modulation and continuous

stimulation in optical imaging to map direction selectivity. Neuroimage 49(2):1416–1431
22. Chemla S, Chavane F (2010) A biophysical cortical column model to study the multi-component origin of the VSDI signal. Neuroimage 53(2):420–438
23. Attwell D, Buchan AM, Charpak S, Lauritzen M, MacVicar BA, Newman EA (2010) Glial and neuronal control of brain blood flow. Nature 468(7321):232–243
24. Haydon PG, Carmignoto G (2006) Astrocyte control of synaptic transmission and neurovascular coupling. Physiol Rev 86(3):1009–1031
25. Henneberger C, Papouin T, Oliet SHR, Rusakov DA (2010) Long-term potentiation depends on release of d-serine from astrocytes. Nature 463(7278):232–236
26. Lee S, Yoon B-E, Berglund K, Oh S-J, Park H, Shin H-S et al (2010) Channel-mediated tonic GABA release from glia. Science 330(6005): 790–796
27. Agulhon C, Fiacco TA, McCarthy KD (2010) Hippocampal short- and long-term plasticity are not modulated by astrocyte Ca2+ signaling. Science 327(5970):1250–1254
28. Shigetomi E, Bowser DN, Sofroniew MV, Khakh BS (2008) Two forms of astrocyte calcium excitability have distinct effects on NMDA receptor-mediated slow inward currents in pyramidal neurons. J Neurosci 28(26): 6659–6663
29. Adesnik H, Bruns W, Taniguchi H, Huang ZJ, Scanziani M (2012) A neural circuit for spatial summation in visual cortex. Nature 490(7419): 226–231
30. Lee S-H, Kwan AC, Zhang S, Phoumthipphavong V, Flannery JG, Masmanidis SC, et al (2012) Activation of specific interneurons improves V1 feature selectivity and visual perception. Nature 488(7411): 379–83. http://www.nature.com/nature/journal/vaop/ncurrent/full/nature11312.html?WT.ec_id=NATURE-20120809. Accessed 9 Aug 2012
31. Kätzel D, Zemelman BV, Buetfering C, Wölfel M, Miesenböck G (2010) The columnar and laminar organization of inhibitory connections to neocortical excitatory cells. Nat Neurosci 14(1):100–107
32. Nauhaus I, Nielsen KJ, Disney AA, Callaway EM (2012) Orthogonal micro-organization of orientation and spatial frequency in primate primary visual cortex. Nat Neurosci 15(12):1683–1690. http://www.nature.com/neuro/journal/vaop/ncurrent/full/nn.3255.html. Accessed 13 Nov 2012
33. Denk W, Delaney KR, Gelperin A, Kleinfeld D, Strowbridge BW, Tank DW et al (1994) Anatomical and functional imaging of neurons using 2-photon laser scanning microscopy. J Neurosci Methods 54(2):151–162
34. Pawley JB (ed) (2006) Handbook of biological confocal microscopy. Springer, Boston. http://www.springerlink.com/content/978-0-387-25921-5/#section=746528&page=1&locus=0. Accessed 24 Aug 2011
35. Zhang Z, Davies K, Prostak J, Fenstermacher J, Chopp M (1999) Quantitation of microvascular plasma perfusion and neuronal microtubule-associated protein in ischemic mouse brain by laser-scanning confocal microscopy. J Cereb Blood Flow Metab 19(1): 68–78
36. Villringer A, Them A, Lindauer U, Einhäupl K, Dirnagl U (1994) Capillary perfusion of the rat brain cortex. An in vivo confocal microscopy study. Circ Res 75(1):55–62
37. Serduc R, Vérant P, Vial J-C, Farion R, Rocas L, Rémy C et al (2006) In vivo two-photon microscopy study of short-term effects of microbeam irradiation on normal mouse brain microvasculature. Int J Radiat Oncol Biol Phys 64(5):1519–1527
38. Schaffer CB, Friedman B, Nishimura N, Schroeder LF, Tsai PS, Ebner FF et al (2006) Two-photon imaging of cortical surface microvessels reveals a robust redistribution in blood flow after vascular occlusion. PLoS Biol 4(2):e22
39. Shih AY, Driscoll JD, Drew PJ, Nishimura N, Schaffer CB, Kleinfeld D (2012) Two-photon microscopy as a tool to study blood flow and neurovascular coupling in the rodent brain. J Cereb Blood Flow Metab 32(7):1277–1309
40. Chaigneau E, Oheim M, Audinat E, Charpak S (2003) Two-photon imaging of capillary blood flow in olfactory bulb glomeruli. Proc Natl Acad Sci U S A 100(22):13081–13086
41. Göbel W, Helmchen F (2007) In vivo calcium imaging of neural network function. Physiology (Bethesda) 22(6):358–365
42. Lütcke H, Helmchen F (2011) Two-photon imaging and analysis of neural network dynamics. Rep Prog Phys 74(8):086602
43. Nimmerjahn A, Kirchhoff F, Kerr JN, Helmchen F (2004) Sulforhodamine 101 as a specific marker of astroglia in the neocortex in vivo. Nat Methods 1(1):31–37
44. Thomas D, Tovey SC, Collins TJ, Bootman MD, Berridge MJ, Lipp P (2000) A comparison of fluorescent Ca2+ indicator properties and their use in measuring elementary and

global Ca2+ signals. Cell Calcium 28(4): 213–223

45. Rochefort NL, Jia H, Konnerth A (2008) Calcium imaging in the living brain: prospects for molecular medicine. Trends Mol Med 14(9):389–399
46. Kerr JND, de Kock CPJ, Greenberg DS, Bruno RM, Sakmann B, Helmchen F (2007) Spatial organization of neuronal population responses in layer 2/3 of rat barrel cortex. J Neurosci 27(48):13316–13328
47. Tian G-F, Takano T, Lin JH-C, Wang X, Bekar L, Nedergaard M (2006) Imaging of cortical astrocytes using 2-photon laser scanning microscopy in the intact mouse brain. Adv Drug Deliv Rev 58(7):773–787
48. Takano T, Han X, Deane R, Zlokovic B, Nedergaard M (2007) Two-photon imaging of astrocytic Ca2+ signaling and the microvasculature in experimental mice models of Alzheimer's disease. Ann N Y Acad Sci 1097: 40–50
49. Riera J, Hatanaka R, Uchida T, Ozaki T, Kawashima R (2011) Quantifying the uncertainty of spontaneous Ca2+ oscillations in astrocytes: particulars of Alzheimer's disease. Biophys J 101(3):554–564
50. Wilson NR, Runyan CA, Wang FL, Sur M (2012) Division and subtraction by distinct cortical inhibitory networks in vivo. Nature 488(7411):343–8. http://www.nature.com/nature/journal/vaop/ncurrent/full/nature11347.html?WT.ec_id=NATURE-20120809. Accessed 9 Aug 2012
51. Boyden ES, Zhang F, Bamberg E, Nagel G, Deisseroth K (2005) Millisecond-timescale, genetically targeted optical control of neural activity. Nat Neurosci 8(9):1263–1268
52. Han X, Boyden ES (2007) Multiple-color optical activation, silencing, and desynchronization of neural activity, with single-spike temporal resolution. PLoS One 2(3):e299
53. Yaseen MA, Srinivasan VJ, Sakadzić S, Wu W, Ruvinskaya S, Vinogradov SA et al (2009) Optical monitoring of oxygen tension in cortical microvessels with confocal microscopy. Opt Express 17(25):22341–22350
54. Huppé-Gourgues F, Bickford ME, Boire D, Ptito M, Casanova C (2006) Distribution, morphology, and synaptic targets of corticothalamic terminals in the cat lateral posterior-pulvinar complex that originate from the posteromedial lateral suprasylvian cortex. J Comp Neurol 497(6):847–863
55. Casanova C (2004) The visual functions of the pulvinar. The visual neurosciences. MIT, Cambridge, pp 592–608
56. Guillery RW, Sherman SM (2002) Thalamic relay functions and their role in corticocortical communication: generalizations from the visual system. Neuron 33(2):163–175
57. Sherman SM, Guillery RW (2011) Distinct functions for direct and transthalamic corticocortical connections. J Neurophysiol 106(3):1068–77. http://jn.physiology.org/content/early/2011/06/10/jn.00429.2011.abstract. Accessed 8 Aug 2011
58. Hamel E (2006) Perivascular nerves and the regulation of cerebrovascular tone. J Appl Physiol 100(3):1059–1064
59. Cauli B, Tong X-K, Rancillac A, Serluca N, Lambolez B, Rossier J et al (2004) Cortical GABA interneurons in neurovascular coupling: relays for subcortical vasoactive pathways. J Neurosci 24(41):8940–8949
60. Nichols AJ, Evans CL (2011) Video-rate scanning confocal microscopy and microendoscopy. J Vis Exp (56). http://www.jove.com/video/3252/video-rate-scanning-confocal-microscopy-and-microendoscopy. Accessed 5 Oct 2012.
61. Pierce M, Yu D, Richards-Kortum R (2011) High-resolution fiber-optic microendoscopy for in situ cellular imaging. J Vis Exp (47). http://www.jove.com/details.php?id=2306. Accessed 9 Aug 2011
62. Vincent P, Maskos U, Charvet I, Bourgeais L, Stoppini L, Leresche N et al (2006) Live imaging of neural structure and function by fibred fluorescence microscopy. EMBO Rep 7(11): 1154–1161
63. Flusberg BA, Cocker ED, Piyawattanametha W, Jung JC, Cheung ELM, Schnitzer MJ (2005) Fiber-optic fluorescence imaging. Nat Methods 2(12):941–950
64. Franceschini MA, Radhakrishnan H, Thakur K, Wu W, Ruvinskaya S, Carp S et al (2010) The effect of different anesthetics on neurovascular coupling. Neuroimage 51(4):1367–1377
65. Garaschuk O, Milos R-I, Konnerth A (2006) Targeted bulk-loading of fluorescent indicators for two-photon brain imaging in vivo. Nat Protoc 1(1):380–386
66. Stosiek C, Garaschuk O, Holthoff K, Konnerth A (2003) In vivo two-photon calcium imaging of neuronal networks. Proc Natl Acad Sci U S A 100(12):7319–7324
67. Sullivan MR, Nimmerjahn A, Sarkisov DV, Helmchen F, Wang SS-H (2005) In vivo calcium imaging of circuit activity in cerebellar cortex. J Neurophysiol 94(2):1636–1644
68. Paxinos G, Watson C (2006) The rat brain in stereotaxic coordinates. Academic Press, Elsevier, London

Chapter 6

Noninvasive Neural Imaging and Tissue Oxygenation in the Visual System

Baowang Li and Ralph D. Freeman

Abstract

Noninvasive functional magnetic resonance imaging (fMRI) has become a primary tool for the measurement of behavior-related neural activity in the human brain. However, the blood oxygen level-dependent (BOLD) signal used in fMRI does not directly measure neural activity. It measures hemodynamic local changes in deoxygenated hemoglobin. To understand the functional significance of these changes, local tissue oxygen concentration may be measured as a way of studying dynamic oxygenation in the brain. Here, we use a dual-purpose sensor to simultaneously measure changes in tissue oxygenation and neural activity in the central visual pathway. We find that this technique can be used reliably in an in vivo section of a functioning visual system. Based on a series of experiments, we have attempted to answer the following questions. First, are there two major components, a small initial dip and a large positive peak, in tissue oxygen response as shown in fMRI and optical imaging studies? If there are two components, is one better coupled with neural activity? If the initial dip in tissue oxygenation is found, we wish to determine if it is unreliable as is the case in fMRI. Second, is tissue oxygen response coupled linearly with neural activity for temporal, spatial, and scaling domains? Third, can neurometabolic coupling be modified by activation of intracortical inhibitory networks? Finally, extracellular neural recordings may be specified in three major categories: single cell, multiple unit activity (MUA), and local field potential (LFP). Is tissue oxygen response better coupled with LFP than the other categories, as shown for fMRI BOLD signals? Our data provide direct evidence regarding the questions above. Results on tissue oxygenation and neurometabolic coupling may be applied to questions concerning human brain mapping with fMRI.

Key words Tissue oxygen, Neural activity, Neurometabolic coupling, Linearity, Excitation, Inhibition, Visual cortex, LGN

1 Background

Noninvasive brain imaging has become a pervasive tool in both basic and applied applications. In functional magnetic resonance imaging (fMRI), blood oxygen level-dependent (BOLD) signals are used to infer neural activity. However, the BOLD signal depends mainly on changes in the relative concentration of oxygenated and deoxygenated blood (see review [1]). Deoxygenated hemoglobin is paramagnetic and influences the local magnetic field

Mingrui Zhao et al. (eds.), *Neurovascular Coupling Methods*, Neuromethods, vol. 88,
DOI 10.1007/978-1-4939-0724-3_6, © Springer Science+Business Media New York 2014

to a greater extent than oxygenated hemoglobin. The presence of deoxygenated hemoglobin increases magnetic susceptibility and reduces the magnetic resonance signal. When neurons in a particular brain region are activated, deoxygenated hemoglobin increases due to oxygen consumption and the BOLD signal is reduced. As local cerebral blood flow (CBF) increases and freshly oxygenated blood is delivered to the active brain region, the concentration of deoxygenated hemoglobin is reduced and the BOLD signal is enhanced [2]. Since the BOLD signal depends primarily on the concentration of deoxygenated hemoglobin, tissue oxygen concentration gives a direct and important estimate of the metabolic state of the local brain region and along with blood flow, is a significant way of arriving at deductions about the state of the neural system.

To consider the implications of oxygen measurements, we first review some factors in the area of neurometabolic coupling. A long lasting dispute concerning cortical metabolism is whether it is aerobic or anerobic. In human positron emission tomography (PET) studies, CBF and cerebral metabolic rate of oxygen consumption ($CMRO_2$) were reported to be coupled when the brain was at rest. However, these two parameters were significantly disassociated, i.e., a 29 % increase in CBF, but only a 5 % rise in $CMRO_2$ were found during somatosensory stimulation [3]. Anerobic glycolysis was assumed to account for this mismatch between CBF and $CMRO_2$ during acute physiological increases in neural activity [4]. However, this assumption of anerobic glycolysis was questioned because substantial $CMRO_2$ responses were observed in later PET and fMRI studies [5–9]. Investigations with optical imaging techniques [10, 11] also showed clear increases in deoxygenated hemoglobin following stimulus presentation. It is possible that the differences in findings could be due to technique limitations in spatial and temporal resolution [12, 13] and uncertainties in data analysis [14, 15]. Because of these factors, the direct measurement of tissue oxygen concentration is a useful approach in the attempts to understand the dynamic changes of oxygenation in the brain.

Local tissue oxygenation in brain tissue can be directly measured with an invasive microelectrode. Clark-style polarographic electrodes have been used successfully for measurements of oxygen partial pressure in bulk solution and tissue oxygen concentration. The polarographic oxygen electrode includes a platinum cathode that is maintained at about −0.6 V and a nonpolarizable anode. The sensor current is proportional to the concentration of oxygen of the solution, and the response time of the electrode is proportional to the size of the exposed platinum surface. The smaller the exposed platinum surface, the smaller the current and the more rapidly the electrode responds to changing oxygen tensions. The field of sensitivity is a sphere of approximately twice the tip diameter of the electrode. This polarographic oxygen electrode was initially developed by Clark and his colleagues to monitor continuously oxygen tension in whole blood [16] and in brain tissue of various animals [17]. Rhythmic fluctuations in oxygen partial pressure were observed in brain tissue in chronically implanted electrodes, and these waves were found to bear no relationship to respiration, heart rate, or fluctuations in arterial or venous pressure [17]. Following the recording of continuous oxygen waves in the brain, multiple electrodes were placed in several other brain areas to measure stimulus-induced tissue oxygenation [18]. Oxygen response was found to spike with stimulus onset and then decrease. This result is not consistent with a mechanism of aerobic metabolism during acute neural activity. However, recent studies which employ direct tissue oxygen measurements show a different pattern of tissue oxygen response. First, there is a small initial negative component followed by a large positive peak [19–29]. Oxygen sensors are now made with a very small cathode and a fine tip. Spatial resolution and response times are such that a rapid localized change in tissue oxygen signal can be reliably detected. This kind of oxygen sensor is suitable for monitoring changes in deoxygenated hemoglobin in brain tissue. It is also possible to combine this measurement with the simultaneous monitoring of neural activity in order to address issues of neurometabolic coupling. This provides a direct assessment of mechanisms that are central to the interpretation of the fMRI BOLD signal.

APOX (Unisense, Denmark) is a dual-purpose sensor that can simultaneously measure co-localized oxygen and neural responses in brain tissue. This double-barrel sensor is a modified Clark-type with an additional platinum electrode extending outside the oxygen permeable membrane as illustrated in Fig. 1. The size of the tip of the APOX sensor is about 30 μm, and the spherical spatial field sensitivity is about 60 μm. The 90 % response time for the oxygen sensor is ~0.5 s. The impedance of the platinum neural electrode is 0.5–2 MΩ. With the APOX sensor, we are able to measure co-localized changes in tissue oxygen and neural activity in the central visual pathway of anesthetized cats. This chapter consists primarily of descriptions of the experimental work we have done using this approach.

2 Aerobic Mechanisms Concerning Acute Physiological Increases in Neural Activity

2.1 Orientation Selectivity

Orientation selectivity is a primary property of neurons in visual cortex [30]. The underlying metabolic activity should also be selective to stimulus orientation. We used the dual-purpose APOX sensor to simultaneously measure changes in tissue oxygenation and single-cell neural activity to visual stimuli at different orientations in striate cortex [26]. Figure 2 illustrates average time courses for both tissue oxygen and neural responses for a representative cell.

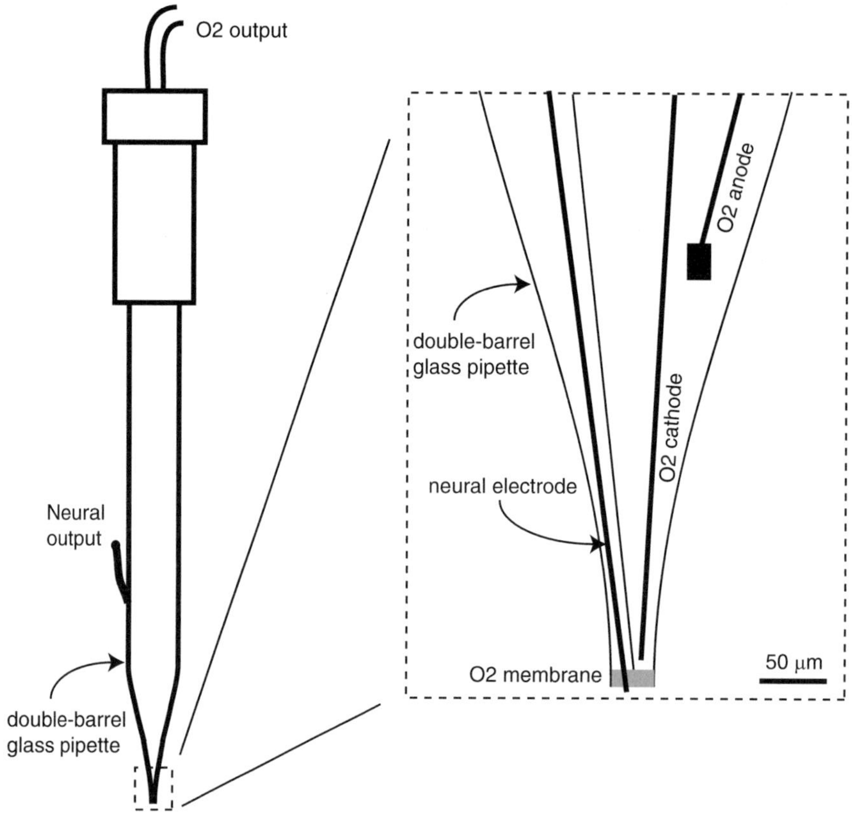

Fig. 1 A schematic plot of the APOX sensor (Unisense, Denmark). A neural microelectrode and an oxygen sensor are enclosed in a double-barrel glass pipette. The tip of the neural electrode extends through the oxygen permeable membrane into the tissue. Both the neural electrode and oxygen cathode are made of platinum. The oxygen anode is a silver–silver chloride reference electrode

Data are averaged across 72 trials. The solid curves represent tissue oxygen responses (dashed lines, ±SEM) and the gray solid areas stand for spike-time histograms. The time course of the oxygen response exhibits an initial dip followed by a positive peak. The amplitudes of the initial dip and positive peak vary with stimulus orientation. For neural response, this cell fired vigorously to visual stimuli at orientations of 270° and 285° but remained silent to those at orientations of 180°, 235°, and 305°, indicating a strong orientation selectivity of neural activity. Optimal orientation conditions, which elicit large neural responses, give rise to largest initial dips and smallest positive peaks.

To further estimate the relationship between oxygen and neural responses, we obtain correlation coefficients from average spike rates and oxygen changes as a function of time. The contour plot (Fig. 3a) shows oxygen response time courses as a function of spike rate for the example cell shown in Fig. 2. Solid and dashed contour plots represent positive and negative changes in

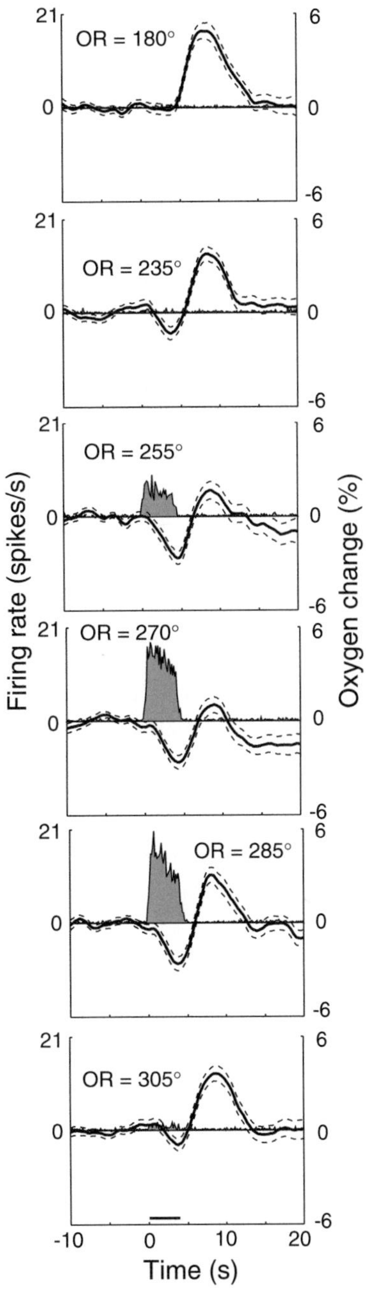

Fig. 2 Orientation selectivity of neural and tissue oxygen responses for a representative cell in area 17. Data are averaged responses across 72 trials. Oxygen responses are quantified as the percentage change from the baseline signals. *Dashed lines* represent ±1 SE of the mean

oxygen response, respectively. High spiking rates of neural response generally correspond to large negative and small positive oxygen responses. In contrast, low firing rates of neural response tend to be associated with small dips and large peaks of oxygen responses.

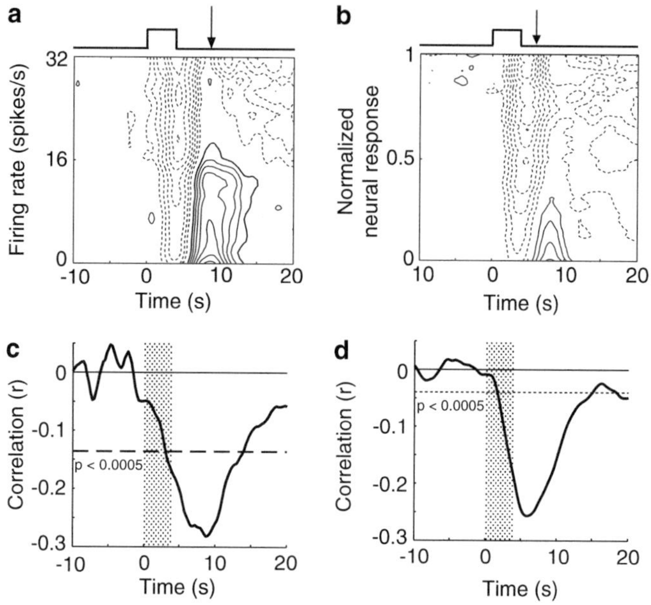

Fig. 3 Correlation between tissue oxygen and neural responses. (**a**) A contour plot of tissue oxygen response, neural firing rate, and time for the example cell in Fig. 2. The line above the contour plot shows stimulus onset and duration, and the *arrow* indicates the maximum correlation time (8.75 s). (**b**) A contour plot for averaged response across a population of 14 cells in area 17. Symbols are identical to (**a**). (**c**, **d**) Correlation coefficients derived from correlating neural and oxygen responses as a function of time for the example cell in (**a**) is shown in (**c**) and a similar fit in (**d**) is from population data in (**b**). Correlation coefficients below the *dotted line* are significantly less than 0 ($p < 0.005$)

The correlation coefficient between spiking activity and oxygen response varies over time (Fig. 3c). Oxygen responses are significantly ($p < 0.0005$) correlated with neural activity between 3.25 and 14.0 s, with the maximum correlation at 8.75 s after the stimulus onset. Consistent with the example cell, data from a group of cells also indicate similar correlations between neural and tissue oxygen responses (Fig. 3b). Significant correlations for the population data begin 1.75 s after the stimulus onset (Fig. 3d). The maximum correlation occurs at 5.75 s after the stimulus onset. These results suggest that a dip in the oxygen response begins early, as a result of increased oxygen consumption by activated cells while a peak in the oxygen response occurs later, presumably as a result of increased blood flow to the activated area of the cortex.

We obtain orientation tuning curves based on average neural and tissue oxygen responses to visual stimuli at different orientations. The average oxygen changes occurring 5.75 s after stimulus onset are chosen to plot the oxygen orientation tuning curve because this time generally gives the best correlation between spike rate and tissue oxygenation for the group data as shown in Fig. 3d.

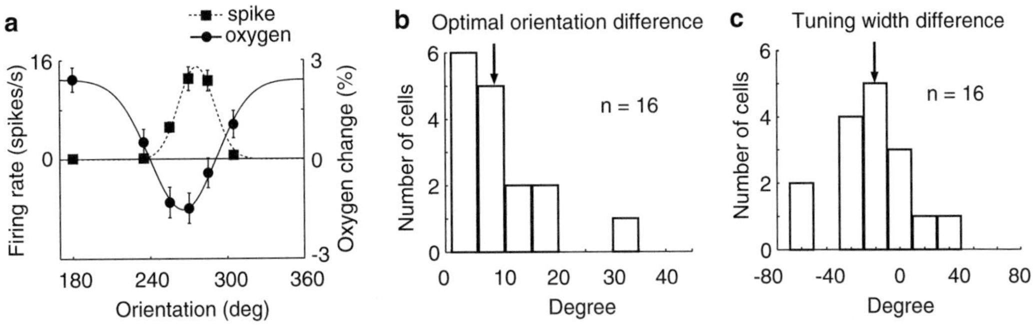

Fig. 4 Comparison of orientation tuning properties between neural and oxygen responses for the example cell shown in Fig. 2 plus other neurons. (**a**) Orientation tuning curves of neural and tissue oxygen responses. The oxygen response at 5.75 s after stimulus onset is determined for each orientation condition. Neural responses are quantified as the average spiking rates different stimulus conditions. (**b**, **c**) Comparison of orientation selectivity between neural and oxygen responses for a population of cells. Distributions of the differences in preferred orientation (**b**) and tuning width (**c**) between neural and oxygen responses

Oxygen and neural orientation tuning curves for the above example cell is shown in Fig. 4a. The oxygen tuning curve (filled circles) is inverted compared to the neural tuning curve (filled squares) because negative changes in the oxygen response corresponded to increases in neural spiking rate, but these two tuning curves give similar estimates of orientation selectivity. Results from our group data ($n = 16$) are consistent with the example cell. The differences between optimal orientations for neural and tissue oxygen responses are plotted in Fig. 4b. For the majority of cells, the optimal orientation differences between neural and oxygen responses are less than 15°. The tuning width differences between these two measurements are shown in Fig. 4c. Oxygen orientation tuning curves tend to be wider than those for neural responses. This difference is significant and suggests that decreases in tissue oxygenation reflect a population of active cells that have a collective tuning width broader than those for individually recorded neurons.

2.2 Ocular Dominance

The visual stream from left and right eyes is first physiologically combined in striate cortex. At this stage, information from both eyes is encoded and transmitted. Based on different degrees of activation from each eye, neurons in striate cortex have been subjectively categorized into varying degrees of ocular dominance [30]. In the original classification system, seven groups were used. Neurons driven only by stimulation from the contralateral eye are defined as group 1 cells while those driven entirely by the ipsilateral eye are termed group 7. Cells with the same degree of eye preference are considered group 4. Groups 2 or 3 and 5 or 6 are intermediate categories [30]. To measure the influence of each of the two eyes in the tissue oxygen and neural responses, optimally oriented stimuli are presented separately to each eye [26].

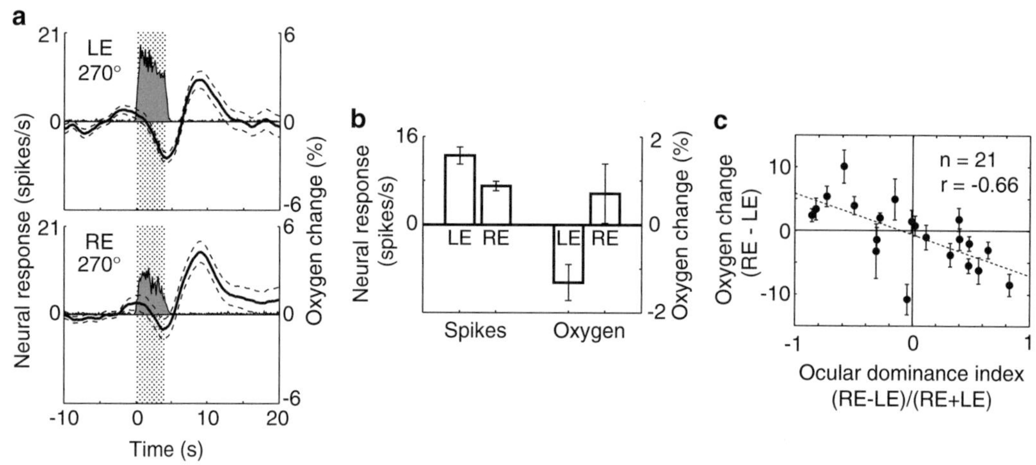

Fig. 5 Comparison of ocular dominance between neural and oxygen responses. Data were obtained from the example cell shown in Fig. 2. Responses are shown for left eye (**a**, *top*) and right eye (**a**, *bottom*) stimulation. (**b**) Bar plots compare the average neural and oxygen responses (5.75 s after the stimulus onset) between the two eyes. (**c**) The difference in tissue oxygen change between the two eyes is plotted for a population of neurons as a function of the ocular dominance index for each cell

Data shown in Fig. 5a illustrate tissue oxygen and neural responses to stimuli in the left and right eyes for the same example cell as shown in Fig. 2. This cell is left eye dominant and exhibits a larger initial dip in response to left eye stimulation (Fig. 5a). Oxygen and neural measurements of ocular dominance exhibit a similar relationship to those for orientation tuning. As the histogram plot shows in Fig. 5b, the stimulus in the left eye activates stronger spiking activity and larger negative oxygen response (5.75 s after stimulus onset). To quantify the relationship between the changes in oxygen and neural responses to the two eyes, the difference in percent oxygen change between the two eyes is plotted as a function of the ocular dominance index for each neuron of our population ($n = 21$). There is a clear inverse relationship between the two measurements (Fig. 5c). This indicates a larger negative percentage of oxygen change in the stronger eye. These data are clearly supportive of aerobic mechanisms for acute neural activation. Moreover, our results suggest that the effect of oxygen consumption is localized within a cortical column, which is consistent with previous findings with optical imaging and fMRI techniques [11, 31, 32].

3 Separate Spatial Scales of the Initial Dip and Positive Peak of Tissue Oxygen Response

As we show above, tissue oxygen response typically consists of two components, a small initial dip followed by a large positive peak. This two-phase response pattern may reflect an initial oxygen dominated mechanism followed by activity-dependent changes in CBF.

An alternative possibility is that CBF and $CMRO_2$ change on similar spatial scales, but there is a delay in the CBF response due to draining veins. To evaluate these two hypotheses, we examine the spatial scales of the two components of tissue oxygen response in the lateral geniculate nucleus (LGN) [27, 28]. The LGN is a visual area of the thalamus with precise retinotopic organization and small receptive fields (RFs) [33]. This allows neural responses to be controlled with greater spatial precision than what is possible in the visual cortex.

Figure 6 illustrates neural and oxygen responses from a representative recording site in the LGN. Visual stimuli comprise drifting sinusoidal gratings with different patterns and sizes.

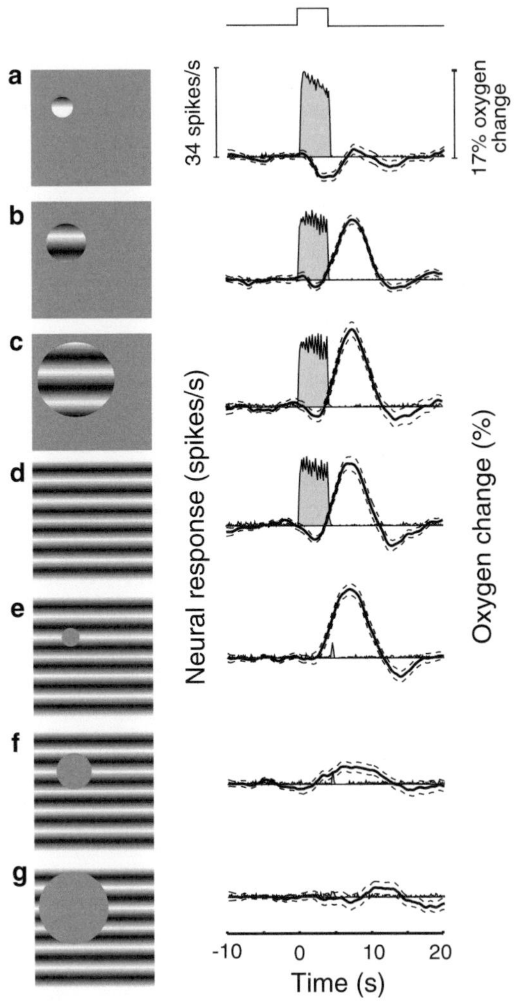

Fig. 6 Tissue oxygen and neural responses in the lateral geniculate nucleus to visual stimuli of different spatial patterns. (**a–c**) Circular stimulus patches of diameters 2° (**a**), 5° (**b**), and 12.7° (**c**). (**d**) A full-field stimulus. (**e–g**) Full-field stimuli masked by blank circular masks of different diameters 2° (**a**), 5° (**b**), and 12.7° (**c**). *Dashed lines* represent ±1 SEM

The first four stimuli (A, B, C, D) are 2°, 5°, 12.7°, and full-screen. Because of the precise retinotopic organization in the LGN, the spatial extent of neural activity is directly related to the stimulus size. Neural firing responses are measured here as multiple unit activity (MUA). The smallest stimulus elicits strongest neural activity in a relatively small population of neurons, and there is a monophasic negative oxygen response which presumably reflects an increase in $CMRO_2$. For the other three stimuli of larger sizes that activate bigger populations of neurons, the amplitudes of the negative oxygen responses are reduced while large positive peaks occur after the initial negative components. The last three stimuli (E, F, G) comprise full-field gratings with blank masks centered over the RFs. The sizes of the blank masks are 2°, 5°, and 12.7°. These stimuli activate neural activity over large populations of neurons. However, neurons that surround the tip of the oxygen sensor are excluded from the stimulation. It is clear that no significant neural responses are observed to these three mask stimuli. The full-field stimulus with the 2° mask elicits a positive oxygen response without an initial negative dip, which suggests that oxygen consumption are highly specific to local changes in neural activity. Compared to the smallest mask stimulus, the two larger mask stimuli elicit much smaller positive oxygen responses, suggesting that activity-induced changes in CBF are also co-localized with neural activity, but at a spatial scale larger than that for neural activity which is detected by the microelectrode.

To examine the spatial specificity of the negative oxygen response, the position of the small-size (2° in diameter) stimulus is moved away from the RF center. Six different locations are included in this experiment as shown in Fig. 7. The first stimulus is centered over the RF of the example recording site in the LGN (Fig. 7a). The other five stimuli are positioned progressively farther away from the center of the RF (Fig. 7b–f). As the stimulus moves farther away from the RF, the initial dip of oxygen response gets smaller. This result confirms the observation that activity-induced oxygen consumption is closely localized to the site of neural activity.

To quantify the spatial scales of the negative and positive components of tissue oxygen response, we compare the full-width at half-maximum (FWHM) of the point spread function (PSF) for each component response as illustrated in the histograms of Fig. 8. The PSFs for the negative and positive components are obtained by deconvolution of tissue oxygen response and neural activity to focal and nonfocal LGN activation, respectively. The distributions of widths for the negative dips and positive responses are shown in Fig. 8a, b, respectively. Mean widths of the PSFs for the two component oxygen responses are 0.15 ± 0.08 and 1.35 ± 0.6 mm, respectively. It is clear that the PSF of the negative response is substantially narrower than that for the positive component.

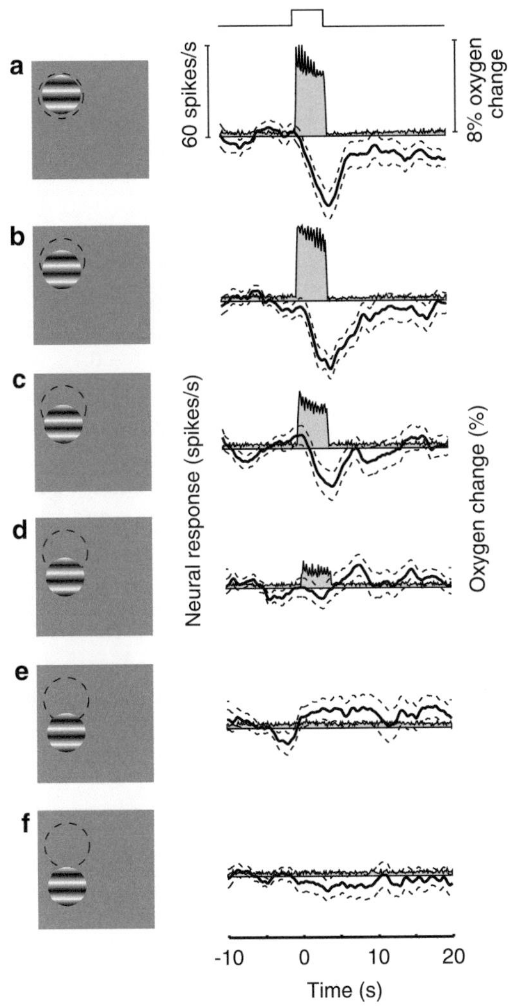

Fig. 7 Position tuning of tissue oxygen and neural responses. The line above the responses indicates stimulus onset and duration. The *dashed circles* indicate the size and position of the RF. Data are averaged across 56 trials. *Dashed lines* represent ±1 SEM. The stimulus size is 1° in diameter. The stimulus position is varied in steps of 0.3°

These data provide direct evidence concerning the assumptions in previous optical imaging and fMRI studies that most spatially specific components of hemodynamic responses are related to early increases in $CMRO_2$ [11, 31, 34]. In addition, our results suggest that the spatial resolution of neuroimaging techniques based on the initial dip may be on the order of a few hundred micrometers, whereas those based on CBF may be limited to a few millimeters. It is necessary to note here that initial dip data in fMRI work is generally weak, variable, and unreliable, or completely absent. However, in our applications, it is usually robust, reliable, and consistent.

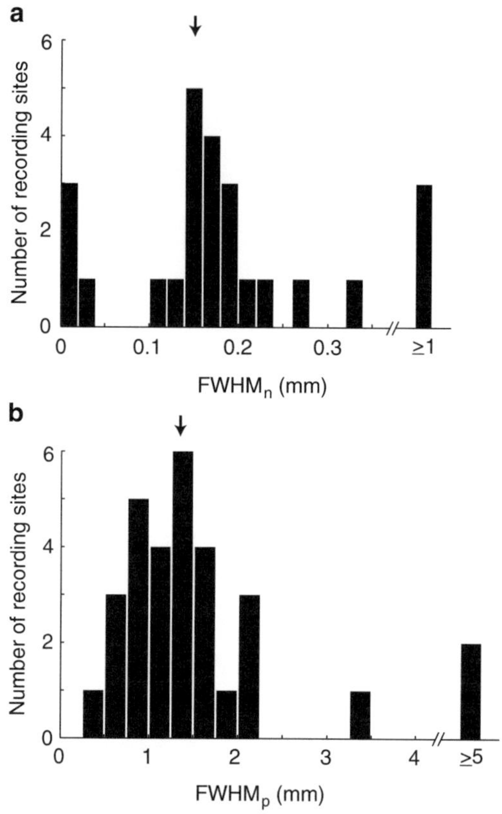

Fig. 8 Comparison of estimated spatial scales of the two components of tissue oxygen responses. Bar plots of the full-width at half-maximum (FWHM) of the point spread function (PSF) are shown for the initial dip (**a**) and the positive peak (**b**) of oxygen response

4 Degree of Linearity of Spatial and Temporal Summation of Tissue Oxygen Signal

In fMRI, BOLD signals are used to infer neural activity with the assumption that the fMRI signal is approximately proportional to local neural activity, averaged over spatial and temporal extents [35, 36]. We have shown that the negative oxygen response is coupled with neural activity and is closely localized to the site of neural activity. A question remains, i.e., is tissue oxygen response linear in spatial and temporal integration? By changing the stimulus size and pattern to control neural response in the LGN, we have observed monophasic negative and positive oxygen responses, which presumably reflect activity-dependent increases in $CMRO_2$ and CBF, respectively. In the following experiments, we evaluate the extent of temporal and spatial linearity of these two component oxygen responses and their coupling with neural activity generated by visual stimuli of different durations and spatial patterns in the LGN.

4.1 Temporal Integration

Two stimulus patterns (small and large) and four stimulus durations (2, 4, 8, and 16 s) are used in the experiment [21]. Figure 9 illustrates average tissue oxygen and neural responses to visual stimuli for a population of recording sites. The small stimulus is a drifting grating centered over the RF of the recording site (Fig. 9a). This stimulus is generally 1–3°, as quantitatively determined by a size tuning function individually for each recording site. Vigorous neural activity and monophasic negative oxygen responses are shown for all four stimulus conditions. The amplitude and width of the negative oxygen responses increase monotonically with stimulus duration. The second stimulus pattern is a full-field drifting grating with a blank mask over the RF (Fig. 9b). Because this stimulus only activates the surround of the RF, minimal neural activity is elicited for all stimuli at four durations. In contrast to the monophasic negative oxygen responses to the small stimulus, the large mask stimulus elicits monophasic positive oxygen responses. Similar to the negative oxygen response, the amplitude and width of the positive oxygen response also increases as the stimulus duration gets longer.

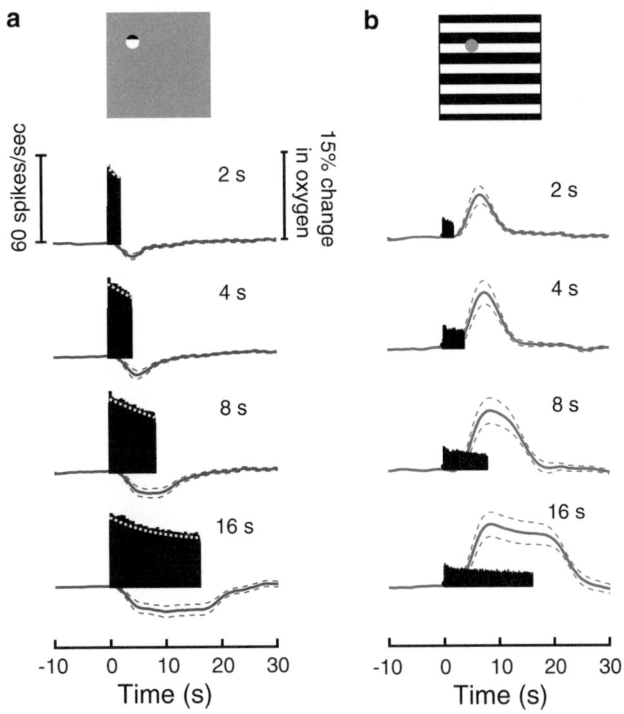

Fig. 9 Neural and oxygen responses to visual stimuli of different durations (2, 4, 8, and 16 s). (**a**) A small stimulus is centered over the RF of the recording site in the LGN. (**b**) A full-field stimulus with a blank mask centered over the RF. *Dashed lines* represent ±1 SEM

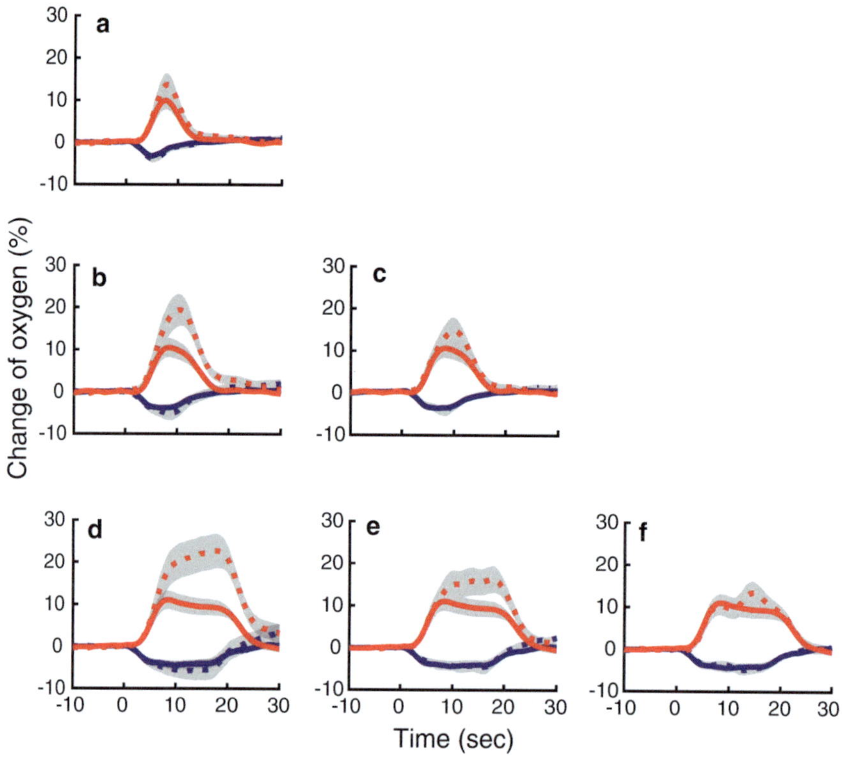

Fig. 10 Estimation of temporal linearity of tissue oxygen response. Predicted responses to long-duration stimuli are generated by replicating, shifting, and summing measured responses to short-duration stimuli. *Blue* and *red* lines indicate responses to the small and large stimuli, respectively, with a mask. (**a**) Measured (*solid lines*) and predicted (*dashed lines*) oxygen responses to 4 s stimuli based on 2 s stimuli. (**b**, **c**) Measured and predicted responses to 8 s stimuli based on 2 s (**b**) and 4 s (**c**) stimuli. (**d–f**) Measured and predicted responses to 16 s stimuli based on 2 s (**d**), 4 s (**e**), and 8 s (**f**) stimuli

To evaluate the temporal linearity of the negative and positive oxygen components, responses to long-duration stimuli are compared with those predicted by summing shifted replicas of the responses to short-duration stimuli to match the durations of longer stimuli. For example, responses to the 4 s stimuli are predicted by summing responses to the 2 s stimuli with 2 s right-shifted responses (Fig. 10a, dashed lines). Blue and red curves represent oxygen responses to the small and large stimuli, respectively. Likewise, responses to the 8 s stimuli are predicted based on those to 2 s (Fig. 10b) and 4 s (Fig. 10c) while responses to the 16 s stimuli are predicted from responses to 2 s (Fig. 10d), 4 s (Fig. 10e), and 8 s (Fig. 10f) stimuli. Measured responses are represented by solid lines. For the negative oxygen response, the predicted and measured responses are approximately consistent with a linear prediction. The approximately linear prediction of the negative oxygen response suggests that metabolic response is likely to be linear in temporal summation. However, this is not the case for the positive oxygen response. Responses to long-duration stimuli tend to linearly predict the responses to longer-duration stimuli, whereas predicted longer-duration responses based on those to short-duration stimuli are generally much larger than those measured. The overestimation of the positive oxygen response is consistent with nonlinear temporal summation of the BOLD signal in previous fMRI studies [37–42]. Because the positive oxygen response presumably reflects activity induced by an increase in CBF, this result suggests that the nonlinearity of the BOLD signal might be due to nonlinear neurovascular coupling.

4.2 Spatial Integration

We have utilized the observation that there is both a classical and nonclassical RF in the surround to evaluate spatial linearity of tissue oxygen response and its coupling with neural activity [24]. The surround is a region outside the classical RF, which has influence on the neural response to a visual stimulus within the classical RF but cannot activate a neuron directly [43]. We have differentially activated these two regions for neurons in the LGN using specifically tailored drifting gratings as shown in Fig. 11. Five stimulus patterns are used in the experiment: a left semicircle (Fig. 11a), a right semicircle (Fig. 11b), a full circle (Fig. 11c), a full-field stimulus with a blank mask over the RF (Fig. 11d), and a full-field stimulus (Fig. 11e). The first three stimuli in Fig. 11a–c have the same diameter, which equals the size of the classical RF that is determined quantitatively by a size tuning function. The summation of the two stimuli in Fig. 11a, b matches that of c, whereas the summation of the stimuli in Fig. 11c, d is equal to that of e. The black curves represent measured tissue oxygen (left column) and neural (right column) responses to these visual stimuli.

We examine the linearity of spatial summation within the classical RF by comparing the measured responses (black curves, Fig. 11c) to the full circle stimulus with the composite responses (gray curves, Fig. 11c) to the two semicircle stimuli individually. The composite oxygen response exhibits a significantly larger negative oxygen response than the measured one (Fig. 11c, left column, 4.1 ± 0.5 % and 2.9 ± 0.4 %, $p = 0.2$, Wilcoxon signed rank test); and the composite neural activity is also significantly stronger than the measured (Fig. 11c, right column, 52.9 ± 6.1 and 42.6 ± 3.9 spikes/s, $p < 0.01$). Similarly, for spatial summation across the classical RF and its surround, we define responses to the full-field as the measured (black curves, Fig. 11e), and the sums of responses to the full-circle stimulus (Fig. 11c) and the full-field stimulus with a blank mask (Fig. 11d) as the composite responses (gray curves, Fig. 11e). The composite oxygen response shows a slightly larger amplitude of the initial dip than that measured (1.9 ± 0.2 and 1.3 ± 0.1 %, $p = 0.04$) but positive peaks are similar (4.8 ± 0.3 and 4.6 ± 0.4 %, $p = 0.5$) (Fig. 11e, left column). The composite neural response is significantly stronger than that measured (43.2 ± 3.8 and 26.6 ± 2.2 spikes/s, $p < 0.01$) (Fig. 11e, right column).

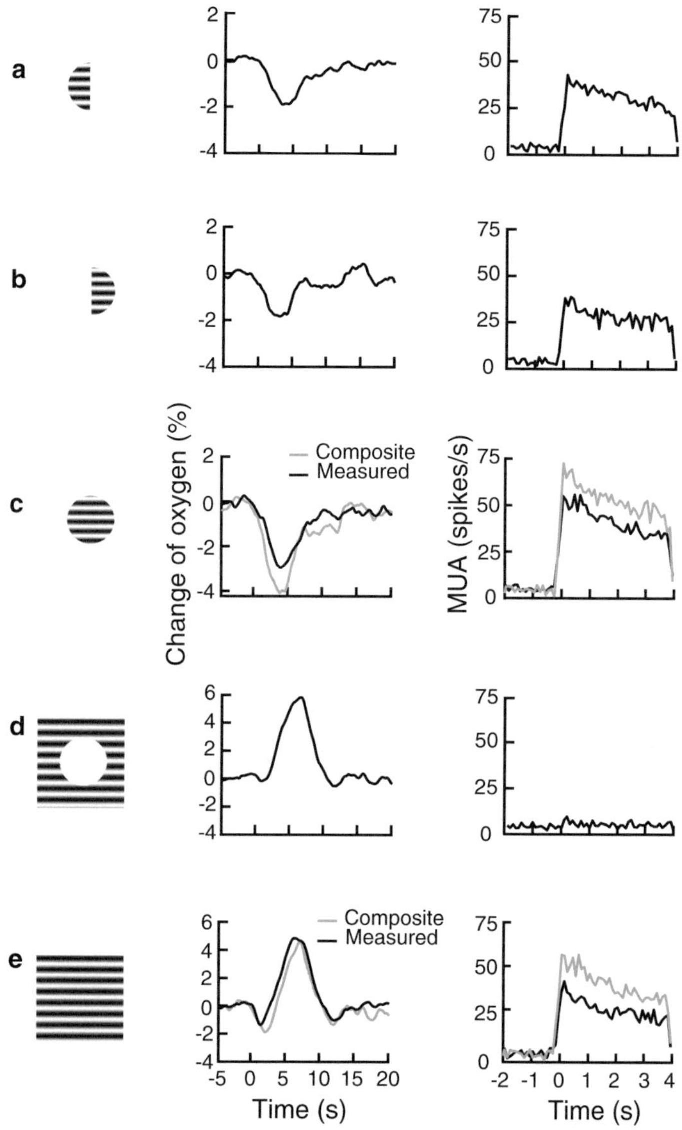

Fig. 11 Tissue oxygen and neural responses to visual stimuli of different spatial patterns for a recording site in the LGN. A left semicircular stimulus (**a**), a right semicircular stimulus (**b**), and a full circular stimulus (**c**) are centered on the RF of a recording site. The size of the stimulus is determined as the optimal size of the RF, and the sum of the stimuli in (**a**) and (**b**) is equal to the stimulus in (**c**). Oxygen and neural responses are shown in the left and right columns, respectively. A full-field stimulus with a blank mask covering the RF (**d**) is used to stimulate the surround of the RF. The summation of the stimuli in (**c**) and (**d**) matches the full-field stimulus (**e**). Data are averaged over 32 trials. *Black* and *gray curves* in (**c**) and (**e**) represent measured and composite oxygen and neural responses to the full circular stimulus (**c**) and the full-field stimulus (**e**), respectively

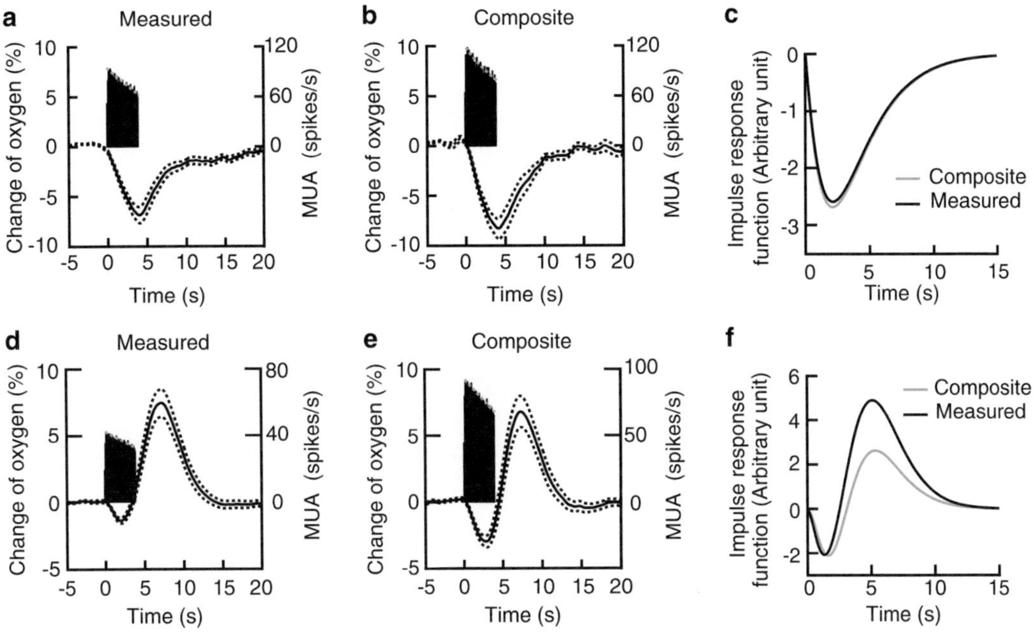

Fig. 12 Linearity of spatial summation of neurometabolic coupling to visual stimuli in the LGN. (**a–c**) Estimation of spatial linearity within the classical RF. Measured and composite responses to the full circular stimulus across the population of recording sites are shown in (**a**) and (**b**), respectively. Measured and composite impulse functions are shown in *black* and *gray curves* in (**c**), respectively. A gamma function is used to fit each impulse response function. (**d**, **e**) Estimation of spatial linearity exceeding the classical RF. Measured and composite responses to the full-field stimulus are shown in (**d**) and (**e**), respectively. Measured and composite impulse response functions are shown in *black* and *gray curves* in (**f**), respectively

Results from our population data confirm the nonlinearity of spatial summation. Figure 12a, b illustrates average measured and composite oxygen and neural responses within the classical RF for a population of recording sites in the LGN. The composite oxygen and neural responses are significantly larger than those measured. To examine if the nonlinearity of oxygen response is due to the nonlinearity of neural activity, we compare impulse response functions which are deconvolved from the average measured and composite oxygen responses and corresponding neural activity. As shown in Fig. 12c, the measured and composite impulse response functions are nearly overlapped ($R^2 = 1.0$), indicating a neural basis of the nonlinear spatial summation of tissue oxygen response within the classical RF. Similarly, for spatial summation across the classical RF and its surround, the population average composite oxygen response (Fig. 12e) tends to show a larger initial dip but a similar positive peak compared to that measured (Fig. 12d). The composite spiking activity (Fig. 12e) is also significantly stronger than that measured (Fig. 12d) for the population data. The measured impulse response function deviates substantially from the composite (Fig. 12f), indicating a nonlinear coupling between tissue oxygen response and neural activity in spatial summation. Note here that the initial dips of the two impulse response functions match closely in amplitude, indicating the nonlinear spatial summation of the initial negative oxygen response might be accounted for by the nonlinearity of neural response when the stimulus size extends beyond the classical RF. Taken together, our results suggest that the initial dip of tissue oxygen response and neural activity in a co-localized brain region are closely coupled in both temporal and spatial domains. The spatial nonlinearity of the BOLD signal reported in previous fMRI studies [44–47] may be due to nonlinear neurovascular coupling.

5 Homogeneity of Tissue Oxygen Signal

A linear system is defined as one for which the input and output are scaled by the same amount. To determine if tissue oxygen changes linearly in amplitude with stimulus-evoked neural activity, we can vary contrast levels of a visual stimulus and measure oxygen responses and neural activity in the LGN [21]. The two stimulus patterns shown at the top of Fig. 9 are used again in this test: a small stimulus that matches the size of the RF site and a full-field with a blank mask over the classical RF. Figure 13 illustrates population-averaged amplitudes of oxygen response and neural activity to the two stimuli at different contrast levels. The blue and red dots represent oxygen responses to the small stimulus and large mask stimulus, respectively (Fig. 13a). Blue and red curves in Fig. 13a represent contrast tuning curves for negative and positive

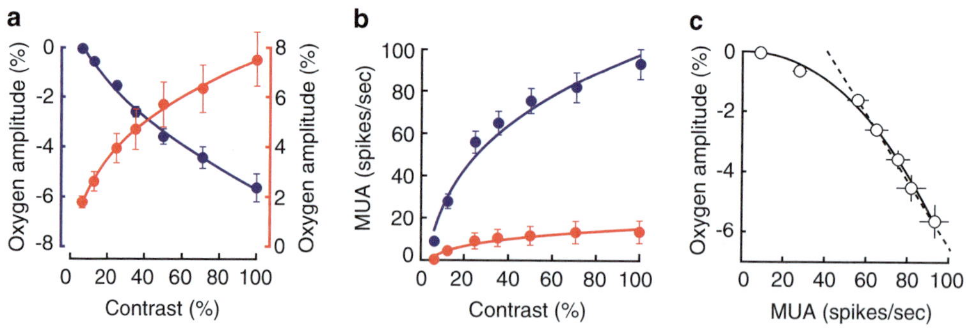

Fig. 13 Contrast nonlinearity of tissue oxygen and neural responses in the LGN. (**a**) Amplitudes of the oxygen responses to the small stimulus (*blue dots*) that only stimulates the RF, and to the large stimulus (*red dots*) that only stimulates the surround of the RF. (**b**) Neural responses to the above two stimuli. The data are fit by a power law function. (**c**) Correlation between the initial dip of oxygen response and neural activity. The amplitude of the initial dip is shown to change monotonically with neural activity, which is best described by a power law function. The *dashed line* indicates a linear coupling between the initial dip and neural activity when neural activity is high. Error bars indicate ±1 SEM

oxygen responses, respectively. Both negative and positive oxygen responses become saturated with increased stimulus contrast and the overall data set is well fit by a power law function ($R^2 = 0.99$ and $R^2 = 1.0$ for the negative and positive oxygen responses, respectively). This result shows a nonlinearity of contrast saturation for tissue oxygen response. Similarly, neural activity also exhibits contrast saturation as shown in Fig. 13b. The neural firing rates to the small stimulus (blue dots, Fig. 13b) and the large mask stimulus (red dots, Fig. 13b) are well fit by a power low function ($R^2 = 0.97$ and $R^2 = 0.95$, respectively).

To examine the coupling between tissue oxygen response and neural activity to the small stimulus, amplitude of the mean negative oxygen response is plotted as a function of neural activity at all contrast levels in Fig. 13c. The data are well fit by a power law function ($R^2 = 0.99$), suggesting nonlinear neurometabolic coupling between oxygen consumption and underlying neural activity. However, if the two low contrast levels are excluded from the plot, oxygen responses and neural activity at the other five contrast levels can be linearly fit ($R^2 = 1.0$). This indicates that neurometabolic coupling is linear when neurons in a local region are strongly activated. A similar relationship has been reported between hemodynamic change and neural activity in optical imaging experiments [48, 49].

6 Effects of Intracortical Inhibition on Neurometabolic Coupling

As we have shown, oxygen consumption and neuronal spiking activity are co-localized, and are coupled linearly in temporal and spatial domains. In visual cortex, excitatory principal neurons and local inhibitory interneurons are densely interconnected and cortical functions depend upon a relative balance of recurrent excitation and inhibition. The detailed functional factors that may influence tissue oxygen responses are not clear. For example, there may be different roles of inhibitory interneurons in neurometabolic coupling. Inhibitory interneurons have been reported to contribute to vascular responses [50, 51] and fMRI BOLD responses [52]. To examine how the recurrent inhibitory drives may be associated with tissue oxygenation, we take advantage of two major suppressive phenomena in visual cortex. One, cross-orientation suppression, occurs within the classical RF when an orthogonally oriented stimulus is superposed on one at optimal orientation. The second, surround suppression, occurs when the size of an optimally oriented stimulus extends beyond the classical RF. Previous neurophysiological studies have shown that cross-orientation suppression can be induced by a relatively high spatial frequency orthogonal stimulus that cannot activate most cortical cells [53–55]. Therefore, the suppression is assumed to be associated with feedforward mechanisms. Surround suppression, on the other hand, is orientation selective and susceptible to contrast adaptation, and it is assumed to be mostly accounted for by intracortical inhibition [56, 57]. By co-localized measurements of neural and tissue oxygen responses in visual cortex, we find different oxygen response patterns when these two types of suppression are activated.

To compare changes in tissue oxygen and neural responses under these two types of suppression, a suppression index (SI) is defined by taking the ratio of the difference and the addition of responses under the suppressed and the control stimulus conditions [23]. A negative SI represents a suppressive effect whereas a positive SI stands for an increase in response. Figure 14 illustrates the SIs of neural and oxygen responses for cross-orientation and surround suppression across a population of recording sites. Negative SIs are shown for neural responses for both surround suppression (Fig. 14a) and cross-orientation suppression (Fig. 14c). The average SIs are −21.6 ± 2.2 and −34.2 ± 3.1 % for neural activity for surround suppression and cross-orientation suppression, respectively. The initial dip of tissue oxygen response, however, exhibits different characteristics for these two types of suppression. The average SIs of the initial dip are 21.2 ± 2.8

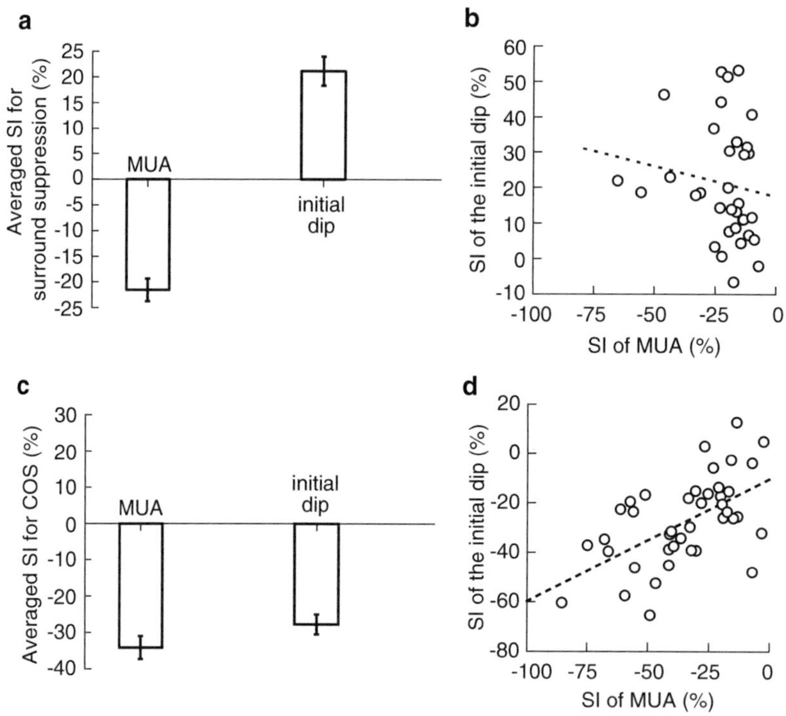

Fig. 14 Comparison of suppression indices (SI) of neural and tissue oxygen responses for the two types of suppression in striate cortex. (**a**, **c**) Average SIs for spiking activity and the initial dip of oxygen response for surround suppression (**a**) and cross-orientation suppression (**c**). (**b**, **d**) Scatter plots of SIs for spiking activity and the initial dip for surround suppression (**b**) and cross-orientation suppression (**d**). Each *open circle* represents a recording site. *Dashed lines* indicate least-squares regression fits

and -27.7 ± 2.7 % for surround suppression and cross-orientation suppression, respectively. In scatter plots in Fig. 14b, d, we compare the SIs for the initial dip and neural response for individual recording sites for these two types of suppression. Consistent with the average results in Fig. 14a, c, an increase in the initial dip is generally elicited for surround suppression whereas a decrease is shown for cross-orientation suppression. Moreover, a significant correlation between SIs of neural activity and the initial dip of oxygen response is shown for cross-orientation suppression (Fig. 14d, $r=0.6$, $p=1.0 \times 10^{-4}$) but not for surround suppression (Fig. 14b, $r=-0.1$, $p=0.4$).

The difference in neurometabolic coupling for surround suppression and cross-orientation suppression may be due to activation of separate neural networks suggested by previous neurophysiological studies [53–55]. Although surround suppression in visual cortex may be partly inherited from the same phenomenon in LGN, it may also be due to intracortical inhibition from long-range horizontal connections and feedback from higher cortical areas [56, 57]. The increase in the initial dip of oxygen response for surround suppression may be due to oxygen consumption by GABA inhibitory interneurons in proximal locations. In contrast, cross-orientation suppression is generally assumed to be based on a feedforward mechanism, which may include contrast nonlinearity in the LGN, synaptic depression in thalamocortical synapses, and the expansive nonlinearity of the spike threshold [53, 58, 59]. Because intracortical inhibition may not be involved in this suppressive process, the reduced initial dip in tissue oxygen response might simply reflect reduced activity-induced oxygen consumption. Similarly, different metabolic responses have been reported when rat primary somatosensory cortex is activated separately by corticocortical and thalamocortical inputs, but no significant difference in neural activity is observed [20]. These results suggest that investigation of neurometabolic coupling may help unfold the neural circuits underlying some neural functions in the brain. The different neurometabolic coupling characteristics for surround and cross-orientation suppression provide additional evidence concerning separate neural mechanisms.

7 Tissue Oxygen Metabolism in Cerebral Cortex Reflects Synaptic More than Spiking Activity

Neural activity can be measured in different ways. Two major components of neural activity are typically included in extracellular recordings from single cells or groups of neurons: single cell or MUA and local field potential (LFP). MUA is assumed to reflect the spiking activity of neurons near the electrode tip while LFP is believed to represent synchronized synaptic activity over a larger volume of tissue. An important issue for the interpretation of fMRI data concerns the nature of the BOLD signal. What type of neural activity is represented? This is a fundamental and controversial issue. In early fMRI studies, a linear relationship was reported between human fMRI response and monkey single-unit spiking rates [60, 61]. This linear relationship was later confirmed by fMRI and spiking data that were collected from human subjects [62]. On the other hand, simultaneous measurements of fMRI response and neural activity in anesthetized monkeys showed that fMRI response was better predicted by LFP compared to MUA measurements [63], suggesting that BOLD fMRI signals reflect synaptic input more strongly than spiking output. In these measurements, LFP accounted for slightly more of the variance in BOLD signals than the MUA. In most cases, spiking activity is highly correlated with the LFP signal. For example, both MUA and LFP increase monotonically with stimulus contrast [63]. Therefore, it is necessary to carefully dissociate these two types of neural activity to address the coupling issues. In addition, because BOLD signals reflect a combined interplay of changes in CBF, CBV, and $CMRO_2$, it is unclear what kind of neural activity is associated with metabolic activity in the brain. To address the neurometabolic coupling from a different approach, we have used our specialized sensor to measure simultaneously neural and tissue oxygen responses to visual stimuli in striate cortex. By manipulation of visual stimulus parameters, we were able to dissociate MUA and LFP responses and found that tissue oxygen metabolism is more clearly associated with synaptic input than spiking output.

We presented small-size drifting sinusoidal gratings at different temporal frequencies and measured tissue oxygen response and neural activity in area 17 as illustrated in Fig. 15. Two temporal frequencies were used in the experiment. A null stimulus condition (blank) was used as a control. The 4-Hz stimulus elicited a large monophasic negative oxygen response and strong MUA (Fig. 15a) and LFP (Fig. 15b) responses. A 30 Hz grating presentation was also used because it is outside the range of cortical response but can activate LGN. The 30-Hz stimulus elicited no change in MUA compared to the null stimulus condition. Tissue oxygen signals, although weak, constituted significant responses ($p = 8.1 \times 10^{-8}$, t-test). Similar to the LFP response to the 4-Hz stimulus, there was a strong initial LFP to the 30-Hz condition. The absence of MUA implies that tissue oxygen metabolism is associated with the concurrent increase in LFP power. These results agree with those from an autoradiography study that show regional glucose metabolism is directly tied to synaptic activity [64]. We note also the fMRI finding that BOLD responses reflect intracortical processing rather than pyramidal cell output activity [63]. The increased metabolism associated with enhanced presynaptic activity might be the primary source that drives fMRI signals.

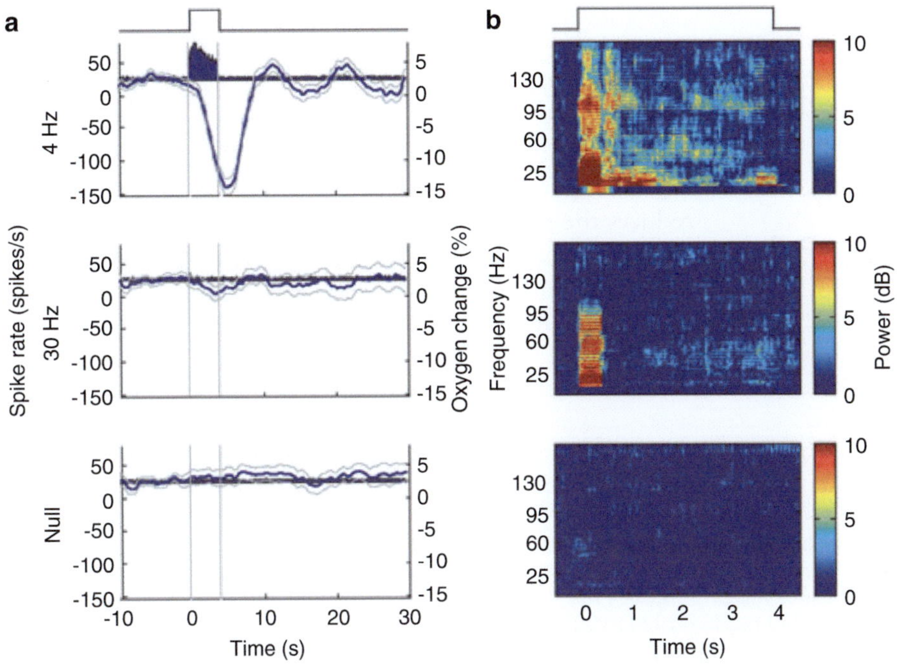

Fig. 15 Neural activity and tissue oxygen response to visual stimuli of different temporal frequencies for a representative recording site in striate cortex. (**a**) Spiking activity (*dark blue histograms*) and tissue oxygen responses (*blue curves*) to 4 and 30 Hz stimuli. Blank screen was used as a null condition. (**b**) LFP spectrograms to the visual stimuli. The size of visual stimuli is 10° in diameter centered on the RF of the recording site, and the spatial frequency is 0.7 cycles/deg. Data are averaged across 48 trials. The lines on the top indicate the onset of visual stimuli and duration

8 Summary

Tissue oxygenation can be reliably measured with a Clark-style polarographic sensor in the brain. The relative high spatial and temporal resolutions of the sensor make it possible to examine changes in tissue oxygen concentration co-localized with neural activity. With the dual-purpose APOX microelectrode, tissue oxygen response and neural activity can be simultaneously measured, which enables us to investigate the coupling between these two measurements. We describe here four major results from experiments in the central visual pathway in anesthetized cats. First, tissue oxygen response typically includes two major components, a small initial dip followed by a large positive peak. Compared to the positive peak, the initial dip is more closely coupled and spatially localized with neural activity. The initial dip of oxygen response presumably reflects activity-induced oxygen consumption while the positive peak reflects activity-induced increase in blood flow. Second, neurometabolic coupling is largely linear in temporal and spatial summation, and is also linear in scaling with high levels of neural activity. Third, neurometabolic coupling can be modified by activation of intracortical inhibitory networks. Finally, tissue oxygen metabolism reflects synchronized synaptic activity more strongly than spiking output. Note that the initial dip and the positive peak of tissue oxygen responses partially overlap in time, and as a result, each component reduces the size of the other. Further studies investigating simultaneous changes in vascular signals and evoked tissue oxygen, neural responses, and other metabolic factors should help elucidate our understanding of neurometabolic coupling.

References

1. Logothetis NK, Wandell BA (2004) Interpreting the BOLD signal. Annu Rev Physiol 66: 735–769
2. Logothetis NK (2002) The neural basis of the blood-oxygen-level-dependent functional magnetic resonance imaging signal. Philos Trans R Soc Lond B Biol Sci 357(1424):1003–1037
3. Fox PT, Raichle ME (1986) Focal physiological uncoupling of cerebral blood flow and oxidative metabolism during somatosensory stimulation in human subjects. Proc Natl Acad Sci U S A 83(4):1140–1144
4. Fox PT, Raichle ME, Mintun MA et al (1988) Nonoxidative glucose consumption during focal physiologic neural activity. Science 241(4864): 462–464
5. Davis TL, Kwong KK, Weisskoff RM et al (1998) Calibrated functional MRI: mapping the dynamics of oxidative metabolism. Proc Natl Acad Sci U S A 95(4):1834–1839
6. Hoge RD, Atkinson J, Gill B et al (1999) Linear coupling between cerebral blood flow and oxygen consumption in activated human cortex. Proc Natl Acad Sci U S A 96(16): 9403–9408
7. Kim SG, Rostrup E, Larsson HB et al (1999) Determination of relative CMRO2 from CBF and BOLD changes: significant increase of oxygen consumption rate during visual stimulation. Magn Reson Med 41(6):1152–1161
8. Roland PE, Eriksson L, Stone-Elander S et al (1987) Does mental activity change the oxidative metabolism of the brain? J Neurosci 7(8): 2373–2389
9. Wey HY, Wang DJ, Duong TQ (2011) Baseline CBF, and BOLD, CBF, and CMRO2 fMRI of visual and vibrotactile stimulations in baboons. J Cereb Blood Flow Metab 31(2):715–724
10. Frostig RD, Lieke EE, Ts'o DY et al (1990) Cortical functional architecture and local coupling between neuronal activity and the microcirculation revealed by in vivo high-resolution optical imaging of intrinsic signals. Proc Natl Acad Sci U S A 87(16):6082–6086
11. Malonek D, Grinvald A (1996) Interactions between electrical activity and cortical microcirculation revealed by imaging spectroscopy: implications for functional brain mapping. Science 272(5261):551–554
12. Kim SG, Richter W, Ugurbil K (1997) Limitations of temporal resolution in functional MRI. Magn Reson Med 37(4):631–636
13. Menon RS, Kim SG (1999) Spatial and temporal limits in cognitive neuroimaging with fMRI. Trends Cogn Sci 3(6):207–216
14. Buxton RB (2001) The elusive initial dip. Neuroimage 13(6 Pt 1):953–958
15. Hathout GM, Varjavand B, Gopi RK (1999) The early response in fMRI: a modeling approach. Magn Reson Med 41(3):550–554
16. Clark LC Jr, Wolf R, Granger D et al (1953) Continuous recording of blood oxygen tensions by polarography. J Appl Physiol 6(3): 189–193
17. Clark LC Jr, Misrahy G, Fox RP (1958) Chronically implanted polarographic electrodes. J Appl Physiol 13(1):85–91
18. Travis RP Jr, Clark LC Jr (1965) Changes in evoked brain oxygen during sensory stimulation and conditioning. Electroencephalogr Clin Neurophysiol 19(5):484–491
19. Ances BM, Wilson DF, Greenberg JH et al (2001) Dynamic changes in cerebral blood flow, O2 tension, and calculated cerebral metabolic rate of O2 during functional activation using oxygen phosphorescence quenching. J Cereb Blood Flow Metab 21(5):511–516
20. Enager P, Piilgaard H, Offenhauser N et al (2009) Pathway-specific variations in neurovascular and neurometabolic coupling in rat primary somatosensory cortex. J Cereb Blood Flow Metab 29(5):976–986

21. Li B, Freeman RD (2007) High-resolution neurometabolic coupling in the lateral geniculate nucleus. J Neurosci 27(38):10223–10229
22. Li B, Freeman RD (2010) Neurometabolic coupling in the lateral geniculate nucleus changes with extended age. J Neurophysiol 104(1):414–425
23. Li B, Freeman RD (2011) Neurometabolic coupling differs for suppression within and beyond the classical receptive field in visual cortex. J Physiol 589(Pt 13):3175–3190
24. Li B, Freeman RD (2012) Spatial summation of neurometabolic coupling in the central visual pathway. Neuroscience 213:112–121
25. Offenhauser N, Thomsen K, Caesar K et al (2005) Activity-induced tissue oxygenation changes in rat cerebellar cortex: interplay of postsynaptic activation and blood flow. J Physiol 565(Pt 1):279–294, Epub 2005 Mar 17
26. Thompson JK, Peterson MR, Freeman RD (2003) Single-neuron activity and tissue oxygenation in the cerebral cortex. Science 299(5609):1070–1072
27. Thompson JK, Peterson MR, Freeman RD (2004) High-resolution neurometabolic coupling revealed by focal activation of visual neurons. Nat Neurosci 7(9):919–920
28. Thompson JK, Peterson MR, Freeman RD (2005) Separate spatial scales determine neural activity-dependent changes in tissue oxygen within central visual pathways. J Neurosci 25(39):9046–9058
29. Viswanathan A, Freeman RD (2007) Neurometabolic coupling in cerebral cortex reflects synaptic more than spiking activity. Nat Neurosci 10(10):1308–1312, Epub 2007 Sep 9
30. Hubel DH, Wiesel TN (1962) Receptive fields, binocular interaction and functional architecture in the cat's visual cortex. J Physiol 160:106–154
31. Kim DS, Duong TQ, Kim SG (2000) High-resolution mapping of iso-orientation columns by fMRI. Nat Neurosci 3(2):164–169
32. Yacoub E, Shmuel A, Pfeuffer J et al (2001) Investigation of the initial dip in fMRI at 7 Tesla. NMR Biomed 14(7–8):408–412
33. Sanderson KJ (1971) The projection of the visual field to the lateral geniculate and medial interlaminar nuclei in the cat. J Comp Neurol 143(1):101–108
34. Duong TQ, Kim DS, Ugurbil K et al (2000) Spatiotemporal dynamics of the BOLD fMRI signals: toward mapping submillimeter cortical columns using the early negative response. Magn Reson Med 44(2):231–242
35. Boynton GM, Engel SA, Glover GH et al (1996) Linear systems analysis of functional magnetic resonance imaging in human V1. J Neurosci 16(13):4207–4221
36. Friston KJ, Jezzard P, Turner R (1994) Analysis of functional MRI time-series. Hum Brain Mapp 1:153–171
37. Birn RM, Saad ZS, Bandettini PA (2001) Spatial heterogeneity of the nonlinear dynamics in the FMRI BOLD response. Neuroimage 14(4):817–826
38. Gu H, Stein EA, Yang Y (2005) Nonlinear responses of cerebral blood volume, blood flow and blood oxygenation signals during visual stimulation. Magn Reson Imaging 23(9):921–928, Epub 2005 Nov 3
39. Liu H, Gao J (2000) An investigation of the impulse functions for the nonlinear BOLD response in functional MRI. Magn Reson Imaging 18(8):931–938
40. Miller KL, Luh WM, Liu TT et al (2001) Nonlinear temporal dynamics of the cerebral blood flow response. Hum Brain Mapp 13(1):1–12
41. Soltysik DA, Peck KK, White KD et al (2004) Comparison of hemodynamic response nonlinearity across primary cortical areas. Neuroimage 22(3):1117–1127
42. Vazquez AL, Noll DC (1998) Nonlinear aspects of the BOLD response in functional MRI. Neuroimage 7(2):108–118
43. Freeman RD, Ohzawa I, Walker G (2001) Beyond the classical receptive field in the visual cortex. Prog Brain Res 134:157–170
44. Nurminen L, Kilpelainen M, Laurinen P et al (2009) Area summation in human visual system: psychophysics, fMRI, and modeling. J Neurophysiol 102(5):2900–2909
45. Press WA, Brewer AA, Dougherty RF et al (2001) Visual areas and spatial summation in human visual cortex. Vision Res 41(10–11):1321–1332
46. Williams AL, Singh KD, Smith AT (2003) Surround modulation measured with functional MRI in the human visual cortex. J Neurophysiol 89(1):525–533
47. Zenger-Landolt B, Heeger DJ (2003) Response suppression in v1 agrees with psychophysics of surround masking. J Neurosci 23(17):6884–6893
48. Devor A, Dunn AK, Andermann ML et al (2003) Coupling of total hemoglobin concentration, oxygenation, and neural activity in rat somatosensory cortex. Neuron 39(2):353–359
49. Sheth SA, Nemoto M, Guiou M et al (2004) Linear and nonlinear relationships between neuronal activity, oxygen metabolism, and hemodynamic responses. Neuron 42(2):347–355

50. Cauli B, Tong XK, Rancillac A et al (2004) Cortical GABA interneurons in neurovascular coupling: relays for subcortical vasoactive pathways. J Neurosci 24(41):8940–8949
51. Kocharyan A, Fernandes P, Tong XK et al (2008) Specific subtypes of cortical GABA interneurons contribute to the neurovascular coupling response to basal forebrain stimulation. J Cereb Blood Flow Metab 28(2):221–231
52. Lee JH, Durand R, Gradinaru V et al (2010) Global and local fMRI signals driven by neurons defined optogenetically by type and wiring. Nature 465(7299):788–792
53. Freeman TC, Durand S, Kiper DC et al (2002) Suppression without inhibition in visual cortex. Neuron 35(4):759–771
54. Li B, Peterson MR, Thompson JK et al (2005) Cross-orientation suppression: monoptic and dichoptic mechanisms are different. J Neurophysiol 94(2):1645–1650
55. Sengpiel F, Vorobyov V (2005) Intracortical origins of interocular suppression in the visual cortex. J Neurosci 25(27):6394–6400
56. DeAngelis GC, Freeman RD, Ohzawa I (1994) Length and width tuning of neurons in the cat's primary visual cortex. J Neurophysiol 71(1):347–374
57. Durand S, Freeman TC, Carandini M (2007) Temporal properties of surround suppression in cat primary visual cortex. Vis Neurosci 24(5):679–690
58. Li B, Thompson JK, Duong T et al (2006) Origins of cross-orientation suppression in the visual cortex. J Neurophysiol 96(4):1755–1764, Epub 2006 Jul 19
59. Priebe NJ, Ferster D (2006) Mechanisms underlying cross-orientation suppression in cat visual cortex. Nat Neurosci 9(4):552–561, Epub 2006 Mar 5
60. Heeger DJ, Huk AC, Geisler WS et al (2000) Spikes versus BOLD: what does neuroimaging tell us about neuronal activity? Nat Neurosci 3(7):631–633
61. Ress D, Backus BT, Heeger DJ (2000) Activity in primary visual cortex predicts performance in a visual detection task. Nat Neurosci 3(9): 940–945
62. Mukamel R, Gelbard H, Arieli A et al (2005) Coupling between neuronal firing, field potentials, and FMRI in human auditory cortex. Science 309(5736):951–954
63. Logothetis NK, Pauls J, Augath M et al (2001) Neurophysiological investigation of the basis of the fMRI signal. Nature 412(6843):150–157
64. Jueptner M, Weiller C (1995) Review: does measurement of regional cerebral blood flow reflect synaptic activity? Implications for PET and fMRI. Neuroimage 2(2):148–156

Chapter 7

High-Resolution Wide-Field Optical Imaging of Microvascular Characteristics: From the Neocortex to the Eye

Ivo Vanzetta, Thomas Deneux, and Amiram Grinvald

Abstract

Measuring microvascular characteristics in cortical tissue and individual microvessels has important applications for functional imaging, biomedical research, and clinical diagnostics. Multiphoton fluorescence microscopy approaches are most effective and allow to reliably record red blood cell (RBC) velocity in individual vessels, but require injecting fluorescent tracers. Moreover, only one or few vessels in a small area can be imaged at a time. Wide-field CCD/CMOS-based optical imaging of intrinsic absorption or reflection changes in macroscopic vascular networks allows to overcome these shortcomings, by recording RBCs' trajectories over several mm^2 of cortical surface. The RBC velocity can then be extracted from these wide-field data using specialized algorithms. Here, we describe two of those, which provide robust RBC velocity estimations that are independent and can thus be used as a control one for another. Although this approach can be used in any part of the body with optically accessible blood vessels, here we show its application in two cases: first the cerebral cortex and then the eye. In this latter application, we go into some more detail in describing the retinal function imager (RFI): a unique, noninvasive multiparameter functional imaging instrument that directly measures hemodynamic parameters such as retinal RBC velocity, oximetric state, and metabolic responses to photic activation. In addition, it allows capillary perfusion mapping without any contrast agent. These parameters of retinal function are degraded by retinal abnormalities. Here, we thus focus on the characterization of microvessels properties. Indeed, clinical studies suggest that knowing these properties should yield multiple clinical applications for early diagnosis of retinal diseases, possible critical guidance of their treatment, as well as implications for vascular diseases of cortex and eye.

Key words Hemodynamics, Functional imaging, Erythrocytes, Vessel segmentation, CBF, CBV blood flow, Intrinsic signal, Metabolic responses, Oximetry, Perfusion maps, Retinal imaging

1 Background

Functional optical imaging based on slow changes in intrinsic properties of active brain tissue has first been applied to the exploration of the neocortex functional architecture. This method has facilitated in vivo visualization of active cortical regions with the unprecedented spatial resolution of 50 μm or less. It reveals

Mingrui Zhao et al. (eds.), *Neurovascular Coupling Methods*, Neuromethods, vol. 88,
DOI 10.1007/978-1-4939-0724-3_7, © Springer Science+Business Media New York 2014

activity-dependent changes in light reflectance recorded using a camera with high spatial and temporal resolution [complementary metal oxide semiconductor (CMOS) or charge-coupled device (CCD)]. The functional signals originate from activity-dependent metabolic, hemodynamic, and fast and slow light-scattering changes. This widespread functional imaging technology has led to major discoveries relating to the functional architecture in neocortical centers for sensory modalities, including visual, auditory, somatosensory, and smell. In addition, the underlying mechanisms have also been extensively studied by the Weizmann Institute group. These studies and studies by other groups that followed were reviewed in [1]. Also functional mapping based on hemodynamic signals has been extensively reviewed [2–6]. Therefore, these topics will not be covered here. Instead, this chapter will focus on blood flow dynamics.

Understanding blood flow dynamics in medium–small-scale vascular networks is important both for clinical applications and, in more basic research, for a deeper knowledge of circulatory physiology. Tissues such as the cerebral cortex and the retina critically rely on an efficient perfusion down to the submillimeter scale. Therefore, being able to reliably measure blood flow over relatively large fields of view (10 mm^2) and at fine spatial scale (~10 μm) is a strong objective for several biomedical imaging applications, such as functional imaging in the retina [7–9] or in the cortex [10, 11]. Such high-resolution measurements are also needed for investigating the spatiotemporal properties of the cortical microcirculation in order to provide critical information necessary to model the coupling between hemodynamic responses and neural activation and to better understand the physiological bases of signals used in modern functional imaging such as fMRI [12–17].

Blood flow can be measured with a laser Doppler probe [18] at one given location, taking advantage of the Doppler shift induced on a laser beam reflected by red blood cells (RBCs) moving in the local (~1 mm^3) vasculature. Although this technique sets the gold standard in measuring local blood flow, it is unable to provide a detailed mapping of blood flow over large territories such as corresponding to, for instance the extent of functional cortical maps (i.e., a few mm^2). More recently, laser speckle imaging [18, 19], which is a truly two-dimensional approach, has made it possible to cover even large fields of view. However, it is characterized by low signal-to-noise ratio (SNR) and published laser speckle images document blood flow changes mostly in the parenchyma, rather than selectively in individual vessels (*see*, e.g., [20, 21], but see also [22]). Lastly, all the above-mentioned methods measure perfusion, that is, the total amount of blood (more precisely hematocrit) traversing the unit tissue volume in unitary time. Therefore, the local density and velocity of RBCs cannot be disentangled in the measurements.

Alternatively, one can take advantage of the strong light absorption of hemoglobin, which creates a contrast between RBCs and the surrounding blood plasma within a vessel. The finding that wide-field imaging without a contrast agent can directly visualize RBC velocity in cortical vessels was first reported by Grinvald and colleagues (Fig. 17 in [3]).

The resulting spatial pattern of contrast and its changes as a function of time allows to directly imaging the movement of RBCs inside the vessel, which is possible whenever it is optically accessible. This approach allows a precise estimation of flow velocity in specific vessels, rather than an average measurement made over different vascular compartments, as given by either laser Doppler or laser speckle techniques.

RBC velocity can be measured accurately and with high SNR using two-photon fluorescence microscopy [23, 24], which has the additional advantage of allowing to image vessels below the cortical surface. However, this approach has two severe limitations. First, it relies on the intravenous injection of extrinsic fluorescent tracers into the bloodstream or fluorescent labelling of the RBCs [11]. Second, it is limited to measuring the movement of RBCs within only one microvessel at a time.

Recently, others and ourselves [3, 6, 9, 25, 26] have proposed a relatively simple and low-cost method to measure RBC velocity in two dimensions in relatively large (i.e., several mm^2) vascular networks that are optically accessible, such as at the surface of the cerebral cortex or the retina [7, 8]. This wide-field optical imaging approach relies on the intrinsic absorption properties of hemoglobin and the strong backscattering of light by biological tissue: in the green part of the spectrum, light absorption by RBCs is much stronger than by the surrounding plasma. RBCs thus appear as dark spots on a bright background. Therefore they can be visualized in images simply taken by a CCD camera (Fig. 1) and their movements can—at least in principle—be reconstructed from time series of images of the imaged vascular network.

The flow of RBCs in the vascular network appears clearly in a high-resolution movie of the cerebral cortex even by visual inspection alone, once low spatial frequency gradients are eliminated from the images. Whereas this can be done relatively easily (e.g., by simple frame-by-frame division by the luminance pattern averaged over the whole movie duration), extracting and quantifying the movement of RBCs (and its changes over time) from the optical data is surprisingly difficult. The main challenges to overcome are the following: (i) Within a vessel, RBCs move with respect to each other and their flow can't thus be reduced to the displacement of a rigid pattern of contrasts long the vessel's curvilinear coordinate. Moreover, (ii) SNR is often low due to shot-noise, which can lead to contrast changes or even the disappearance/reappearance of individual BDCs from one frame to another. Shot-noise is a

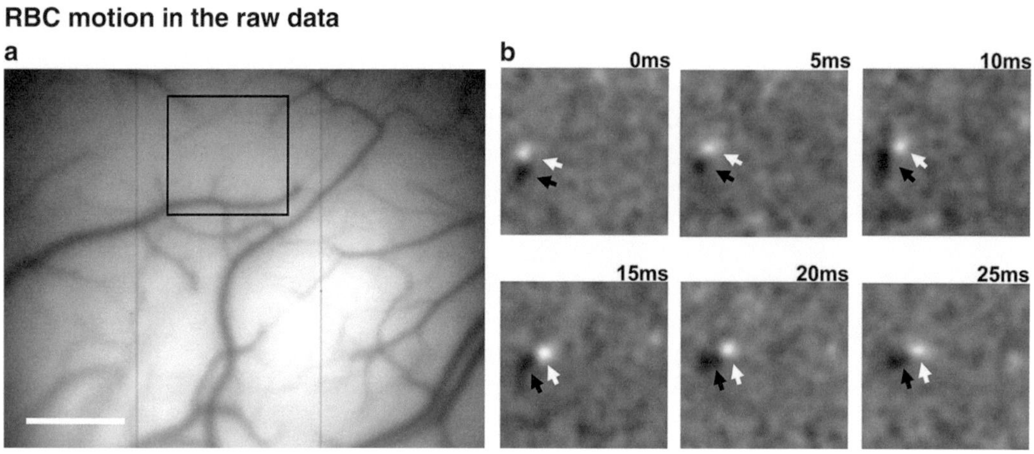

Fig. 1 RBC motion as seen by a CCD: (**a**) the reference frame, with the rectangle enlarged in (**b**) and re-displayed in Fig. 2. Scale bar: 1 mm. (**b**) Illustration of the RBC motion to be estimated from the images after elimination of low spatial frequency gradients: here, each frame acquired at time *t* has been divided by the frame acquired at time *t*−1; the *black* and *white* "dots" are a "cluster" of RBCs, that moved from its old (*t*−1) position (*white*) to its new (*t*) one (*black*) in a left-to-right motion along a small vessel. The same small vessel can be seen more clearly in the last panel of Fig. 2, where it has been selected by a semiautomatic segmentation algorithm. Reprinted from [29], Copyright (2012), with permission from Elsevier

particularly important limiting factor in applications where conditions of high illumination are difficult to obtain, such as the primate retina, because of the limited aperture of the pupil and the low reflectivity/backscattering of the ocular fundus. To deal with such complex and noisy data, a sophisticated and efficient image processing algorithm is thus needed. In our hands, the two algorithms that perform best are based on a Radon transform [27] or on a variational approach (the "fasttrack" algorithm) that we recently proposed [28] and compared to the first one [29].

Importantly, these algorithms require several key preprocessing stages. First, it is necessary to spatially coregister all data frames to correct for movement artifacts that are unavoidable during in vivo experiments due to heartbeat, respiration, and—especially when recording from awake subjects—small displacements of the sample. Second, vessels must be selected from the images—a task that can be very tedious and time consuming—especially in the case of wide-field images. Next, data from each vessel have to be analyzed and stored separately after spatial and temporal filtering. Altogether, these different processes result in a very large dataset. Typical experiments can contain up to several hundred trials, each trial containing more than 1,000 frames at 512 × 512 pixels. It is thus not uncommon that data volumes reach 100 GB or more, requiring an adequate data handling and management.

We have therefore developed a complete analysis suite (free for download, *see* [29]), which includes these various steps, starting from organizing the raw data, tackling coregistration and semiautomatic assisted blood vessel selection, and ending up with a two-dimensional RBC velocity field displayed over the network of blood vessels that have been selected for analysis. The user can decide which one of the two algorithms mentioned above to use, i.e., the Radon or the "fasttrack"-based one. Both have user-selectable optimization parameters. In addition to RBC velocity, our analysis suite also allows to estimate the blood vessels' diameters and its changes, as well as changes in backscattered light intensity, i.e., absorption.

Some care is however required when extrapolating the results obtained with our method to those obtained with other blood flow measurement techniques, in particular PET and MRI. Indeed, the definition of blood flow varies among methods. Some, such as MRI and PET, are sensitive mostly to blood plasma flow and its changes. Others, such as all the other methods discussed herein, measure the flow of particles within the blood. Our method is sensitive exquisitely to RBC flow. If all these different definitions of flow are certainly strongly correlated, they are not necessarily identical, especially in small vessels and/or when it comes to measuring small changes [23, 24, 30].

2 Hardware Requirements and Applicability Limits

In wide-field intrinsic optical imaging, a camera takes images over time from the surface of the tissue under examination, such as cortex [31, 32] or the retina [33]. To allow tracking the RBCs in their displacements even in arterial flow regimes (10 mm/s and faster), the camera used for imaging must be able to provide high frame rates. On the other hand, the resulting short exposure times call for considerable illumination intensity, which is critical for maximizing the SNR: contrast grows linearly with light intensity, whereas shot-noise increases as its square root. It also has to be considered that to maximize the RBC's contrast, green light (540–570 nm) is used for the illumination because of the strong hemoglobin absorption in this part of the visible spectrum. However, for the same reason, a large amount of light is absorbed by the tissue and only a small part is scattered back, therefore demanding for large illumination power. Finally, this imposes a limit on the maximal measurable RBC velocity. As a rule of thumb, very reliable estimations are obtained for speeds up to 3 pixels/frame (i.e., 3 mm/s at 100 Hz and 10 μm/pixel and up to 6 mm/s at 200 Hz), reliable ones up to 10 pixels/frame. If the noise (especially the jitter in RBC speeds) is low, RBC speed estimations are possible up to 20 pixels/frame. Similarly, the thinnest vessels our method can deal with (~2 pixels in diameter, provided RBCs can still be resolved) is also set by a pixel-based spatial scale, which sets the lower limit for the used magnification and therefore the maximal field of view for a given detector.

Conversely, large vessels pose a problem that is independent of magnification: due to the long optical path through the vessel, contrast differences resulting from individual RBCs are difficult if not impossible to detect. As a result, vessels larger than ~0.1 mm cannot be treated by the present approach. It might, nevertheless, be possible to extract RBC speed from the border of thicker vessels, because there the vessel is thinner than in its middle, although the detected RBC speed might not be representative for the whole vessel thickness due to inhomogeneous flow regime.

3 Data Preprocessing

3.1 Coregistration

Residual motions during in vivo recordings are unavoidable and will affect data. In particular, they induce displacements of blood vessels across frames. Therefore, the RBC trajectories as seen by a camera-centered reference frame need to be transformed into vessel-centered coordinates. To do so, it is essential to compensate for any full image motion by aligning all frames acquired during a given experimental session to a given reference frame. This operation produces a new set of images, in which vessels are correctly aligned over time (Fig. 2). For details, *see* [29].

Ideally, this estimation should be performed for each frame. However, in order to spare computation time, it is thus often sufficient to correct only for slow drifts (in particular in anesthetized preparations) by estimating the motion parameters once for every trial.

3.2 Vessel Segmentation

Tracking RBCs within a two-dimensional image is a highly complex problem. However, once the spatial extent of a vessel is known, estimating RBC motion along the vessel axis becomes a one-dimensional problem that is much easier to solve.

Selecting manually the subset of the image corresponding to a blood vessel is a long and fastidious operation that can introduce considerable errors, in particular for small vessels and low contrast image parts (e.g., the central vessel in Fig. 2b–d). Therefore, we have implemented a semiautomatic vessel segmentation algorithm that produces more "objective" results. Figure 2f, g shows an example of vessel segmentation. The user has only to specify the starting and ending points (respectively, points a and b) of a desired vessel through an interactive interface, and eventually a few intermediate points (point c) if the curvature is too strong. This user input is necessary because the segmentation algorithm is not yet able to deal with vessel bifurcations or crossings between two vessels. Once these points have been selected, the segmentation is performed automatically. First, a mask of the vasculature is computed

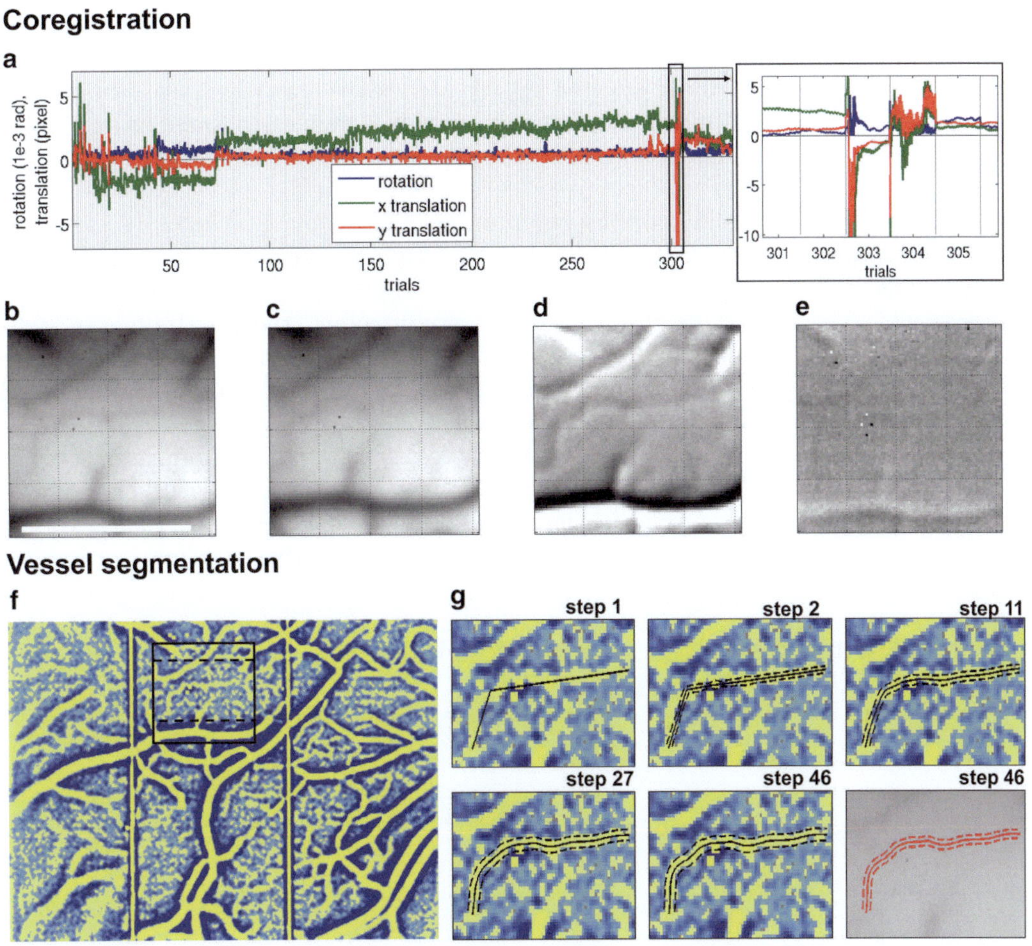

Fig. 2 Preprocessing: coregistration. (**a**) Parameters of the coregistration stage, estimated for each frame in each trial, for a total of 1,200 frames/trial × 330 trials recorded from visual area V4 of an awake macaque. The zoom to the right shows that, in noisy trials (in trial **#**303 the monkey made a sharp body movement), estimation parameters can vary considerably from one frame to the next (note that frames are not numbered on the abscissa) (**b**, **c**) A region shown in two frames taken at different times. Scale bar: 1 mm. (**d**) Dividing (**b**) by (**c**) emphasizes the motion artifact. (**e**) The same division after coregistration, showing a much better match. Vessel segmentation: (**f**) the vasculature mask (Laplacian) of the reference image (Fig. 1a). The *dashed rectangle* corresponds to the enlargement shown in (**g**) to exemplify the segmentation process. (**g**) Successive steps of the semiautomatic vessel segmentation algorithm. A few points (**a**–**c**) have been selected by the user at step 1; the rest is automatic. Note that the algorithm succeeds despite of the very low contrast (the vessel is barely visible in (**b**)). Reprinted from [29], Copyright (2012), with permission from Elsevier

as the image Laplacian, which is an estimate of the local curvature. Pixels lying inside the valleys of the image (positive curvature) are therefore likely to belong to the blood vessel (low intensity). Next, a vessel is defined by two snake functions, the "skeleton" $l(s)$ and the half-diameter $h(s)$, such that the interior of the vessel consists of all pixels located between the two lines $(s+h)$ and $(s-h)$. Intuitively, the vessel's skeleton can be thought of as its axis of cylindrical symmetry. We then define an energy consisting of this Laplacian and three regularization terms enforcing the endpoints to be close to those selected by the user, as well as vessel smoothness. The solution minimizing this energy is then easily found using a simple gradient descent method. For details, *see* [29].

3.3 Creation of "Trajectories Images" and RBC Signal Enhancement

As in previous two-photon microscopy approaches [23, 24], we next cast the RBCs' spatiotemporal trajectories into a "trajectories image" for each segmented vessel (Fig. 3). Intuitively, this process can be thought of as "straightening" the vessel along a spatial coordinate and determining, for each point in space, the intensity values as a function of time. We call the resulting spatiotemporal function "trajectories image" $I(s,t)$.

More in detail, a trajectories image is a two-dimensional matrix $I(s,t)$ containing the intensity values of a given vessel at all times t of a given trial, and for all discrete value s of the curvilinear coordinate along the vessel. To discretize the curvilinear coordinate

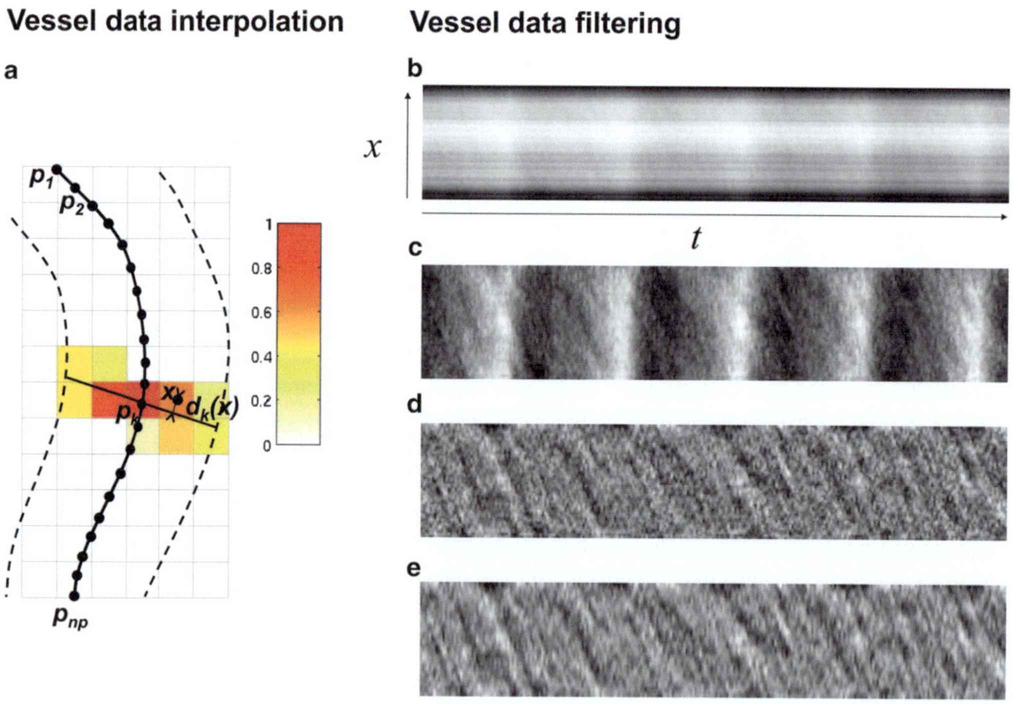

Fig. 3 Production of "Trajectories images" $I(x,t)$: (**a**) Illustration of the local convolution filter (weights are color-coded) used to interpolate the data along the vessel's width (the distance between p_k and p_{k+1} is equal to $1/\sqrt{2}$ times the original pixel size) used to create the raw trajectories image: in the calculation of the signal at point p_k, along the vessel's skeleton, the weight assigned to a given pixel (e.g., pixel x) is related to the distance $d_k(x)$ of the pixel center to the segment normal to the skeleton and passing through p_k. Filtering. (**b–e**) Trajectories image (x-axis = time, y-axis = space: the faster the RBCs, the more the lines tend to be vertical): raw (**b**), after temporal normalization and high-pass (**c**), after spatial high-pass filtering (**d**), and finally after a slight spatial low-pass filtering (**e**). Reprinted from [29], Copyright (2012), with permission from Elsevier

s along the vessel, we first defined a set of equidistant points p_k (Fig. 3a) along the vessel's skeleton, separated by $d/\sqrt{2}$ (d being the original pixel size). The reason for this interpolation is that otherwise information might be lost in case of vessels running obliquely with respect to the pixels' orientation. To optimize the SNR of the value I at that particular point (s,t) in space and time, we take into account the data along the whole thickness of the vessel for each point p_k on the vessel's skeleton, rather than only at p_k itself. For convenience of notation, in the following we shall use x and t (rather than s and t) to denote the spatial and temporal coordinates.

In the raw trajectories image $I(x,t)$, RBCs' positions and their changes over time are difficult to appreciate because of large intensity gradients along vessels. This is illustrated in Fig. 3b, where different intensity values between the vessel's extremities and its midpoint generate strong spatial gradients. Moreover, periodic luminance changes induced by the cardiac pulsation can also be seen. Those changes in light absorption are nearly simultaneous over the whole imaged region and thus remain roughly constant over the entire spatial dimension. To enhance the SNR, three filtering steps are applied successively to the raw trajectories images $I(x,t)$. First, to enhance local contrasts corresponding to RBCs' positions at a given point in time t, the trajectories image at time t, $I(x,t)$ is divided by a reference image $\hat{I}(x)$, which is equivalent to a subtraction and a normalization. This reference image is taken as a local time average, centered at t and long enough to "fill in" the whole vessel length with temporally smeared RBCs, but short enough to keep the difference between $I(x,t)$ and $\hat{I}(x)$ small. As a consequence, constant or slowly varying offsets are eliminated and so are constant or slowly varying differences between nearby spatial locations. Moreover, the fractional signal $I(x,t)/\hat{I}(x)$ remains unaffected by slow drifts (Fig. 3c). Next, the resulting spatiotemporal images are high-pass filtered along the spatial dimension (Fig. 3d) to remove signals that are (quasi-) simultaneous at many different pixel locations within the vessel. To reduce shot-noise, which considerably affects the algorithms' performance [28], we finally applied a weak smoothing along the spatial dimension (Fig. 3e). Estimating the local speed of an RBC at a certain time is equivalent to estimating the slope of the lines in the trajectory image in position (x,t).

Importantly, the first filtering operation described above can be performed as well to the whole three-dimensional data (rather than on two-dimensional trajectories images), after coregistration of the time series of two-dimensional frames. The result is a movie of the vascular network with strongly enhanced RBC's contrast. When displayed at near-to-real speed, it allows visualizing RBCs moving within the vascular network, yielding a tool for the straightforward determination of the blood flow direction in each vessel. This tool is also very useful for the correct selection of a given vessel's starting and endpoint, to check the good continuity of the selected vessels, and for the discrimination between bifurcations and crossing points.

4 Algorithms for the Extraction of RBC Movements

In early studies, the slope of the lines in the trajectories images was estimated manually [23]. Since then, several automated methods have been proposed. The most straightforward approaches [26] consider small blocks of data (a few pixels large) and estimate the speed of RBCs from how much they should be shifted in time for a given shift in space in order to maximize the correlation between original and time-shifted data. However, such strategy is very demanding in terms of computing time. More sophisticated methods have been proposed and are based on specific image processing techniques in order to estimate the RBC translation speed from the orientation of the trajectories in the image, that is, the angle between these oblique lines and the horizontal axis. For instance, we have previously implemented two variants of this orientation-based approach, based on either the structure tensor or oriented Gabor filters [26]. Kleinfeld and colleagues have proposed two other approaches. The image is first cut into small blocks; then, the main orientation of the trajectories in each block is estimated independently. The first method applies a rotation to the block before evaluating the horizontality through singular-value decomposition [10]. The second technique is much faster and relies on a block-by-block Radon transform [27]. As this last technique seems to perform the best for locally estimating RBC velocity, we implemented and customized it to match the constraints of wide-field optical imaging (*see* Box 1). To our knowledge, to date, RBC

Box 1
The Radon Transform-Based Approach

The method developed by [27] is based on the Radon transform. It estimates the velocity of RBCs in (x,t) as the slope of the trajectories in $I(x,t)$. In practice, a sub-patch of the trajectories image (Fig. 4a) that is small enough for trajectories to be considered straight therein is centered on (x,t) and projected onto lines of different orientations passing through (x,t) (Fig. 4b).

The local RBC velocity is then given by the cotangent of the orientation that maximizes the projection's variance (Fig. 4c). The operation is then repeated for every (x,t). In order to optimize this technique for wide-field imaging, we developed several techniques that improved the robustness of the method with respect to its original version. First, instead of using a square-shaped patch, we implemented a windowing scheme by multiplying the original image by a symmetric Gaussian function. Second, a local sliding averaging of the estimates is performed within a rectangular, rather than square, window, which can be chosen according to the local characteristics of the image. Finally, both the number and the choice of the orientations used in the Radon transform are optimized such as to limit computational cost as well as to improve estimation robustness [28]. An example of RBC speed determination using this method is shown in Fig. 4d.

(continued)

Box 1
(continued)

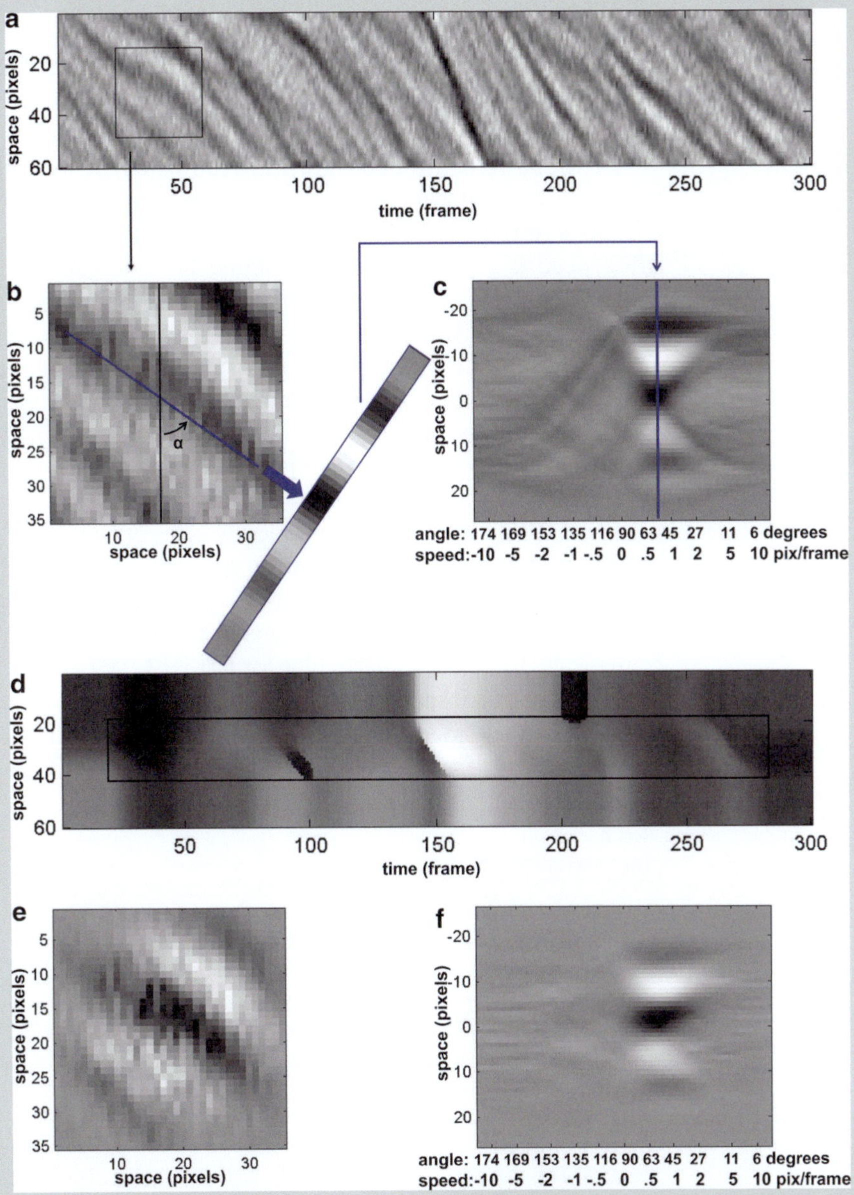

Fig. 4 RBC velocity estimation based on the Radon transform. (**a**) Trajectories image. A *square* shows the sub-block centered on the point where to estimate the trajectories direction. (**b**) Projection of the content of the square parallel to the direction of RBC trajectories α. (**c**) The Radon transform of the square is the set of all such projections: each column is the projection for a different orientation. The *x* axis shows the angles and the corresponding speeds. The preferred orientation of the RBC trajectories in the square is shown by a *blue vertical*: it is the column with maximal variance and allows to estimate the velocity as $v = \text{cotan}(\alpha)$. (**d**) Velocity can be estimated in each point of the image which is distant of the sides of the image by more than half the size of the square (points inside the *black rectangle*). Velocities for points close to the sides are extrapolated. (**e**) Improvement of the Radon-based method: masking the square content by a two-dimensional Gaussian. (**f**) This prevents artifacts induced by the edges of the image and yields less variance in the Radon transform for orientations other than those of the RBC trajectories. © 2011 IEEE. Reprinted, with permission, from [28]

velocity can be best estimated from the trajectories images $I(x,t)$ using this latter algorithm or the "fasttrack" approach, a novel algorithm that was developed by ourselves. The underpinnings and performances of both algorithms have been tested both on simulated and real data and are discussed extensively in ([28, 29], where they are also downloadable). Here, we provide only a short description of their principles.

All the aforementioned algorithms estimate RBC velocity *locally*. Velocity is calculated in (x,t) at a given time, by considering a small sub-block of the trajectories image I centered on (x,t), within which RBC speed is assumed to be constant. The local RBC velocity is then estimated from this block either directly, or by first measuring the orientation of the trajectories inside the block. This operation is then repeated for all time points or a subset thereof, depending on the required time resolution. One critical problem of this approach is related to the criterion used to choose block sizes. When they are too small, estimations are strongly affected by noise. Conversely, with larger sizes, the assumption that a given block would display only local RBC velocity (i.e., the velocity value at (x,t)), breaks down as RBC velocities will fluctuate both in space and time. By consequence, a local estimate will lack both temporal and spatial resolutions in the best case. In the worst conditions, accuracy will be dramatically lowered, in particular for fast RBC velocities [28].

We have proposed a novel RBC tracking algorithm, thereafter called "fasttrack" (*see* Box 2), which allows overcoming these

Box 2
The "Fasttrack" Variational Approach

The details of the "fasttrack" variational approach have been described extensively in [28]. In a nutshell, the main idea is that we estimate the "actual" trajectories $J(x,t)$—underlying the measured trajectories image $I(x,t)$ (shown in Fig. 5a) and the velocity field $V(x,t)$, as they would appear in a noise-free situation. This framework is based on minimizing a cost function (energy) consisting of several terms, each one coding a specific requirement to be satisfied by the solution. These requirements are: (i) the estimated trajectories image $J(x,t)$ must be close to the measured one, $I(x,t)$; (ii) the optical flow equation $J(x+V(x,t),t+1)=J(x,t)$ must be satisfied; and (iii) the estimated trajectories image $J(x,t)$ and velocity field $V(x,t)$ should not vary abruptly. An example of RBC velocity field estimation can be seen in Fig. 5b. The red arrows in Fig. 5c confirm that velocity field $V(x,t)$ estimated via this cost function is indeed an excellent estimate of the tangent to the trajectories in $J(x,t)$.

Minimizing the cost function is an optimization problem that can be conveniently solved iteratively, with the interior-reflective Newton method [35], implemented in the Matlab function *fmincon*. To avoid misinterpreting local minima as the solution, it is advised to repeat the first iteration several times with different initial conditions for V and check the obtained energy values. One can then continue with the rest of the minimization using the initial conditions leading to the minimum energy after the first iteration (for more details, *see* [28, 29]).

Box 2 (continued)

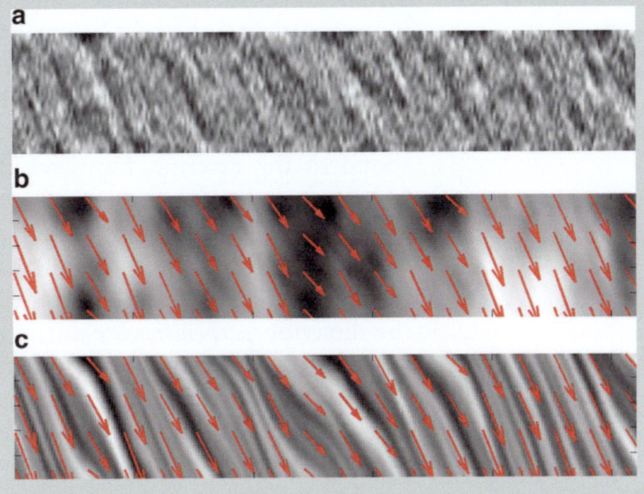

Fig. 5 Velocity estimation through the "fasttrack" variational approach. (**a**) The trajectories image. (**b**) The estimated velocity field (grayscale codes for RBC velocity). The vertical component of the *red arrows* is proportional to the RBC velocity, which is calculated here on a regular grid for illustration purposes (the *horizontal component* represents the time step, which is constant). The same grid is then superimposed onto the reconstructed trajectory images (**c**), to show the link between velocity field and computed trajectories: the *arrows* approximate very closely the local tangents to the RBC trajectories. This match (or mismatch) provides an easy test on whether (or not) the relative weighting of the different terms in the cost function for RBC velocity estimation had been appropriate (or not). Reprinted from [29], Copyright (2012), with permission from Elsevier

shortcomings [28]. By contrast with all other previous methods, it performs a *global* estimation of RBC velocities using a variational approach [34]. Rather than estimating speed values independently in numerous subsets of the data, the fasttrack algorithm estimates the whole spatiotemporal velocity field at once. Its distortions (i.e., fluctuations of the velocity as a function of time and space) are allowed and controlled through regularization parameters. Moreover, this algorithm has the capacity to deal with low SNR by estimating also the real trajectories, unlike the previously described methods that estimate only velocities. This property allows separating, in the measured trajectories image, features resulting from RBC motion from features that result from noise (e.g., photonic noise, contamination from other vessels...) and that should thus be discarded.

4.1 Blood Vessel Section Estimation and Calculation of Blood Flow

In the cortex, the two-dimensional images acquired with the CCD allow also to estimate the vessels' calibers and their changes. If we assume the cross sections of a blood vessel is assumed to be round, it can be calculated as pi/4 times the square of its diameter, which is obtained by tracing a straight line across the vessel perpendicular to its axis and evaluating its gray levels. The low values are then identified with the vessel's interior, whereas the high ones correspond to the exterior. However, at typical magnifications, most vessels are only a few pixels thick and the accuracy of such a rough estimate is much too low to warrant detection of small activity-evoked diameter changes. To increase it, we thus use an interpolation of the gray level values along the line, following—this time in the direction *perpendicular* to the vessel's axis—a procedure similar to before (*see* Fig. 3a: interpolation *parallel* to vessel's curvilinear coordinate). Finally, the multiplication of RBC velocity (mm/s) and vessel cross section (mm^2) then allows calculating blood flow (mm^3/s). All further details of the procedure can be found in [29].

5 Examples of Cortical Blood Flow Measurements

Different objectives can underlie the need of measuring RBC velocity in cortex (i) establishing the stationary blood flow in a vascular network, and its repartition between the various individual vascular elements, eventually comparing the healthy and the pathological case. (ii) The study of massive changes in local blood flow, as can be induced by a local pharmacological or mechanical agent, laser coagulation [36] and the like. Alternatively, one can use RBC velocity (and vessel caliber) measurements to assess the much smaller changes evoked by sensory stimulation in (iii) anesthetized and (iv) awake animal models.

To illustrate points (i) and (ii) above, we show the example of a spreading depression experiment conducted in an anesthetized rat. The passage of a spreading depression over a given cortical location is characterized by large RBC speed responses that yield high SNR recordings, allowing to clearly detecting the response and its spatiotemporal characteristics at the single trial level. We also show how an overall picture of the baseline perfusion through the local network can be obtained in terms of mean arterial and venous RBC speed, blood flow, etc.

The second example we show illustrates the most challenging type of measurement—at least in cortex—namely the RBC speed, blood flow and volume responses elicited in the visual cortex (extrastriate area V4) by visual stimulation in an awake macaque monkey. Despite both low SNR and large signal variability, these responses could indeed be visualized. For this case, however, extensive trial averaging ($n \sim 300$) was necessary. As a comparison, much less averaging was needed to perform the same measurement in the anesthetized cat ($n \sim 16$), due to the reduced movement artifacts and a lower variability of the responses (not shown here, *see* [29]).

5.1 Spreading Depression in Rat Barrel Cortex

A spreading depression (SD) is a massive collective neuronal phenomenon consisting in a wave of neuronal depolarization followed by a period of hyperpolarization. It can be initiated by applying KCl onto the cortex [37] or by mechanical trauma [38]. It then spreads across the cortex at a speed of a few millimeters per minute and is known to be accompanied by large hemodynamic responses (*see* e.g., [39]). The variability in both its spatial and temporal structure is large from one realization ("trial") to the other, which makes averaging over trials useless. Therefore, we will show only one representative recording of our experiments. We evoked SDs in the barrel cortex of an anesthetized rat by gently touching the cortical surface with the tip of a needle, following a protocol described by [38]. Data were acquired as one single trial that lasted several minutes, including a ~100 s before the event, the response to the SD, and the return to baseline, for a total of more than 7 min. Figure 6 illustrates the results.

We illuminated the cortex with green light (540 ± 15 nm) and recorded a movie of the rat's cortex at 100 Hz and a pixel size of 9×9 μm. Hardware limitations imposed us to group camera frames into many, successive 1 s long data files (instead of into one single, large file) before storing them to disk. Since over the time span of one such file (1 s) movements could be neglected, to spare computation time images were coregistered on a file-by-file rather than on a frame-by-frame basis. Next, vessels were selected and segmented (Fig. 6a), and the RBC speed was estimated. Incidentally, the RBC flow direction obtained from the analysis confirmed the classification of arteries and veins that had been identified by their color and pulsatility. Figure 6b illustrates in an example vessel (white arrow in a) the large increase of RBC velocity due to spreading depression wave front, followed by a somewhat smaller temporary reduction. Figure 6c shows that the vessel's width also first increased and then deceased, with a temporal dynamics similar but not identical to that of RBC speed. RBC speed and vessel diameter can then be used to calculate blood flow and its dynamics (not shown).

Figure 6f and g show the full spatiotemporal dynamics of the hemodynamic response to the SD. Note that, as a consequence of the SD, the increase of all hemodynamic variables propagate similarly to a wave sweeping over the cortex from the upper left of the image (~120 s) towards the lower right (~180 s). To illustrate this, Fig. 6d and e detail the temporal sequence of the RBC speed changes in 43 vessels. A gradual increase in response latency can be observed when moving from the upper left towards the lower right of the image (vessels and corresponding time courses are identified by the same color-code).

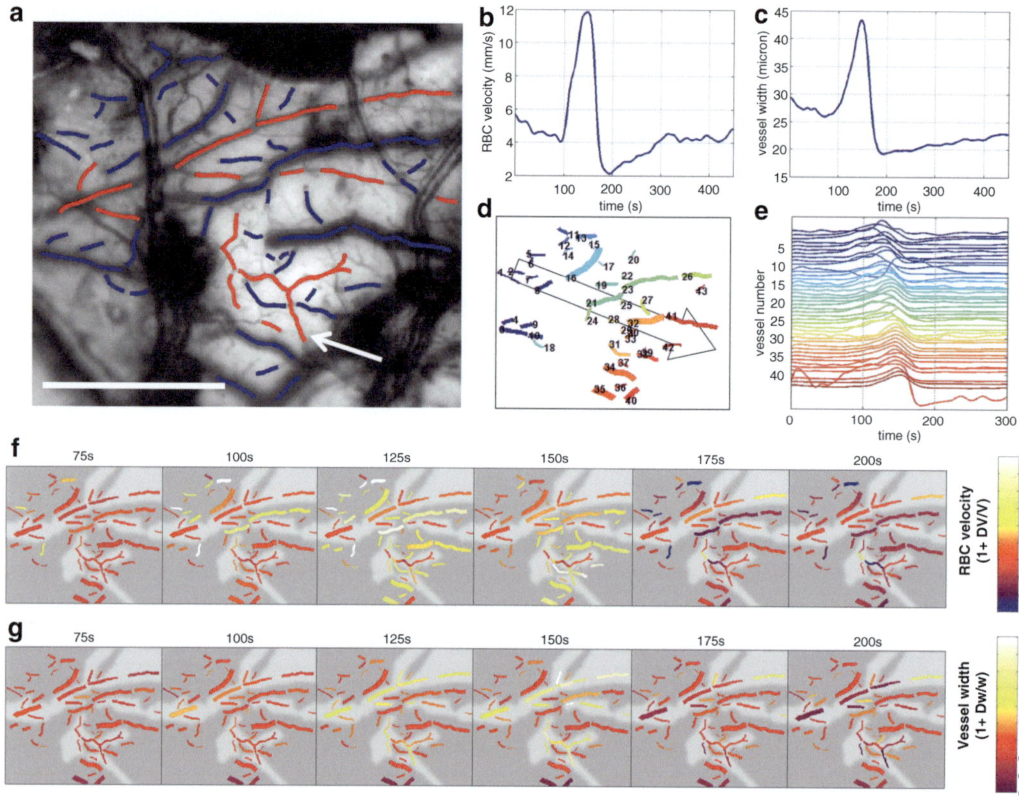

Fig. 6 Estimation results on a spreading depression in anesthetized rat barrel cortex. The cortex was touched in a location not visible here (~5 mm left of the leftmost border of the image) with the tip of a needle, causing a massive wave of neuronal depolarization, followed by a period of hyperpolarization. (**a**) Vasculature surface and segmented vessels. Scale bar: 1 mm. In the subset of surface vessels that were selected, veins appear in *blue* and arteries in *red*. For the vessel designated by the *arrow* in (**a**), *panels* (**b–d**) show plots across time of RBC velocity (**b**) and vessel diameter (**c**). The traces in (**b**, **c**) were smoothed by convolution with a Gaussian (sigma = 50 ms). When considering the full vascular network, the RBC speed increase propagates over the cortex by sweeping over the image from the upper left to the lower right. The vessels selected in (**d**) were numbered (from 1 to 43) and colored (from *blue* to *red*) according to the details of their responses, which are shown in (**e**). *Panels* (**f**, **g**) also show the spatial mapping over time of the increases in RBC velocity (**f**) and vessel width (**g**), with respect to the baseline. Vessels lying on bright and dark background are arteries and veins, respectively. Reprinted from [29], Copyright (2012), with permission from Elsevier

Finally, the histograms in Fig. 7 show RBC velocity, vessel diameter, and blood flow (calculated as the product of RBC velocity and vessel cross section), as well as their changes, for the same population of 43 vessels classified into arteries and veins (Fig. 6a).

As expected, RBC speeds were in the average larger in arteries. Arteries also displayed the larger fractional dilations, in agreement with previously reported results [30].

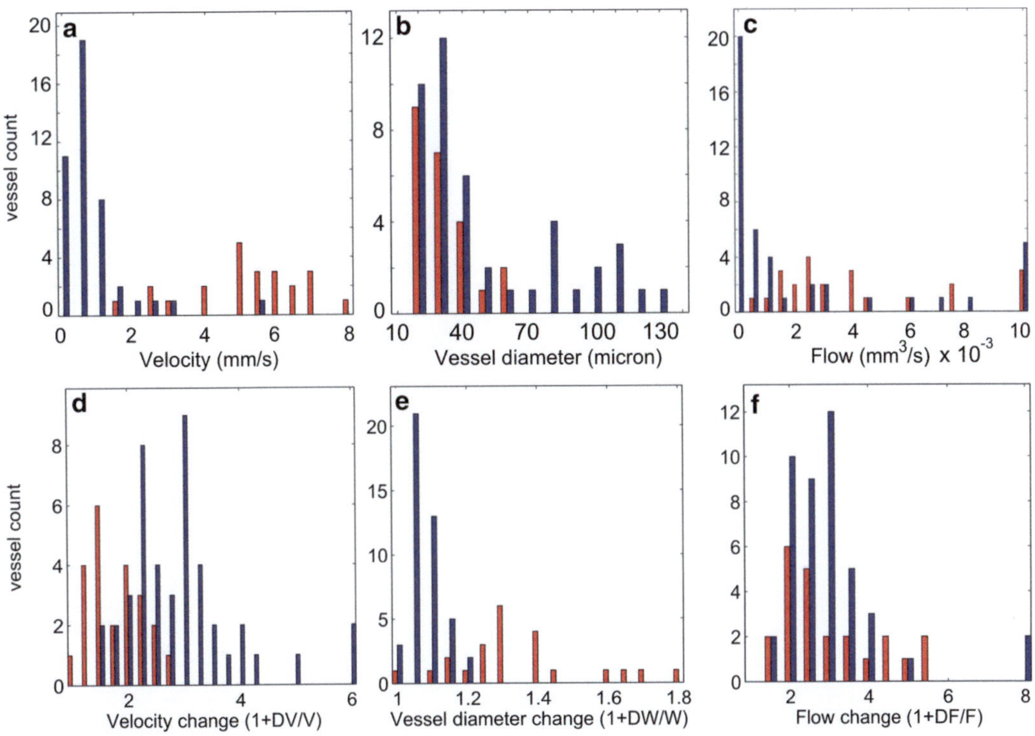

Fig. 7 Spreading depression in anesthetized rat barrel cortex: cumulative data. Histograms were produced for RBC speed (**a**), vessel width (**b**) and flow (**c**), as well as for their changes (**d**–**f**). Veins and arteries are represented separately, respectively, as *blue* and *red bars*. Reprinted from [29], Copyright (2012), with permission from Elsevier

5.2 Sensory-Evoked Hemodynamic Response in Visual Area V4 of the Awake Macaque

High-resolution optical imaging also allows exploring in detail the sensory-evoked hemodynamic responses in awake, behaving monkeys. In the shown example, an awake macaque monkey, trained to fixate a small fixation dot, was presented with a 3 s duration visual stimulus (a drifting grating of optimal spatial and temporal frequency). Hemodynamic responses were imaged in visual area V4, which is an extrastriate visual cortical area, using an isosbestic wavelength of 570 nm.

A clear, overall darkening of the cortex is visible, peaking some 2.5 s after stimulus onset (Fig. 8e). This phenomenon is well known and is due to an increase of blood volume due to activity-evoked vasodilation and perhaps also capillary recruitment [31]. Interestingly, some vessels had a positive and some had a negative RBC velocity response (Fig. 8f). A similar result was obtained also upon sensory stimulation in the anesthetized cat (not shown here, *see* [29]), but not in the anesthetized rat upon in the case of an SD. The measurement and analysis method we present here provide the resolution and SNR that would be required to investigate in greater detail this difference, as well as the precise interactions between the various hemodynamic variables (blood flow, velocity, volume).

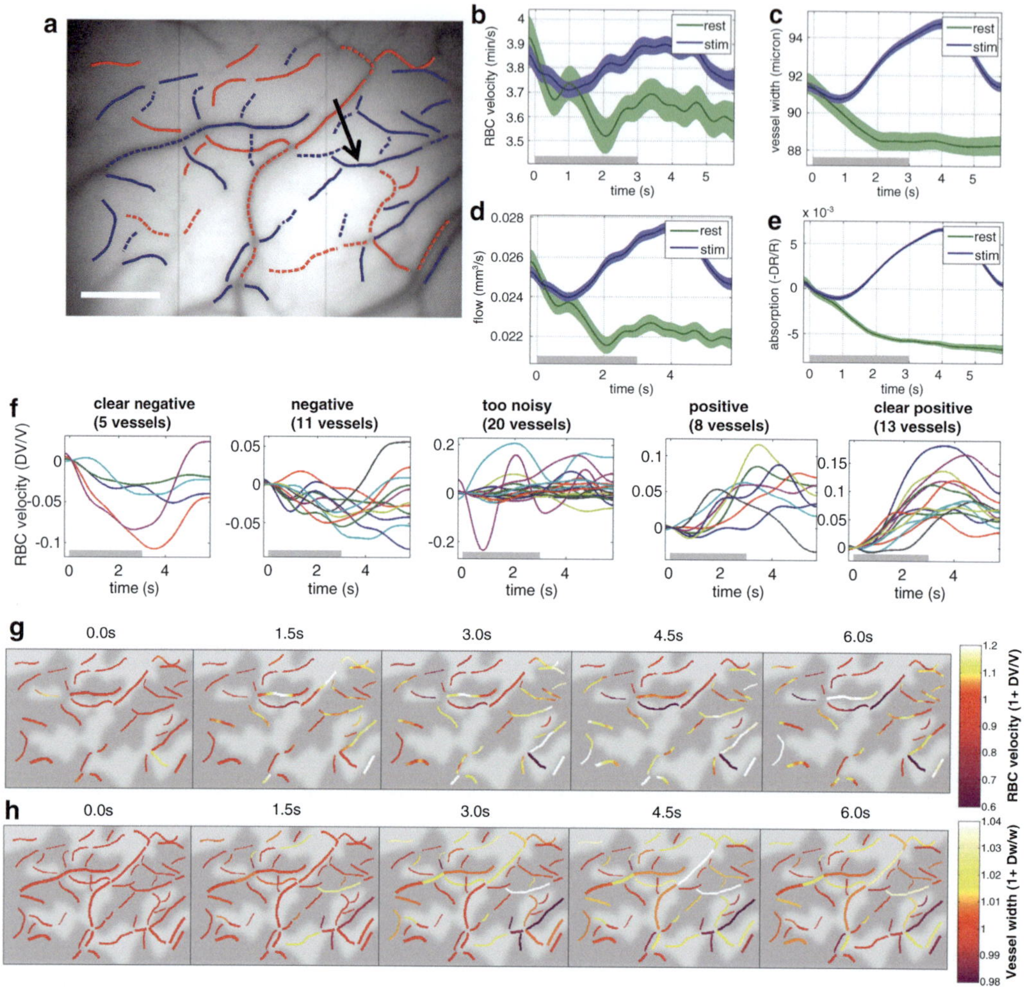

Fig. 8 Responses to a visual stimulation in awake monkey visual area V4. (**a**) Vasculature surface and segmented vessels. Scale bar: 1 mm. Veins and arteries, respectively, appear in *blue* and *red*. In the dotted vessels, RBC velocity could not be estimated due to low SNR. (**b–e**) show the time courses of the various hemodynamic variables for the vessel designated by the *arrow* in (**a**). (**b**) The RBC velocity; (**c**) the vessel diameter; (**d**) the blood flow; (**e**) the absorption signal (~blood volume). In all cases, both the rest ("rest") and the stimulated ("stim") conditions are displayed for comparison. (**f**) shows the estimated velocity responses (stim/rest) for a large set of vessels, grouped into five categories, according to their RBC velocity response. The last two subfigures, respectively, show the mapping over time of the fractional changes in: RBC velocity (**g**) and vessel diameter (**h**); the background behind vessels has been artificially brightened or darkened, to allow their identification as arteries and veins, respectively. Reprinted from [29], Copyright (2012), with permission from Elsevier

We stress that data acquired in awake animals represent the biggest challenge in terms of data analysis because of the multiple additional sources of noise that strongly affect the SNR of the effects of interest. As a result, to obtain the same SEM of the responses as in the anesthetized case, about ten times more signal averaging becomes necessary. The responses recorded from the awake preparation also show a much larger variability as compared to the anesthetized one (vessels with responses classified as "too noisy" in Fig. 8f). For a more detailed comparison between the anesthetized and the awake case, *see* [29]. Figure 8g and h show the full spatio-temporal dynamics of the various hemodynamic variables.

5.3 Blood Flow/RBC Velocity Estimation

The RBC speed estimation method presented here should be seen as complementary to the other existing techniques for blood flow measurement, rather than as an exclusive alternative. For instance, in spite of its lack of spatial resolution, classical laser Doppler measurements provide excellent SNR and are comparatively simple to implement. Two-photon fluorescence microscopy has access to below-surface vessels. Moreover, its ultra-high resolution allows distinguishing different flow speeds within the same vessel (inhomogeneous flow) or even to measure changes in RBC density and vessel size [11, 23, 24]. However, the measurements being limited to very few vessels at a time make imaging vascular network dynamics difficult. This shortcoming is not shared by the approach set forth here, which allows estimating RBC speed in large vascular networks simply starting from wide-field optical data. Most important, it does not require any extrinsic tracer injection or fluorescent RBC labelling [11]. Its application in humans [6–9] is therefore relatively straightforward.

The other wide-field technique, laser speckle imaging is able to spatially map blood flow and its changes in tissue (most likely imaging capillary perfusion) devoid of visible vessels, but its sensitivity to blood flow in visible vessels appears to be surprisingly low [20, 21]. Moreover, the two named Doppler Effect-based techniques measure blood flow in the sense of perfusion, and disentangling RBC speed from density of flowing RBCs is difficult.

The main difficulty with extracting RBC motion from such wide-field data is low SNR, as a result of large shot-noise, some intrinsic ambiguity in the data due to the fact that all RBCs are identical, and inter-trial variability. Recently, we have proposed an original algorithm able to satisfactory deal with these difficulties and adapted it to such wide-field data [28], together with another algorithm based on the Radon transform, originally developed for two photon fluorescence microscopy data [27]. Here we describe the whole processing flow that integrates this approach, and apply it to three radically different datasets. We show—to our knowledge for the first time—that RBC speed changes can be detected, reliably, in large vascular networks, not only as a result of large, non-physiological stimulation, but also as a response to neuronal activation, evoked by physiological sensory stimuli in the awake primate.

We stress that, like all methods, also the one presented here has a limited applicability range. Yet, it can be somewhat adjusted by adequately choosing magnification, frame rate, and illumination intensity. Typically, RBC speed estimation is reliable up to 10 pixels/frame. If frame rate and SNR are high, this limit can be doubled, but speed estimation above 20 pixels/frame fails systematically—at least with current equipment and/or analyses. Similarly, the thinnest vessels our method can deal with (~2 pixels in diameter) can be set by choosing the adequate magnification. Conversely, large vessels pose an intrinsic problem, independent of magnification: due to the long optical path, contrast differences resulting from individual RBCs are difficult if not impossible to detect. As a result, the approach proposed here fails, in practice, in vessels larger than ~0.1 mm. It might, nevertheless, be possible to extract RBC speed from the border of thicker vessels, because there the vessel is thinner than in its middle. However, it shall be kept in mind that, in that case, the detected RBC speed values are unlikely to be representative for the whole vessel thickness due to the presence of an inhomogeneous flow regime.

5.4 Blood Volume Dynamics: Vessel Diameter Estimation

Vessel dilation is an important component of hemodynamic responses in both physiological and pathological conditions as, for instance, following venule occlusion or strokes (e.g., [40]). Our analysis suite allows the estimation of the individual blood vessels' diameters and their changes as a function of time, therefore enabling to study the dynamics of blood volume changes at the scale of a vascular network. For instance, we have shown that, during a spreading depression, increases (and decreases) in vessel diameters are concomitant with an increase (decrease) in RBC velocity. This result appears to be in contradiction with a recent study, carried out with confocal microscopy [41]. There, no local changes in vessel diameters where observed together with the redistribution in RBCs induced by a spreading depression. Clearly, further work is needed to fully explain these differences. Yet, our data suggest that vessel diameter changes may be missed when using a too small field of view, underscoring the importance of measuring hemodynamic responses at the level of comparatively large vascular networks.

As opposed to confocal microscopy (e.g., [42]), our wide field of view allows to track these changes over a large (several mm^2) cortical territory, from large-sized vessels (for diameter estimation, vessel thickness is not limited by the need to resolve individual RBCs), down to the level of small capillaries. Finally, estimating both RBC velocity and vessel diameter allows to calculate local blood flow and its variations.

5.5 Oximetry

In its current version, the method presented here does not provide information about local oximetric changes. However, the latter could be obtained, simultaneously with the blood flow and volume measurements by a relatively small modification. Indeed, [43] have shown that two-dimensional high-resolution optical imaging spectroscopy is possible by using several multiplexed wavelengths and a fast filter-wheel. If all these wavelengths are chosen from the green part of the spectrum, their RBC contrast can then be relatively easily adjusted to similar values, thus allowing using all of them for the RBC speed measurement. In this way one would still have the full frame rate available for the RBC velocity measurement, rather than the slower frame rate characterizing each individual wavelength alone.

6 Imaging the Retina

Progress in the battle against blindness depends on the development of new technologies that can facilitate early detection and accurate monitoring of retinal pathologies such as age-related macular degeneration (AMD), diabetic retinopathy, and glaucoma. New technologies should improve our ability to provide early diagnosis, facilitating preventive treatment and better understanding of retinal pathogenesis, which can lead directly to the development of novel treatments by either drugs or surgery. Treatment follow-up with such technologies fall in the category of personal medicine receiving a growing recognition. In recent years, several attempts were made to develop new retinal imaging technologies [44–49]. However, most systems available for clinical use monitor structural changes in the retina and do not provide direct information about the retina's functioning. The RFI is a novel device [8, 9, 33]. It offers a noninvasive diagnostic approach to retinal function assessment and provides easy, direct, qualitative, and quantitative imaging of four parameters: (1) blood flow velocity, (2) capillary perfusion maps, (3) blood oximetry [9], and (4) metabolic intrinsic state [9, 33]. All of these parameters directly assess retinal function and are thought to be modified in the course of retinal diseases. Most importantly, as retinal functional changes precede structural changes, quantitative imaging of these four parameters across the retinal surface should ultimately provide a useful tool for early detection of blinding diseases.

The RFI's imaging approach is similar to the one first used in the brain. Adapting brain functional imaging techniques to the retina provides imaging of functional parameters directly related to the vitality and viability of the retina superior to most current powerful technological methods, which mostly evaluate structural changes in the retina.

The RFI (*see* Box 3) is capable of directly determining blood flow velocity in the capillaries. Since many retinal diseases affect small vessels, information regarding blood flow, particularly in the capillaries, is applicable for diagnosing retinal abnormalities. Moreover, by analysis of where in the retina blood motion occurs, microvascular maps are obtained that complement the functional

Box 3
Principles and Recording Protocols of the Retinal Function Imager

The retinal function imager (RFI) system combines digital fundus imaging with functional optical imaging, providing a turnkey system with flexible parameter settings and open standard data storage. In order to measure the previously described parameters, the following enhancements have been incorporated into the fundus camera:

1. *Rapid sequential imaging*. A fast 1,024 × 1,024 pixel CCD camera (such as Dalsa 1M60P) in conjunction with fast stroboscopic illumination (2 ms flashes) at green wavelength (548 ± 8 nm) takes snapshots of the retina at rates high enough (50–100 Hz) to reduce interframe retinal motion and to follow erythrocytes moving at up to 10 mm/s. This provides high-resolution imaging of flow information obtained from small vessels, down to the level of a single RBC moving in a single capillary.
2. *Multispectral imaging*. A fast filter wheel has been integrated, capable of rapidly switching up to four illumination wavelengths, allowing multiple wavelength image acquisition with little or no global eye movement.
3. *Stimulus generator*. An LED-based stimulus generator can stimulate the retina with a visual patterned stimulus with a specified pattern, frequency, and duration. This is useful for functional signal imaging.

Figure 9 summarizes the principles the RFI uses to measure blood flow velocity parameter. Sequences of up to eight frames each are obtained from humans (following pupil dilation) by methods familiar to operators of standard fundus cameras. High optical magnification (standard 20° setting on the fundus camera optics) is preferred for imaging modalities that rely on erythrocyte motion; for all other types of imaging, both narrow- and wide-angle settings are appropriate. A subject may sit for 5–15 min per session, and image sequences and initial results can be quickly reviewed during the acquisition session, allowing reacquisition whenever needed.

The raw data obtained in a single sequence resemble familiar digital black-and-white red-free images, except that they were taken only 10–20 ms apart and, depending on requirements, at different wavelengths (e.g., for qualitative oximetry, choroidal visualization). Automatic algorithms perform subpixel registration of the images within a sequence, and then apply differential analysis and other algorithms to reveal the minute differences among them. This analysis reveals image differences due to the phenomena described above (*see* Fig. 10b and c). Differences are subject to further analysis and interpretation as described below.

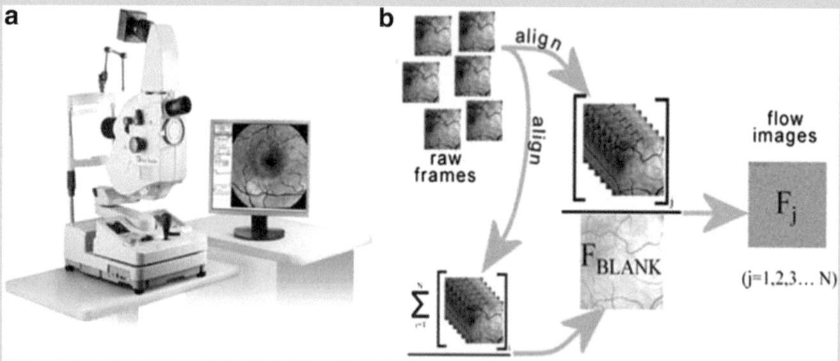

Fig. 9 The retinal function imager (RFI). (**a**) The RFI is based on a standard fundus camera with some modifications. (**b**) The RFI can acquire sets of up to eight sequential retinal images from a human subject at 55–100 Hz, using flash illumination (548 ± 8 nm). Reprinted from [8], Copyright (2009), with permission from Springer

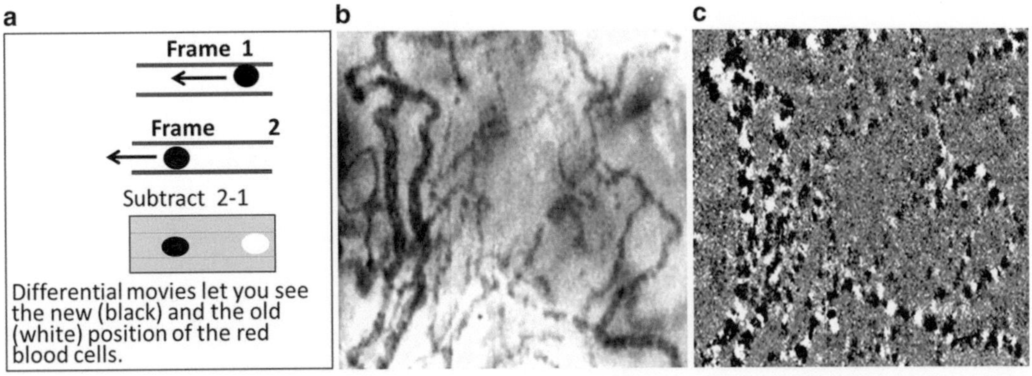

Fig. 10 Direct visualization of blood flow in a diabetic patient. (**a**) (*Left*) a single moving erythrocyte in a single vessel is depicted in the first snapshot (frame 1). In frame 2 one sees the same erythrocyte in its new position 20 ms later. Subtracting the two frames from each other eliminate any detail which does not move and largely enhance the present and past position of the red blood cell. (**b**) (*Middle*) The *middle panel* shows the retinal image. (**c**) (*Right*) One frame from a flow movie showing individual red cells clusters (dark) spot

maps with microvascular information. From measurement of how reflected light varies with wavelength, the relative concentration of hemoglobin chromophores can be assessed, both in vessels and in capillary background, for information about intravascular oxygen content related to tissue vitality [50, 51]. Finally, the response of the retina to photic stimulation is visible with the functional intrinsic signal imaging capability of the RFI. This chapter describes the imaging of all four parameters in detail and provides results from pilot clinical studies.

6.1 Retinal Blood Flow Velocity

Under green illumination, hemoglobin within the erythrocytes in the bloodstream provides a natural, high-contrast chromophore for tracking blood flow. Figure 10c (right) illustrates the contrast thus observed upon differential imaging without any additional processing.

Cross-correlation matching of moving erythrocyte patterns within an image series gives a direct measure of velocity. Re-registration and differential processing of a series of images taken at 55 Hz produces a "flow movie," in which it is possible to follow the motion of individual clusters of RBCs—or even single RBCs—in vessels as small as capillaries.

Figure 11 shows a blood velocity map and a frame from a differential movie. Cross-correlation matching determines the relative offset of path segments in sequential images that contain approximately the same pattern of moving blood cells [9]. The size of this offset, suitably calibrated and multiplied by the frame rate, gives blood flow velocity in units of mm/s in the small vessels including capillaries (Fig. 11b).

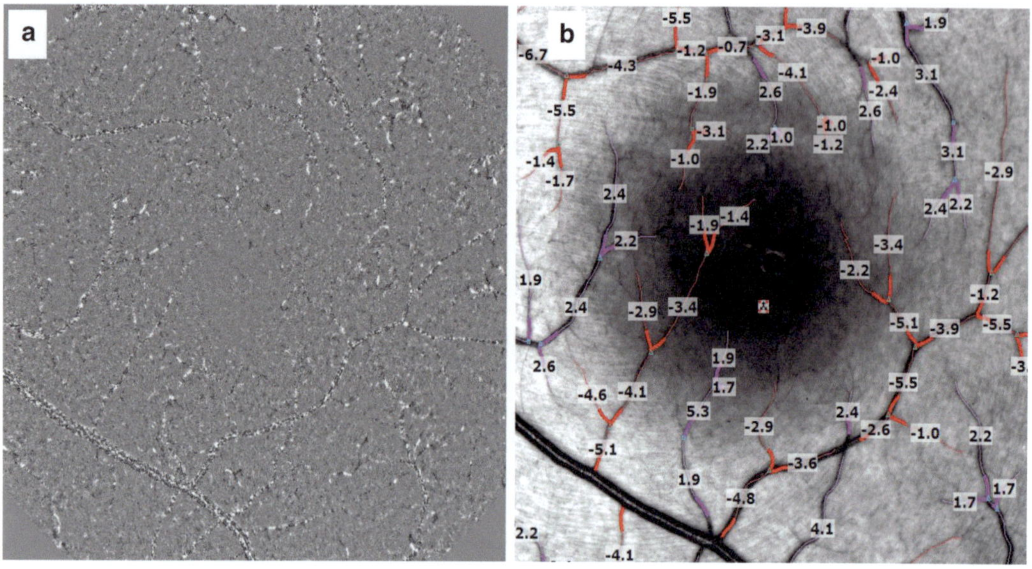

Fig. 11 Quantification of retinal blood flow velocity in the human retina. (**a**) Single frame of a flow movie from the same human retina. Erythrocyte clusters (*black*) and gaps (*white*) move through the small vessels shown here at an average velocity rate of 2 to 3 mm/s. (**b**) RFI image showing calculated velocity in mm/s, obtained automatically. The image shows several arteries (*reddish*) and veins (*pink*) that were manually selected for quantification. A negative value indicates blood flow away from the heart, whereas a positive value indicates blood flow towards the heart. Reprinted from [8], Copyright (2009), with permission from Springer

Flow movies provide a tool for qualitative assessment of retinal blood flow rates (to view blood flow movies, see electronic supplementary material). In order to chart a complete map of retinal blood velocity, it is necessary to quantify the velocities inside each microvessel displayed in each movie in a direct and objective manner. Such quantified measures assist in detecting and monitoring clinical diseases, in investigating the effects of various physiological challenges such as flicker stimuli on blood flow velocity, and for longitudinal follow-up of the effects of drugs. Currently, the RFI provides a velocity map; conversion of the velocity units to flow units has not yet been implemented. To accomplish this goal, the precise vessel diameter must be established, a task already accomplished by [52]. Retinal abnormalities can also provide an indication of the effects of systemic diseases. For example, retinal lesions with vascular characteristics such as vessel narrowing [53] and nonvascular features such as hemorrhages, microaneurism, and other features [54] are important indicators of serious systemic diseases such as diabetes or hypertension. Thus, an instrument that can robustly measure blood flow velocity might provide a better diagnostic tool for increased cardiovascular morbidity and mortality in either diabetic [55] or hypertensive patients [56, 57]. The functional results obtained by such a method can also help to differentiate between diseases whose structural or anatomic effects, albeit pronounced, may be open to multiple interpretations.

The first step in using the RFI for clinical diagnostics based on blood velocity is to establish the baseline for normal subjects. Two pilot studies have recently reported their results [58, 59]. A recent pilot clinical study with the RFI show a significant velocity decrease in arteries and veins of patients with nonproliferative diabetic retinopathy (mean age, 63.3 ± 9.1 years; male/female, 22/15) compared with healthy controls (mean age, 59.73 ± 10.0 years; male/female, 16/8) [7]. On the other hand, they detected increased blood flow velocity compared with controls in retinal arteries and veins of patients with early diabetes mellitus without diabetic retinopathy (mean age, 58 and 61, for diabetes mellitus patients and healthy volunteers, respectively) [60]. In a different study, the RFI detected reduced blood flow velocity in exudative AMD eyes compared with fellow dry AMD eyes (in all sizes of veins and in small arteries; mean age, 78.5 ± 4.5 years) [61]. Central serous chorioretinopathy shows reduced retinal circulation in retinal function imaging was reported [62]. Retinal blood flow velocity in retinitis pigmentosa was found to be lower than normal [63].

Drug effects on the retina have also been studied with the RFI. Following intravitreal bevacizumab (Avastin) injections, a distinctive pattern of change in patient retinal blood flow velocity was noted in responders versus nonresponders [61]. The effect of topical nipradilol on retinal microcirculation was evaluated by RFI [64]. In a different study [65], compared venous blood flow velocity with central retinal volume in patients with various ocular diseases, and observed a strong correlation between venous retinal blood flow velocity and central retinal volume. The RFI may thus be a useful tool for evaluating changes in retinal thickening. Landa and his colleagues also reported cilioretinal arteries in diabetic eyes are associated with increased retinal blood flow velocity and occurrence of diabetic macular edema [64].

6.2 Capillary Perfusion Map

The ability to directly detect the motion of RBC aggregates provides a tool to map the surface retinal vasculature in detail. Following image registration, pixel value distribution parameters are analyzed to locate blood motion, tracing microvasculature based on motion contrast rather than on total reflectance contrast. The RFI's noninvasive capillary perfusion map (nCPM) shows vessels that are indistinct or invisible in even sharp red-free images. Indeed, small vessels often appear as clearly as larger ones on these maps because the red cell clusters moving in these vessels are more distinct and less motion blurring occurs (Fig. 12a, b) [9, 66].

Recently, in studies with an improved algorithm, as much and often more microvascular detail could be seen on the nCPM than on corresponding fluorescein angiography (FA) images [66]. The RFI can assist in cases where adverse reactions to fluorescein injection have been observed [67]. The most significant improvements of nCPM maps over FA are (1) they are noninvasive; (2) far more

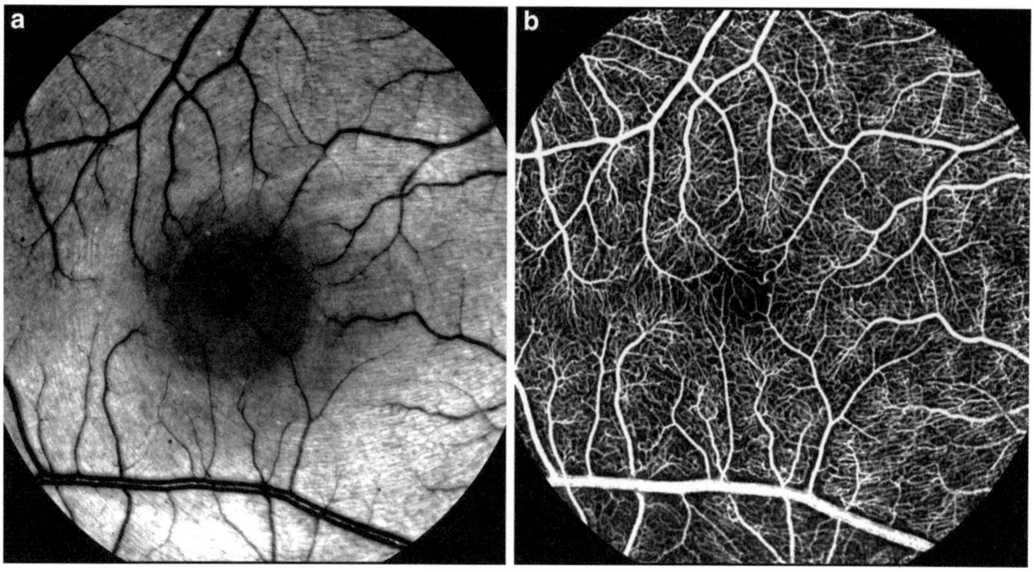

Fig. 12 Invisible vasculature (in red-free images) revealed noninvasively by the RFI. (**a**) Red-free image of a healthy subject. (**b**) The RFI's capillary perfusion map showing fine microvasculature detail (scale bar = 1 mm)

Table 1
Comparison of measurements from the retinal function imager capillary perfusion maps and measurements from intravenous fluorescein angiograms from the same eye

Microvascular measurements	IVFA	RFI	*P* value
Vessel loops per eye	0.970 (±1.503)	4.424 (±3.580)	<0.001
Eyes with 0 loops	20 (61%)	8 (24%)	
Eyes with 1–3 loops	10 (30%)	7 (21%)	
Eyes with >3 loops	3 (9%)	18 (55%)	
Vertical anastomoses per eye	0.424 (±0.540)	1.333 (±1.514)	<0.001
Eyes with 0 anastomoses	23 (70%)	11 (33%)	
Eyes with 1–2 anastomoses	8 (24%)	14 (42%)	
Eyes with >2 anastomoses	2 (6%)	8 (24%)	
Order of vessel branching	4.576 (±0.708)	5.182 (±0.638)	<0.001
Fourth order visible	18 (55%)	4 (12%)	
Fifth order visible	11 (33%)	19 (58%)	
Sixth order visible	4 (12%)	10 (30%)	
Foveal avascular zone (mm^2)	0.753 (±0.757)	0.352 (±0.181)	0.017

IVFA intravenous fluorescein angiogram, *RFI* retinal function imager detailed; (3) allow follow-up at any frequency; and (4) show where RBCs are moving at approximately normal speed. In the latter context, variations between normal capillaries and capillaries exhibiting reduced RBC velocity cannot be detected with FA. At present, one cannot infer from nCPM maps regions of leakage readily detected with FA. To detect leakage, we used the RFI to obtain nCPM just after oral administration of fluorescein dissolved in juice. By the end of RFI session, one fluorescence imaging of the fluorescein yield a late phase FA revealing the leaks. Recent comparative study of FA and nCpM by an independent group [68] confirmed that the nCPM maps are indeed far more informative than the gold standard FA as seen in Table 1.

6.3 Multispectral Imaging for Retinal Oximetry

The balance of oxygen supply and demand in the retina is closely regulated to maintain the processes of visual perception. Alterations in either oxygen supply or consumption might directly indicate the early onset of retinal abnormalities [69–72]. Thus, it is important to have a tool for qualitative and quantitative assessment of oxygen utilization in the tissue.

The difference between the absorption spectra of oxyhemoglobin and deoxyhemoglobin can be used to determine the oxygenation of blood with multispectral imaging methods. In multispectral imaging mode, the RFI can perform spectroscopic decomposition to qualitatively assess the oximetric state of the retina. For example, Fig. 13 shows a qualitative oximetry map of a healthy volunteer. The oximetry image was obtained by acquiring retinal images at two wavelengths (oximetric, 575 ± 5 nm; and isosbestic, 569 ± 5 nm). Differential decomposition analysis was used to generate the oximetry image and will be described in detail elsewhere.

6.4 Functional Metabolic Signal

Retinal reflectance changes in response to photic stimulation carry information about metabolic processes underlying light responses in the retina. High-resolution, contrast agent-free optical imaging

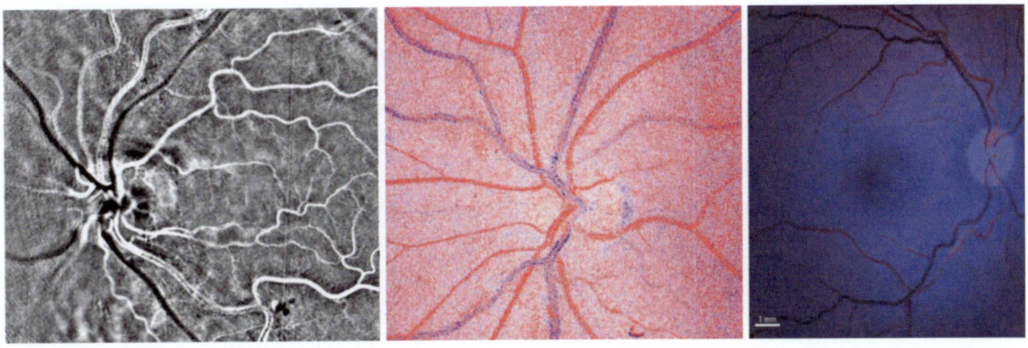

Fig. 13 Three qualitative oximetric images obtained with the RFI from the retina of healthy volunteers. Arteries, veins, and capillaries appear in different colors because of the different oxygen saturation levels of arterial, venous, and capillary blood

based on intrinsic signals [6, 73] in vivo has significantly contributed to understanding of the functional architecture of the neocortex and the mechanisms underlying its development [3, 4]. It reveals activity-dependent changes in light reflectance, recorded using a digital camera with high spatial and temporal resolution. Such functional signals are usually small, originating from activity-dependent metabolic, hemodynamic, and fast and slow light-scattering changes [31, 74–77].

The RFI is capable of imaging outside the absorption range of photoreceptors under near-infrared light (750–840 nm) and can be used to optically monitor retinal activity in response to a well-defined visual stimulus (562 ± 20 nm). The difference between the photo-stimulated and prestimulated images is used to determine the metabolic state of the retinal compartments.

Figure 14 shows how the RFI optics was set to be able to image the functional signal.

Figure 15 illustrates the pioneering imaging of the functional signal in the cat retina [9, 33]. It shows changes in light reflectance in response to a visual stimulus (542 ± 20 nm) flashing at 100 Hz for 1 s during recording in the near-infrared (750 ± 20 nm).

More recent experiments with more complex patterns evaluated the effects of other visual stimulus parameters as well as the retinal status. Some unpublished results from monkeys are depicted in Fig. 16.

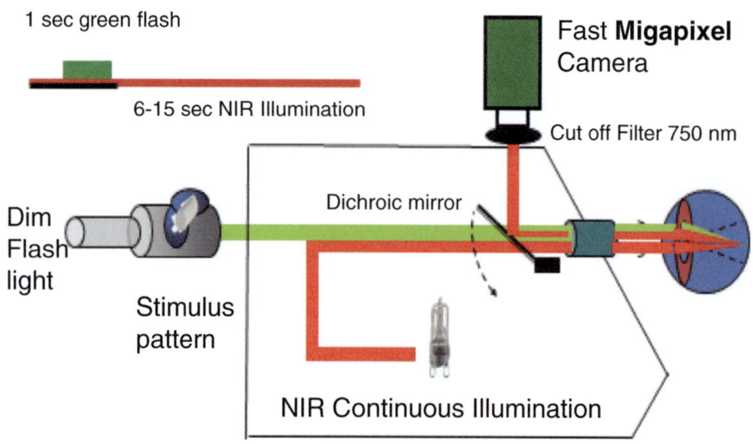

Fig. 14 Modification of the RFI to be able to stimulate with visible light and image the metabolic signals in the near IR. One of the image plans was used as a photo-stimulation port. Whatever image was produced falls directly on the retina (*green path*). Here the image was produced by a thin disk containing a cross or a ring shaped holes. Images on the retina can thus be produced also by computer controlled projector. The retinal was also illumined with near-IR light (*red pathway*). Changes in near-infrared reflections corresponding to metabolism reached the camera but the green stimulation light was blocked by a long pass filter

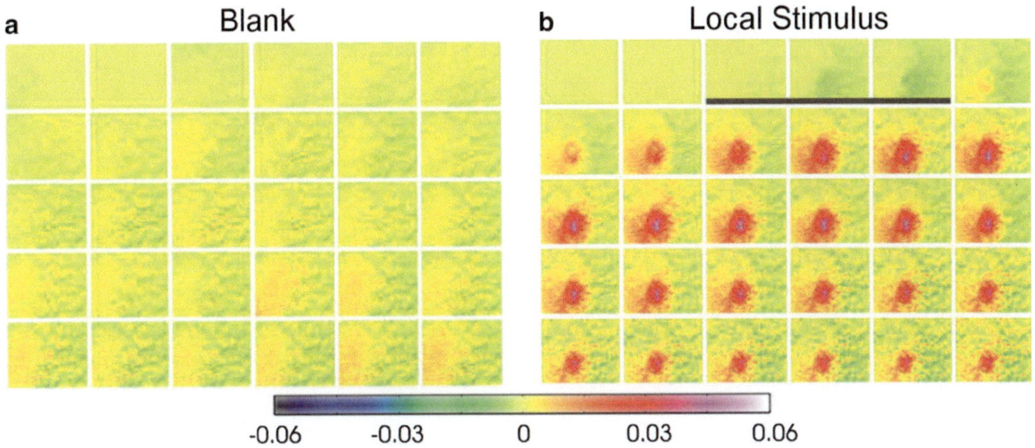

Fig. 15 Time course image series of the functional signal from the cat retina. (**a**) Control time series without photic stimulation. (**b**) Near-infrared local changes in the retinal reflection of a cat eye, evoked by weak local photic stimulation in the visible wavelength range. While a 1-s local 100-Hz train of green flashes was delivered to the eye, video images of the feline retina were taken in the near-infrared at 400 ms/frame. Time increases from left to right, then from the top to the bottom row. The stimulus period is marked by a *black bar*. The very slow time course of the signal and its wavelength dependency (not shown) indicates that it is not the well-known hemodynamic response, nor it is rapid enough to be a signal directly due to neural activity or bleaching. The *color scale* indicates the fractional change in the reflected near-infrared light. The width of each image is 8 mm. Reprinted from [8], Copyright (2009), with permission from Springer

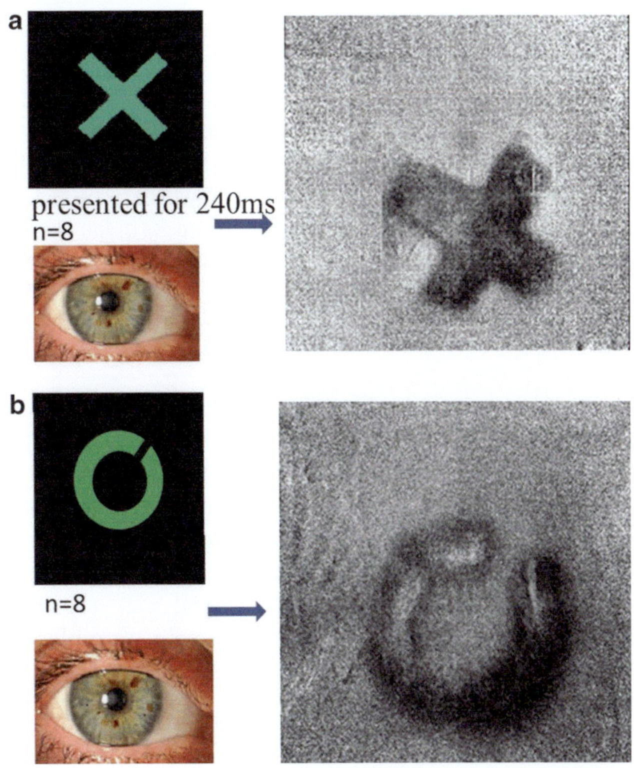

Fig. 16 Imaging the metabolic responses in response to a cross or a ring. (**a**) Retinal image 3 s after flashing the *green cross* for quarter of 1 s. (**b**) Retinal image 3 s after flashing the *green ring* for quarter of 1 s

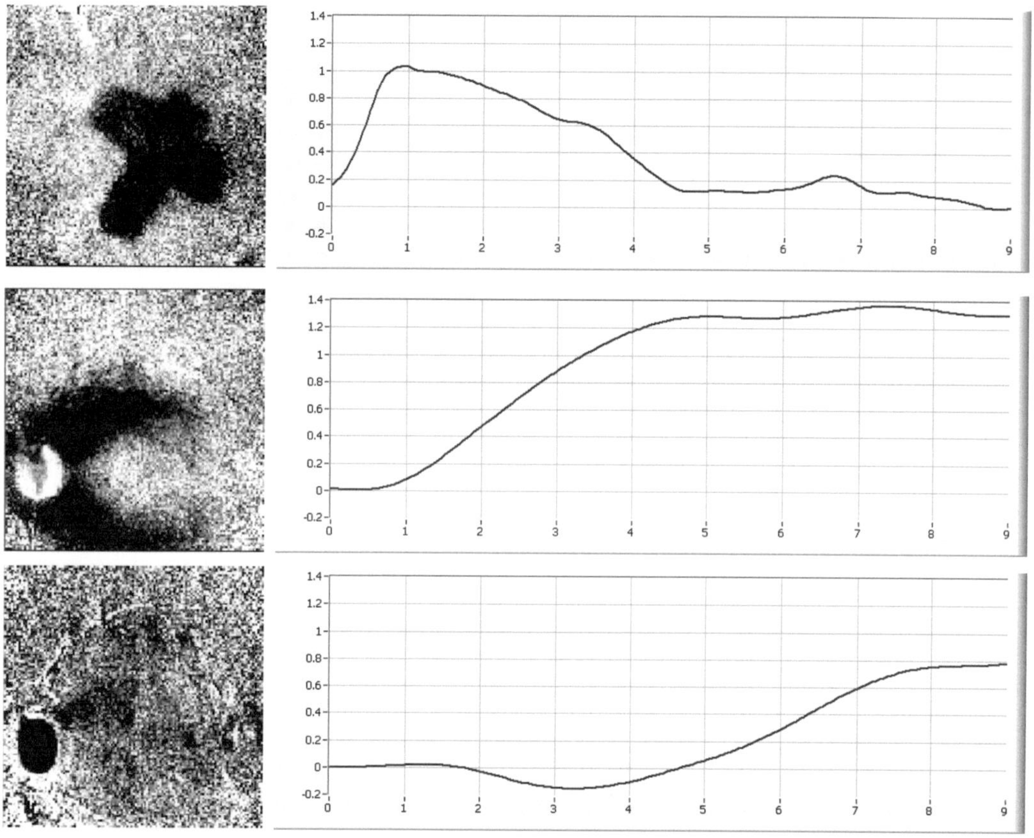

Fig. 17 The three components of the metabolic signal revealed by ICA analysis. The first component includes the projected image of the cross. The time course at the right shows that this response has a relatively fast time course peaking at about 1 s and it decays slowly to baseline in 4 s. The second component has a slower rise and decay and it revealed the activated axonal trucks of the stimulated ganglion cells. The third component coincides with the blood vessels-rich optic nerve. Its time course resemble the two component time course of the oximetric signal in the neocortex

The signals reported here originated from multiple mechanisms which are different at different retinal sites. Figure 17 shows time course and the shape of individual component in the response to a briefly projected cross separated by ICA analysis.

Additional experiments were also carried out on a monkey model of glaucoma with pressure IOP increase in one eye only. The functional signal was very different in each case (D. Nelson and D. Yitzhaky et al., unpublished results). Similar experiments conducted on cats and monkeys [78–83] have provided spectacular functional maps resulting from photic pattern activation and are described in detail in other reviews in a special issue of Japan J Ophthalmology edited by K. Tsunoda in 2009.

7 Summary

Detection of functional parameter abnormalities in the retina may permit diagnosis of diseases and disease progression before anatomic abnormalities become evident, allowing treatment intervention before irreversible retinal damage occurs. Furthermore, treatment can be improved by noninvasive follow-up. The RFI's multimodal approach to functional imaging also opens research and drug development opportunities respecting a wide range of retinal diseases, beyond the capabilities of structural imaging. The ongoing advances in the treatment of retinal disease, in conjunction with better diagnostic tools, will improve outcomes for patients suffering from some of the most prevalent ophthalmic diseases. The same optical approaches can be applied to almost any part of the body via endoscopy or minimally invasive micro-endoscopy. Importantly, Jiang and her colleague have recently shown that human conjunctival microvasculature status could be assessed with an RFI [84]. This example is significant because the conjunctiva receive the same circulation as the brain. Therefore it offers optical monitoring of brain microcirculation even without shining light on the retina.

Acknowledgements

This work was supported by CNRS and Ministère de la Recherche (ACI Neuroscience), by a CNRS PICS grant to IV, by the Foundation NRJ. Part of this work was supported by the European Union (Integrated Project BrainScales, IST-FET-2010-269921), by an ANR grant to IV (French-Hungarian international project "Multiscalefunim"), and by a grant from the Grodetsky and Dominic centers to AG. The work on the retina was supported by the Weizmann Institute of Science and Optical Imaging Ltd. grants to AG. Sub-parts of the text have been reprinted from [29], Copyright (2012), with permission from Elsevier, and from [8] copyright (2009) with permission from Springer.

8 Appendix: Surgery, Optical Imaging Data Acquisition Stimulation Parameters

8.1 Surgery

Details for the surgeries in all three preparations (rat, cat, monkey) can be found elsewhere [29, 30, 85]. Briefly: in all three cases, a recording chamber was implanted on the skull—either in an acute procedure (rat, cat) or chronically (monkey). A cranial window was opened above the cortical area of interest and the dura was either left in place (rat) or removed (cat, monkey), and in the awake monkey case, replaced by an artificial transparent dural substitute

[86]. The chamber was filled with saline (rat), silicon oil (cat), or agar (monkey) to stabilize the cortex against movements (heartbeat, respiration…) and closed with a cover-glass to provide a flat optical interface and, in the case of the monkey, a sterile seal. Anesthesia was urethane in the rat, sodium pentobarbital for the cat and isofluorane for the monkey (in this case for surgery only). All surgical and maintenance procedures were in agreement with NIH guidelines.

8.2 Stimulation Protocol

In the rat case, a spreading depression was elicited by carefully touching the cortex with a needle about 2 mm away from the border of the imaged area (for details, see [29]). In the alert monkey case, the experimental paradigm was a simple fixation task. A trial started when the monkey began fixating on a fixation point (0.1°), displayed on a CRT screen. After 200 ms, a high luminance contrast drifting square grating (0.125–0.25 cycles/deg, 8–32°/s) appeared for 3 s, except in the "blank" conditions (no grating). The stimulus was then turned off, and the monkey had to continue to fixate until the fixation point disappeared, for a total fixation period of 6.5 s. An isoluminant, uniformly gray-screen inter-trial interval of ~10 s followed. Data from trials where the monkey broke fixation were rejected.

8.3 Optical Imaging

The light source was a tungsten-halogen lamp (100 or 150 W, Zeiss, Germany) from which specific wavelengths can be selected using adequate interference filters (Omega Optical, Brattleboro, USA). The cortex was illuminated using light guides. To maximize RBC contrast and thus blood vessel contrast in general, we used green light, since, in this part of the spectrum, hemoglobin (Hb) absorption is large.

To reach a satisfactory illumination level despite small exposure times corresponding to high frame-rate collection (100 Hz in the rat, 200 Hz in the monkey) and high magnifications, we used an interference filter with a rather large transmission window (30 nm), centered at 540 nm in the case of the rat, and a more narrow one (10 nm) centered at the isosbestic point of 570 nm in the case of the monkey. Images were acquired using a commercial CCD-based imaging system (Imager 3001, Optical Imaging Inc., Germantown, USA). A macroscope [87] was mounted onto the camera. It was composed of a 50 mm lens and an inverted 135 mm lens, yielding a magnification of 2.7, such that our spatial resolution was 9×9 μm/pixel, in the case of the rat. In the case of the monkey, we used an upper lens of 100 mm, yielding a magnification of 2 and a spatial resolution of ~12×12 μm/pixel. To obtain an optimal identification of vascular activity, the optical system was focused onto the cortical vasculature.

In the case of the monkey, data acquisition lasted the time of a trial (~6 s). Data were written onto the hard disk after each trial.

In the rat case, data acquisition lasted 7.5 min and was split into 1 s long periods of continuous 100 Hz acquisition (100 frames), interleaved with 20 ms pauses necessary to store the data onto the hard disk, as required by the data acquisition software we used for this purpose (Optical Imaging Inc., Germantown, USA).

Veins and arteries were first identified by visual inspection (morphology, pulsation, color of vessels...) through a microscope and through the camera. This classification was then integrated with oximetric information obtained from an "oximetric image" of the vasculature [30]: Two images of the cortical surface were taken, one at 560 and the other at 540 nm illumination (peaks of deoxy- and oxyhemoglobin absorption, respectively). One image was then divided by the other and the gray level scale of the resulting "oximetric image" was shifted, such that the parenchyma, which contains mainly capillaries, appeared to be in the middle of it (gray). With this procedure, the various vessel types in the image were thus automatically shifted to the bright or the dark half of the grayscale according to their different degree of oxygenation, allowing for a further identification of veins and venules (dark), as well as of arteries and arterioles (bright).

References

1. Vanzetta I, Grinvald A (2008) Coupling between neuronal activity and microcirculation: implications for functional brain imaging. HFSP J 2:79–98
2. Bonhoeffer T, Grinvald A (1996) Optical imaging based on intrinsic signals: the methodology. In: Toga AW, Mazziotta JC (eds) Brain mapping; the methods. Academic, San Diego, pp 55–97
3. Grinvald A, Shoham D, Shmuel A, Glaser D, Vanzetta I, Shtoyerman E, Shlovin H, Wijnbergen C, Hildesheim R, Arieli A (1999) In-vivo optical imaging of cortical architecture and dynamics. In: Windhorst U, Johansson H (eds) Modern techniques in neuroscience research. Springer, Heidelberg, pp 893–970, Chapter 34
4. Grinvald A, Malonek D, Shmuel A, Glaser D, Vanzetta I, Shtoyerman E, Shoham D, Arieli A (1999) Intrinsic signal imaging in the neocortex. In: Imaging of neuronal activity. Cold Spring Harbor Laboratory, Cold Spring Harbor, NY, pp 1–17, Chapter 45
5. Grinvald A, Sharon D, Omer DB, Vanzetta I (2011) Imaging the neocortex functional architecture based on multiple intrinsic signals; implications for hemodynamic based functional imaging. In: Helmchen F, Konnerth A, Yuste R (eds) Imaging in neuroscience. Cold Spring Harbor Laboratory, Cold Spring Harbor, NY, pp 907–926, Chapter 83
6. Grinvald A, Lieke E, Frostig RD, Gilbert CD, Wiesel TN (1986) Functional architecture of cortex revealed by optical imaging of intrinsic signals. Nature 324:361–364
7. Burgansky-Eliash Z, Nelson DA, Bar-Tal OP, Lowenstein A, Grinvald A, Barak A (2010) Reduced retinal blood flow velocity in diabetic retinopathy. Retina 30:765–773
8. Izhaky D, Nelson DA, Burgansky-Eliash Z, Grinvald A (2009) Functional imaging using the retinal function imager: direct imaging of blood velocity, achieving fluorescein angiography-like images without any contrast agent, qualitative oximetry, and functional metabolic signals. Jpn J Ophthalmol 53:345–351
9. Nelson DA, Krupsky S, Pollack A, Aloni E, Belkin M, Vanzetta I, Rosner M, Grinvald A (2005) Special report: noninvasive multiparameter functional optical imaging of the eye. Ophthalmic Surg Lasers Imaging 36: 57–66
10. Schaffer CB, Friedman B, Nishimura N, Schroeder LF, Tsai PS, Ebner FF, Lyden PD, Kleinfeld D (2006) Two-photon imaging of cortical surface microvessels reveals a robust redistribution in blood flow after vascular occlusion. PLoS Biol 4:e22

11. Kamoun WS, Chae SS, Lacorre DA, Tyrrell JA, Mitre M, Gillissen MA, Fukumura D, Jain RK, Munn LL (2010) Simultaneous measurement of RBC velocity, flux, hematocrit and shear rate in vascular networks. Nat Methods 7:655–660
12. Buxton RB, Uludag K, Dubowitz DJ, Liu TT (2004) Modelling the hemodynamic response to brain activation. Neuroimage 23:220–233
13. Vazquez AL, Cohen ER, Gulani V, Hernandez-Garcia L, Zheng Y, Lee GR, Kim SG, Grotberg JB, Noll DC (2006) Vascular dynamics and BOLD fMRI: CBF level effects and analysis considerations. Neuroimage 32:1642–1655
14. Wei HL, Zheng Y, Pan Y, Coca D, Li LM, Mayhew JE, Billings SA (2009) Model estimation of cerebral hemodynamics between blood flow and volume changes: a data-based modeling approach. IEEE Trans Biomed Eng 56:1606–1616
15. Mullinger KJ, Mayhew SD, Bagshaw AP, Bowtell R, Francis ST (2013) Poststimulus undershoots in cerebral blood flow and BOLD fMRI responses are modulated by poststimulus neuronal activity. Proc Natl Acad Sci U S A 110:13636–13641
16. Okell TW, Chappell MA, Kelly ME, Jezzard P, Piilgaard H, Lauritzen M (2013) Contribution of somatosensory cortex to evoked cerebellar blood flow responses. Cerebral blood flow quantification using vessel-encoded arterial spin labeling. J Cereb Blood Flow Metab 33: 1716–1724
17. Guibert R, Fonta C, Estève F, Plouraboué F (2013) On the normalization of cerebral blood flow. J Cereb Blood Flow Metab 33:669–672
18. Briers JD (2001) Laser Doppler, speckle and related techniques for blood perfusion mapping and imaging. Physiol Meas 22:35–66
19. Briers JD, Richards G, He XW (1999) Capillary blood flow monitoring using laser speckle contrast analysis. J Biomed Opt 4: 164–175
20. Dunn AK, Bolay H, Moskovitz MA, Boas DA (2001) Dynamic imaging of cerebral blood flow using laser speckle. J Cereb Blood Flow Metab 21:195–201
21. Royl G, Leithner C, Sellien H, Müller JP, Megow D, Offenhauser N, Steinbrink J, Kohl-Bareis M, Dirnagl U, Lindauer U (2006) Functional imaging with laser speckle contrast analysis: vascular compartment analysis and correlation with laser Doppler flowmetry and somatosensory evoked potentials. Brain Res 1121:95–103
22. Winship IR (2013) Improved cerebral blood flow measurement with multiexposure speckle imaging. J Cereb Blood Flow Metab 33:797
23. Kleinfeld D, Mitra PP, Helmchen F, Denk W (1998) Fluctuations and stimulus-induced changes in blood flow observed in individual capillaries in layers 2 through 4 of rat neocortex. Proc Natl Acad Sci U S A 95:15741–15746
24. Chaigneau E, Oheim M, Audinat E, Charpak S (2003) Two-photon imaging of capillary blood flow in olfactory bulb glomeruli. Proc Natl Acad Sci U S A 100:13081–13086
25. Vanzetta I, Deneux T, Masson GS, Faugeras O (2006) Cerebral blood flow recorded at high sensitivity in two dimensions using high resolution optical imaging. In: Proceedings of IEEE international symposium on biomedical imaging, pp 1264–1267
26. Hillman EM, Devor A, Bouchard MB, Dunn AK, Krauss GW, Skoch J, Bacskai BJ, Dale AM, Boas DA (2007) Depth-resolved optical imaging and microscopy of vascular compartment dynamics during somatosensory stimulation. Neuroimage 35:89–104
27. Drew PJ, Blinder P, Cauwenberghs G, Shih AY, Kleinfeld D (2010) Rapid determination of particle velocity from space-time images using the Radon transform. J Comput Neurosci 29:5–11
28. Deneux T, Faugeras O, Takerkart S, Masson GS, Vanzetta I (2011) A new variational method for erythrocyte velocity estimation in wide-field imaging in-vivo. IEEE Trans Med Imaging 30:1527–1545
29. Deneux T, Takerkart S, Grinvald A, Masson GS, Vanzetta I (2012) A processing work-flow for measuring erythrocytes velocity in extended vascular networks from wide field high-resolution optical imaging data. Neuroimage 59:2569–2588
30. Vanzetta I, Hildesheim R, Grinvald A (2005) Compartment-resolved imaging of activity-dependent dynamics of cortical blood volume and oximetry. J Neurosci 25:2233–2244
31. Frostig RD, Lieke EE, Ts'o DY, Grinvald A (1990) Cortical functional architecture and local coupling between neuronal activity and the microcirculation revealed by in vivo high-resolution optical imaging of intrinsic signals. Proc Natl Acad Sci U S A 87:6082–6086
32. Ts'o DY, Frostig RD, Lieke EE, Grinvald A (1990) Functional organization of primate visual cortex revealed by high resolution optical imaging. Science 249:417–420
33. Grinvald A, Bonhoeffer T, Vanzetta I, Pollack A, Aloni E, Ofri R, Nelson D (2004) High-resolution functional optical imaging: From the neocortex to the eye. Ophthalmol Clin North Am 17:53–67
34. Horn BKP (1986) Robot vision. MIT, Cambridge, MA

35. Coleman TF, Li Y (1994) On the convergence of reflective Newton methods for large-scale nonlinear minimization subject to bounds. Math Program 67:189–224
36. Nishimura N, Schaffer CB, Friedman B, Tsai PS, Lyden PD, Kleinfeld D (2006) Targeted insult to subsurface cortical blood vessels using ultrashort laser pulses: three models of stroke. Nat Methods 3:99–108
37. Ayata C, Shin HK, Salomone S, Ozdemir-Gursoy Y, Boas DA, Dunn AK, Moskowitz MA (2004) Pronounced hypoperfusion during spreading depression in mouse cortex. J Cereb Blood Flow Metab 24:1172–1182
38. Piilgaard H, Lauritzen M (2004) Contribution of somatosensory cortex to evoked cerebellar blood flow responses. Neuroreport 15: 695–698
39. Kohl M, Lindauer U, Dirnagl U, Villringer A (1998) Separation of changes in light scattering and chromophore concentrations during cortical spreading depression in rats. Opt Lett 23:555–557
40. Nguyen J, Nishimura N, Fetcho RN, Iadecola C, Schaffer CB (2011) Occlusion of cortical ascending venules causes blood flow decreases, reversals in flow direction, and vessel dilation in upstream capillaries. J Cereb Blood Flow Metab 31:2243–2254
41. Tomita M, Tomita M, Unekawa M, Toriumi N, Suzuki N (2011) Oscillating neuro-capillary coupling during cortical spreading depression as observed by tracking FITC-labeled RBCs in single capillaries. Neuroimage 56:1001–1010
42. Seylaz J, Charbonné R, Nanri K, Von Euw D, Borredon J, Kacem K, Méric P, Pinard E (1999) Dynamic in vivo measurement of erythrocyte velocity and flow in capillaries and of microvessel diameter in the rat brain by confocal laser microscopy. J Cereb Blood Flow Metab 19:863–870
43. Berwick J, Johnston D, Jones M, Martindale J, Redgrave P, McLoughlin N, Schiessl I, Mayhew JE (2005) Neurovascular coupling investigated with two-dimensional optical imaging spectroscopy in rat whisker barrel cortex. Eur J Neurosci 22:1655–1666
44. Fujimoto JG (2003) Optical coherence tomography for ultrahigh resolution in vivo imaging. Nat Biotechnol 21:1361–1367
45. Trick GL, Calotti FY, Skarf B (2006) Advances in imaging of the optic disc and retinal nerve fiber layer. J Neuroophthalmol 26:284–295
46. Singh R, Kaiser PK (2007) Advances in AMD imaging. Int Ophthalmol Clin 47:65–74
47. Drexler W, Fujimoto JG (2008) State-of-the-art retinal optical coherence tomography. Prog Retin Eye Res 27:45–88
48. Schmitz-Valckenberg S, Holz FG, Bird AC, Spaide RF (2008) Fundus autofluorescence imaging: review and perspectives. Retina 28: 385–409
49. Podoleanu AG, Rosen RB (2008) Combinations of techniques in imaging the retina with high resolution. Prog Retin Eye Res 27:464–499
50. Denninghoff KR, Smith MH, Hillman L (2000) Retinal imaging techniques in diabetes. Diabetes Technol Ther 2:111–113
51. Harris A, Dinn RB, Kagemann L, Rechtman E (2003) A review of methods for human retinal oximetry. Ophthalmic Surg Lasers Imaging 34:152–164
52. Blum M, Bachmann K, Wintzer D, Riemer T, Vilser W, Strobel J (1999) Noninvasive measurement of the Bayliss effect in retinal autoregulation. Graefes Arch Clin Exp Ophthalmol 237:296–300
53. Hubbard LD, Brothers RJ, King WN et al (1999) Methods for evaluation of retinal microvascular abnormalities associated with hypertension/sclerosis in the Atherosclerosis Risk in Communities Study. Ophthalmology 106:2269–2280
54. Ege BM, Hejlesen OK, Larsen OV, Møller K, Jennings B, Kerr D, Cavan DA (2000) Screening for diabetic retinopathy using computer based image analysis and statistical classification. Comput Methods Programs Biomed 62:165–175
55. Van Hecke MV, Dekker JM, Nijpels G, Stolk RP, Henry RM, Heine RJ, Bouter LM, Stehouwer CD, Polak BC (2006) Are retinal microvascular abnormalities associated with large artery endothelial dysfunction and intima-media thickness? The Hoorn Study. Clin Sci (Lond) 110:597–604
56. Stanton AV, Wasan B, Cerutti A et al (1995) Vascular network changes in the retina with age and hypertension. J Hypertens 13(12 Pt 2):1724–1728
57. Van den Born BJ, Hulsman CA, Hoekstra JB, Schlingemann RO, van Montfrans GA (2005) Value of routine funduscopy in patients with hypertension: systematic review. BMJ 331:73
58. Landa G, Jangi AA, Garcia PM, Rosen RB (2012) Initial report of quantification of retinal blood flow velocity in normal human subjects using the retinal functional imager (RFI). Int Ophthalmol 32:211–215
59. Burgansky-Eliash Z, Lowenstein A, Neuderfer M, Kesler A, Barash H, Nelson DA, Grinvald A, Barak A (2013) Retinal function imager measurements of retinal blood flow velocity and their relationship to various physiological parameters. Ophthalmic Surg Lasers Imaging Retina 44:51–58

60. Burgansky-Eliash Z, Barash H, Nelson DA et al (2012) Retinal function imager measurements of retinal blood flow velocity in patients with early diabetes mellitus. Retina 32:112–119
61. Barak A, Burgansky-Eliash Z, Barash H, Nelson DA, Izhaky D, Barak A, Lowenstein A, Neuderfer M, Kesler A, Grinvald A (2012) The effect of intravitreal bevacizumab (Avastin) injection on retinal blood flow velocity in patients with choroidal neovascularization. Eur J Opthtalmol 22:423–430
62. Beutelspacher SC, Serbecic N, Barash H, Burgansky-Eliash Z, Grinvald A, Jonas JB (2011) Central serous chorioretinopathy shows reduced retinal circulation in retinal function imaging (RFI). Acta Ophthalmol 23:1755–3768
63. Beutelspacher SC, Barash H, Burgansky-Eliash Z, Grinvald A, Serbecic N, Jost J (2011) Retinal blood flow velocity measured by retinal function imaging in retinitis pigmentosa. Graefes Arch Clin Exp Ophthalmol 249: 1855–1858
64. Landa G, Amde W, Haileselassie Y, Rosen RB (2011) Cilioretinal arteries in diabetic eyes are associated with increased retinal blood flow velocity and occurrence of diabetic macular edema. Retina 31:304–311
65. Landa G, Garcia PM, Rosen RB (2009) Correlation between retina blood flow velocity assessed by retinal function imager and retina thickness estimated by scanning laser ophthalmoscopy/optical coherence tomography. Ophthalmologica 223:155–161
66. Nelson DA, Amit R, Oaknin J, Burgansky-Eliash Z, Barash H, Izhaky D, Lowenstein A, Barak A, Bartov E, Rock T, Grinvald A (2011) Wide-field high-resolution imaging of perfused capillaries without the use of contrast agent. Clin Ophthalmol 5:1095–1106
67. Kwan AS, Barry C, McAllister IL, Constable I (2006) Fluorescein angiography and adverse drug reactions revisited: the Lions Eye experience. Clin Experiment Ophthalmol 34:33–38
68. Watkin AJ, Alshareef RA, Rezeq SS, Sampat KM, Chhablani J, Bartsch DU, Freeman WR, Haller JA, Garg SJ (2012) Comparative analysis of the retinal microvasculature visualized with fluorescein angiography and the retinal function imager. Am J Opthtalmol 154:901–907
69. Stefansson E, Landers MB 3rd, Wolbarsht ML (1983) Oxygenation and vasodilatation in relation to diabetic and other proliferative retinopathies. Ophthalmic Surg 14:209–226
70. Stefansson E, Machemer R, de Juan E, Jr MCBW 2nd, Peterson J (1992) Retinal oxygenation and laser treatment in patients with diabetic retinopathy. Am J Ophthalmol 113:36–38
71. Tiedeman JS, Kirk SE, Srinivas S, Beach JM (1998) Retinal oxygen consumption during hyperglycemia in patients with diabetes without retinopathy. Ophthalmology 105:31–36
72. Sebag J, Delori FC, Feke GT, Weiter JJ (1989) Effects of optic atrophy on retinal blood flow and oxygen saturation in humans. Arch Ophthalmol 107:222–226
73. Grinvald A, Sharon D, Slovin H, Vanzetta I (2005) Intrinsic signal imaging in the neocortex; implications for hemodynamic based functional imaging. In: Yuste R, Konnerth A (eds) Imaging in neuroscience and development: a laboratory manual. Cold Spring Harbor Laboratory, Cold Spring Harbor, NY, Chapter 89
74. Hill DK, Keynes RD (1949) Opacity changes in stimulated nerve. J Physiol 108:278–281
75. Cohen LB, Keynes RD, Hille B (1968) Light scattering and birefringence changes during nerve activity. Nature 218:438–441
76. Cohen LB (1973) Changes in neuron structure during action potential propagation and synaptic transmission. Physiol Rev 53: 373–418
77. Malonek D, Grinvald A (1996) Interactions between electrical activity and cortical microcirculation revealed by imaging spectroscopy: implications for functional brain mapping. Science 272:551–554
78. Tsunoda K, Oguchi Y, Hanazono G, Tanifuji M (2004) Mapping cone- and rod-induced retinal responsiveness in macaque retina by optical imaging. Invest Ophthalmol Vis Sci 45: 3820–3826
79. Abramoff MD, Kwon YH, Ts'o D, Soliz P, Zimmerman B, Pokorny J, Kardon R (2006) Visual stimulus-induced changes in human near-infrared fundus reflectance. Invest Ophthalmol Vis Sci 47:715–721
80. Hanazono G, Tsunoda K, Shinoda K, Tsubota K, Miyake Y, Tanifuji M (2007) Intrinsic signal imaging in macaque retina reveals different types of flash-induced light reflectance changes of different origins. Invest Ophthalmol Vis Sci 48:2903–2912
81. Hanazono G, Tsunoda K, Kazato Y, Tsubota K, Tanifuji M (2008) Evaluating neural activity of retinal ganglion cells by flash-evoked intrinsic signal imaging in macaque retina. Invest Ophthalmol Vis Sci 49:4655–4663
82. Grieve K, Roorda A (2008) Intrinsic signals from human cone photoreceptors. Invest Ophthalmol Vis Sci 49:713–719

83. Srinivasan VJ, Chen Y, Duker JS, Fujimoto JG (2009) In vivo functional imaging of intrinsic scattering changes in the human retina with high-speed ultrahigh resolution OCT. Opt Express 17:3861–3877
84. Jiang H, Ye Y, DeBuc DC, Lam BL, Rundek T, Tao A, Shao Y, Wang J (2013) Human conjunctival microvasculature assessed with a retinal function imager (RFI). Microvasc Res 85: 134–137
85. Shtoyerman E, Arieli A, Slovin H, Vanzetta I, Grinvald A (2000) Long-term optical imaging and spectroscopy reveal mechanisms underlying the intrinsic signal and stability of cortical maps in V1 of behaving monkeys. J Neurosci 20:8111–8121
86. Arieli A, Grinvald A, Slovin H (2002) Dural substitute for long-term imaging of cortical activity in behaving monkeys and its clinical implications. J Neurosci Methods 114:119–133
87. Ratzlaff EH, Grinvald A (1991) A tandem-lens epifluorescence macroscope: hundred-fold brightness advantage for wide-field imaging. J Neurosci Methods 36:127–137

Chapter 8

Optical Intrinsic Signal Imaging for Elucidating Functional Structures in Higher Visual Area

Takayuki Sato and Manabu Tanifuji

Abstract

To understand neural mechanisms of object recognition, representation of visual object images has been studied in inferior temporal (IT) cortex, which is located at the final stage of the ventral object recognition pathway. Optical intrinsic signal imaging (OISI) is a powerful imaging technique to visualize spatial patterns of object representation in IT cortex. In this chapter, we will describe technical procedures of OISI particularly required when applying to IT cortex, and also recent advancement with OISI in understanding object representation and functional organization of IT cortex.

Key words Object recognition, Column, Cortical mosaic, Population code, Functional organization, Feature column, fMRI, fOCT, Inferotemporal, Inferior temporal

1 Introduction

Visual psychology and neuroscientific studies suggest that primates have a unique capability in object recognition. For example, we recognize faces easily regardless of views, positions, photometric conditions, and some deformation in their shapes such as different facial expressions. Early studies with macaque monkeys showed that there are neurons specifically responding to faces in IT cortex [1, 2]. These neurons may consist of a component of face processing. However, activities of individual face-selective neurons cannot fully explain the above invariant face recognition. For example, face neurons typically respond to a particular view of faces but not to the entire view of faces [1–3]. Freiwald and Tsao reported that population of neurons in anterior IT cortex can show view-invariant representation of faces [3]. Thus, a combination of multiple neuronal activities seems to be essential for the view-invariant recognition of faces.

Specific response to certain objects is not the general property of IT neurons except for faces and some other behaviorally important

Mingrui Zhao et al. (eds.), *Neurovascular Coupling Methods*, Neuromethods, vol. 88,
DOI 10.1007/978-1-4939-0724-3_8, © Springer Science+Business Media New York 2014

classes of objects, such as bodies. Most of IT neurons respond to visual features that are less complex than object images [2, 4]. Some neurons prefer local features, such as asymmetric curvatures, protrusions, and some other neurons respond to specific spatial arrangements of local features. Therefore, object images are not represented by single neurons, but are represented by a combination of multiple neuronal activities in IT cortex [5, 6]. Altogether, to explore object representation, we need techniques that allow us to elucidate population activities of neurons, and we believe that optical intrinsic signal imaging (OISI) is the most suitable tool for measuring the large population of neural activities.

2 Target Functional Structures for OISI in IT Cortex

Unlike functional imaging with two-photon laser-scanning microscopy, intrinsic signal imaging with conventional CCD cameras does not have a single cellular level resolution. OISI is an imaging technique to visualize functional structures larger than or equal to cortical columns. The columnar organization first discovered in primary visual cortex (V1) is the general organization principle in cerebral cortices. In V1, for example, neurons responding to a line segment with a particular orientation and location in visual field are clustered in a columnar region and works as a functional unit that encodes local orientation in the visual field. Similarly in IT cortex, Fujita and his colleagues found that neurons responding to similar visual features are clustered in a columnar region in IT cortex [7]. IT columns (feature columns) are considered to be functional units that encode local or global visual features of object images [5, 6]. Spatial patterns of columnar organizations in cortices including IT cortex have been successfully visualized by OISI [5, 6, 8–12].

fMRI studies with macaque monkeys revealed that there are category-specific regions in IT cortex such as body-specific regions and face patches [3, 13–17]. These regions are much larger in size than columns in IT cortex typically spanning several millimeters wide [16]. If object images are represented by feature columns randomly distributed in IT cortex, we cannot expect such category-specific regions, and thus regional activation revealed by fMRI such as face patches are seemingly inconsistent with IT columnar organization. As shown below, however, we identified mosaic structures in IT cortex where feature columns related to a particular category are clustered together into category-specific regions [18]. OISI gives us the opportunity to visualize such lager activation areas corresponding to category-specific regions shown by fMRI and feature columns by choosing an appropriate spatial filter range.

3 Principle and Spatiotemporal Patterns of Intrinsic Signals

Functional imaging of intrinsic signals is a technique to visualize reflection changes in cortical surface when sensory stimuli activate cortical neurons. Historically, intrinsic signals were first found as a component of absorption changes associated with voltage-sensitive dye signals when the cortex was stained extrinsically with the dye [8, 9]. Following studies addressing sources of absorption changes have revealed that multiple intrinsic signals are involved: increase of deoxygenated hemoglobin [19], blood volume changes in capillaries [20, 21], and changes in tissue structure that cause changes in light scattering properties [22, 23]. The relative contribution of these three components to the intrinsic signals depends on the wavelength of light used to illuminate the cortical surface. The most commonly used wavelength of light is from 600 to 620 nm, where increase of deoxygenated hemoglobin is the dominant component.

For example at 610 nm, neurons in cat primary visual cortex activated by grating stimuli consume more oxygen than neurons under the resting state, which in turn causes an increase in deoxyhemoglobin at the activated region. The local increase of deoxyhemoglobin leads to an increase in the absorption of the illuminated light, which is ultimately detected as darkening of the cortex, since absorption coefficient of deoxyhemoglobin is higher than the coefficient of oxyhemoglobin. This absorption starts to increase around 0.5 s after the onset of visual stimulation, while spiking activity starts to increase only 50 ms after the onset (Fig. 1a). Latency is different because intrinsic signals do not directly reflect neural activities but consequent metabolic changes. After 1–2 s from the stimulus onset, the absorption increase reaches its maximum and begins to decrease, eventually crossing the baseline absorption level around two-second mark. The signal then overshoots the baseline between 2 and 4 s and stays in the positive range, forming a new maximum. This second peak corresponds to an increase in blood volume due to excess of blood flow. The blood volume increase is associated with the replacement of deoxyhemoglobin with oxygenated hemoglobin and causes an increase in brightness. The temporal patterns of intrinsic signals in IT cortex are very similar to those in cat primary visual cortex (Fig. 1b).

To visualize spatial patterns of neural activation, we temporally average intrinsic signals for the 0.5–2.5 s period after the stimulus onset. This period corresponds to the phase of deoxyhemoglobin increase (Fig. 1a, b). The reason for focusing this period is that the following phase includes the replacement of deoxyhemoglobin with oxygenated hemoglobin in draining vessels running along cortical surface as well as the change in neural tissue. In cat primary visual cortex, spatial patterns of intrinsic signals evoked by orthogonal

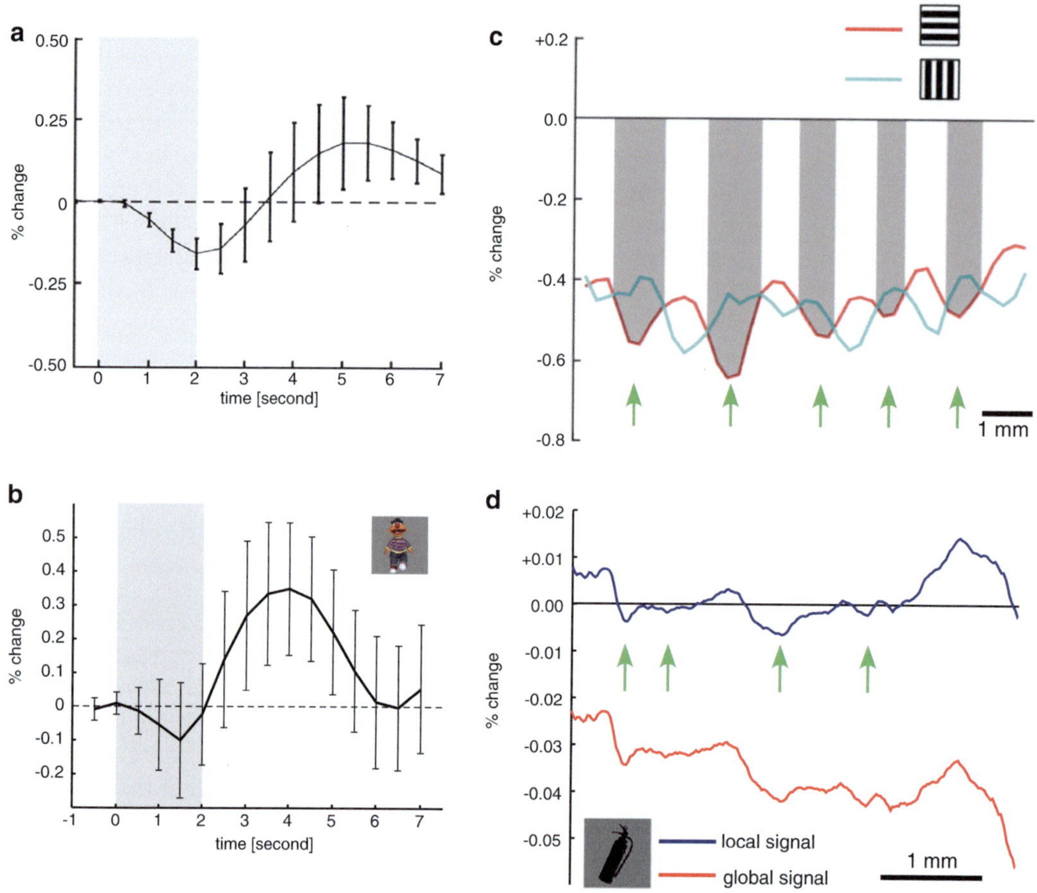

Fig. 1 Time courses (**a**, **b**) and spatial patterns of intrinsic signals (**c**, **d**) in cat primary visual cortex (**a**, **c**) and macaque monkey IT cortex (**b**, **d**). In (**a**, **b**), the horizontal axes represent time from stimulus onset, and the vertical axes represent percentage of reflection intensity change relative to reflection intensity before the stimulus onset. Intensities of all pixels except on vessels were averaged. Error bars represent the standard deviation obtained from 14 cats (**a**) and trial by trail standard deviation (**b**). The shaded regions indicate the duration of stimulus presentation. The *horizontal broken lines* indicate the baseline corresponding to no absorption changes. In (**c**, **d**), signal intensities along *solid lines* in Fig. 2 (**c**) and along a *solid line* in Fig. 3a (**d**). The *horizontal lines* indicate the baseline corresponding to the absence of absorption change. In (**c**), spatial patterns with orthogonal orientation stimuli are plotted in different colors, and shaded regions indicate the portions showing larger responses to a grating at 0° orientation (modified from [21]). In (**d**), signal intensities before (*red*) and after (*blue*) spatial high-pass filtering are plotted. These spatial patterns were obtained after temporally average responses for the early components (see text). *Arrows* in (**c**, **d**) and those in Figs. 2 and 3 represent spatial locations of local darkening evoked by horizontal grating (**c**) and by an object stimulus (**d**)

whole-field grating patterns reveals global darkening irrespective of stimulus selectivity (Fig. 1c) [21]. The global darkening is spatially modulated in a stimulus-specific manner (Fig. 1c, arrows). The local regions specific to each orientation (local signal component) correspond to columnar activation and were extracted by subtracting signals with one orientation from the signals with orthogonal

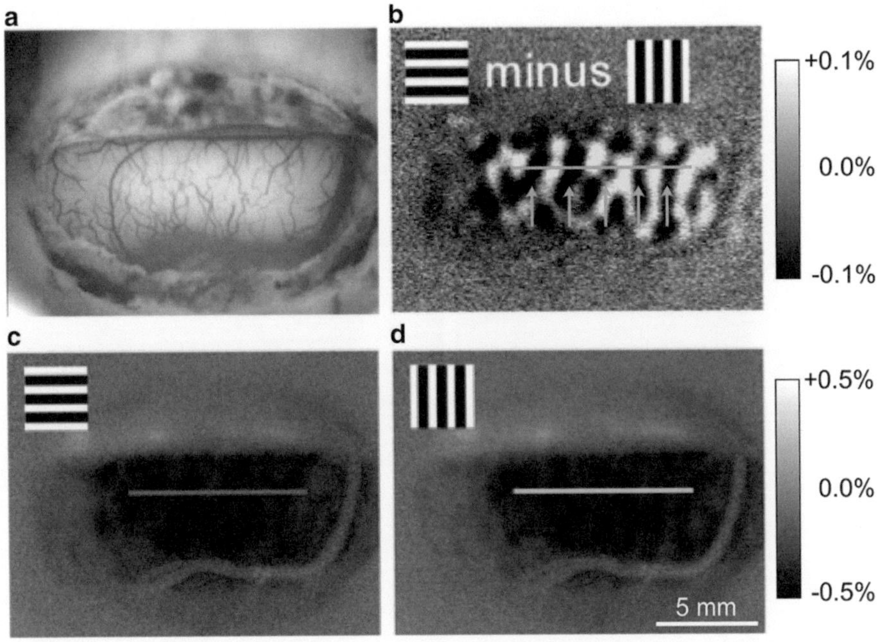

Fig. 2 Spatial patterns of intrinsic signals evoked by orthogonal grating stimuli in cat primary visual cortex. (**a**) Image of exposed cortex. (**b**) Spatial pattern of local darkening obtained by subtraction between intrinsic signals elicited by orthogonal orientations. (**c**, **d**) Global signal components of intrinsic signals that were evoked by horizontal grating (**c**) and vertical grating (**d**). Gray scale in (**b**–**d**), percentage of reflection intensity change relative to reflection intensity before the stimulus onset. Intensity along the horizontal line in (**b**–**d**) is plotted in Fig. 1c

orientation (Fig. 2), by subtracting signals with one orientation from average responses for all other orientations, or by processing signals through high-pass spatial filtering. Spatial patterns of intrinsic signals in IT cortex also consist of global and local signal components (Fig. 1d). Since it is difficult to identify a stimulus corresponding to orthogonal orientation in V1 and it is also difficult to obtained average responses to all possible stimuli in IT cortex, columnar activation patterns (arrows in Fig. 1d) evoked by object stimuli are typically extracted by a high-pass spatial filtering (Fig. 3a, b). The width of the spatial filter is selected according to the size of columns (~500 μm in diameter), which is estimated by electrophysiological [7, 24] and anatomical [25] studies.

4 Critical Procedures for OISI in IT Cortex

The experimental procedures for OISI in IT cortex are basically the same as those in cat primary visual cortex. However, increases in spiking activity evoked by visual stimuli in IT cortex are about five to ten times smaller than those in primary visual cortex and thus local components of intrinsic signals are also about five to ten

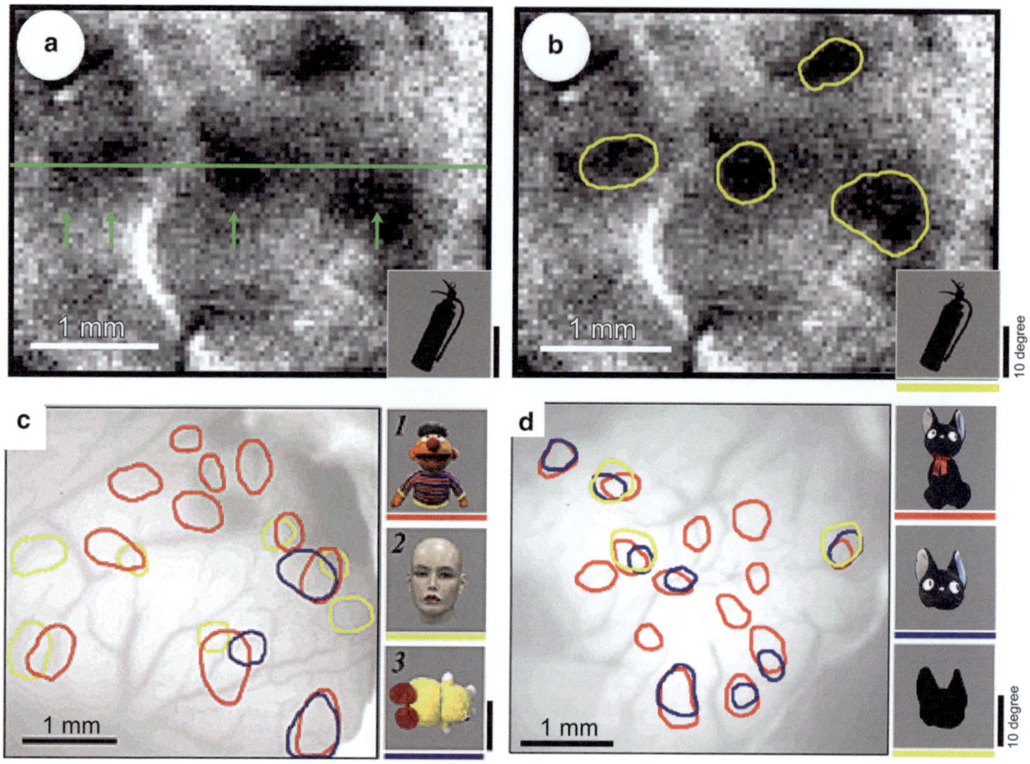

Fig. 3 Spatial patterns of intrinsic signals evoked by object stimuli in monkey IT cortex. (**a**) Spatial patterns of local component of intrinsic signals evoked by the stimulus given in lower right corner. Intensity along the horizontal line is plotted in Fig. 1d. (**b**) The same as (**a**) except for extracted spots outlined by connecting pixels with half the peak absorption value. (**c**) Spatial distributions of activated columns elicited by three different object images. (**d**) Spatial distributions of activated columns elicited by original black cat, head of the cat, and silhouette of the head. Activated columns are contoured by corresponding colors below each image (modified from [5])

times smaller (*see* vertical scale in Fig. 1c, d). Thus, OISI is more difficult in IT cortex than in V1. Here, we will describe some necessary procedures for successful imaging in IT cortex (for general procedures in conventional OISI, *see* [26]).

Recording chamber is placed over IT cortex (Fig. 4a, b). We have to remove thick and nontransparent dura mater. It is critical not to mechanically stimulate the cortex since reliable OISI highly depends on biological conditions of the exposed cortex. For example, we should avoid touching cortex with cotton tips; we should also avoid spraying saline directly onto the cortex with syringes. One has to keep in mind that the cellular mechanisms to generate intrinsic signals associated with neural activity are much easily damaged than neurons.

Control of brain pressure is also essential during duratomy. If one fails, the cortex will develop herniation, and it rarely recovers from herniation. To avoid such a situation, we (1) raise the head of the monkey above the body and slightly rotated the head to make the IT cortex facing upward rather than downward,

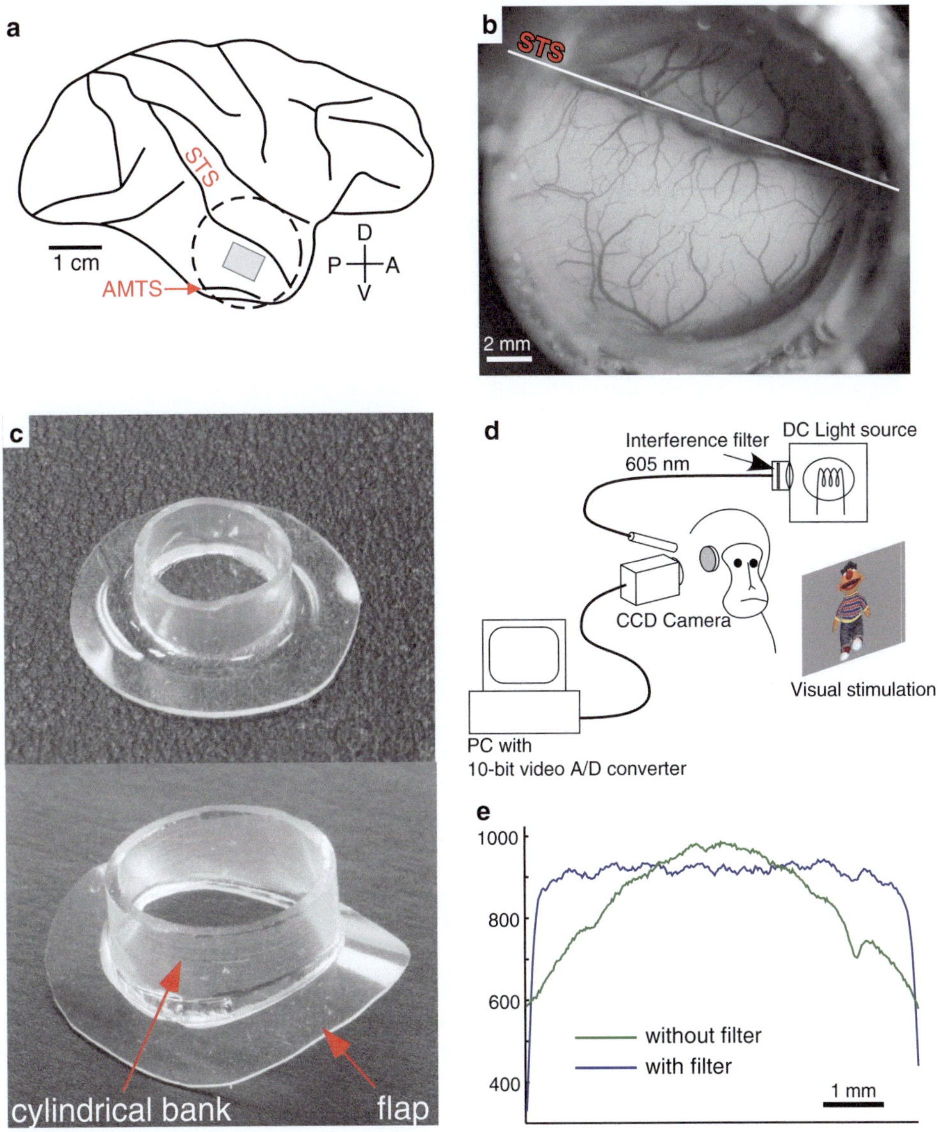

Fig. 4 Technical details of OISI from IT cortex. (**a**) The lateral view of the postmortem brain where the location of the chamber (*circle* in *broken line*) and imaging area within the chamber (*shaded area*) are indicated. *STS* superior temporal sulcus, *AMTS* anterior middle temporal sulcus. (**b**) A picture of exposed IT cortex. Position of STS is shown in *white line*. (**c**) Artificial dura made of silicone for V1 (*upper*) and IT cortex (*bottom*). For IT cortex, *bottom* part was curved to fit the cortical shape. (**d**) Schematic drawing of OISI setup. (**e**) spatial profile of intensity of light reflected back from the cortex before (*green*) and after (*blue*) inserting neutral density optical filter in the light pass (see text)

(2) intravenously infuse hypertonic solution, (3) keep deep anesthesia, and (4) maintain hyperventilation condition.

After removing the dura, we insert an artificial dura (Fig. 4c) [27]. The artificial dura consists of a flap and a cylindrical bank, and both are made of silicone. The flap is circular and its diameter is larger than the hole made on the dura. We insert the flap between the cortex and the remaining dura mater. In this way, the cylindrical bank protects the exposed cortical surface from invasion of regrowing tissue typically starting from the edge of remaining dura mater. Since IT cortex has a curved surface, the flap is also curved. Furthermore, the flap is not extended inside the cylindrical bank so that the cortex inside the cylindrical bank is not covered by a sheet of silicone. In this way, we prevent the exposed cortex from mechanical stimulation when the operated monkey is in his home cage. This is critical for IT recordings since the cortex is facing downward. Finally, we cover the exposed cortex with heavy silicone oil (KF-96-1000CS, Shin-Etsu Chemical, Japan) and put a transparent glass cover slip on top of the entire surgical site. We especially select high viscosity of silicone oil because infiltration of silicone oil to the cortex may cause damage.

As for the conventional OISI in cat primary visual cortex, the exposed cortex is illuminated with 605 nm light (Fig. 4d) and the reflected light from the cortex is detected by a CCD camera and digitized. The curved surface of IT cortex creates inhomogeneity in reflected light intensity. The inhomogeneity often causes concentric circular-shaped artifacts, which are related to nonlinearity of A/D converter. To avoid the artifacts, the reflected light intensity is flattened by inserting a spatially inhomogeneous neutral density filter to the light pass (Fig. 4e) [28]. We make this filter by printing a B/W inverted image of the exposed cortex on a transparent sheet.

5 Combinatorial Code of Object Images in IT Cortex

In the following sections, we will describe how OISI contributes to our understanding of functional structures in IT cortex. Wang and colleagues [29] first introduced OISI in IT cortex. They identified features of neurons before exposing cortical surface by extracellular recordings. Then, they exposed cortical surface and conducted OISI with identified features as visual stimuli. They found that the electrophysiological recording sites overlapped with spots of local darkening, providing supporting evidence for the columnar organization in IT cortex with respect to visual features [7]. Tsunoda and colleagues [5] used OISI to investigate object representation with these columns (feature columns) in monkey IT cortex. OISI showed that object images activated multiple spots in IT cortex (Fig. 3a, b). They found that activated patterns of columns were different from object to object (Fig. 3c). Some spots were co-activated by multiple objects but some spots were activated only by one of them, suggesting that each spot represents a visual feature within object images. This proposal was confirmed by comparison of activation patterns with an object and simplified version of the object. They found some spots disappeared when some features of the object are removed by simplification (Fig. 3d). Subsequently, Yamane and colleagues [6] found that there were spots sensitive to spatial relationship among parts but were not sensitive to changes in local features within each part. These studies revealed that object images are represented by local features and features representing spatial relationship among local features. In such ways, OISI has contributed to establish a concept of combinatorial codes of object representation in IT cortex. One aspect of OISI in IT cortex is that we use this technique to find target sites with specific response properties and to make further characterization of neurons with extracellular recordings. In fact, in the above two studies [5, 6], critical analyses were made by electrophysiological recordings from the spots to characterize visual features represented by these spots. Since intrinsic signals reflect metabolic changes evoked by neural activity, the combination with extracellular recordings is critical to conduct reliable and fine investigations of electrophysiological properties of neurons.

6 Subcolumnar Structures Revealed by OISI in IT Cortex

As OISI provides fine spatial organization of orientation columns in the primary visual cortex of cats and monkeys [30, 31], OISI is also suitable to characterize finer functional structures in IT cortex. In the study of Wang and colleagues [29, 32], they found that some spots activated by faces were continuously shifted against viewing angle of faces, and the length of the overall region along the direction of movement was around 1 mm (Fig. 5). These results suggest that columns in IT cortex may have substructures where neurons are systematically arranged against certain parameters such as viewing angle of faces. Unfortunately, there was no confirmation with extracellular recording at the time. Recently, Lin and colleagues [33] conducted dense electrophysiological recordings from face-selective regions in IT cortex, and obtained supporting evidence for systematic arrangement of face-selective neurons with respect to preferred viewing angles. It is still an open question whether this substructures within a column is common among other feature columns. Future studies are required.

7 Mosaic Structures in IT Cortex

As mentioned earlier, monkey fMRI studies have suggested the existence of larger functional structures than columns [13, 15–17]. For example, Tsao and colleagues revealed six face patches along temporal lobe [16]. The face patches are spanning ~5 mm wide that are much larger in size than columns in IT cortex. Since one of the patches was found at about the same location of anterior IT

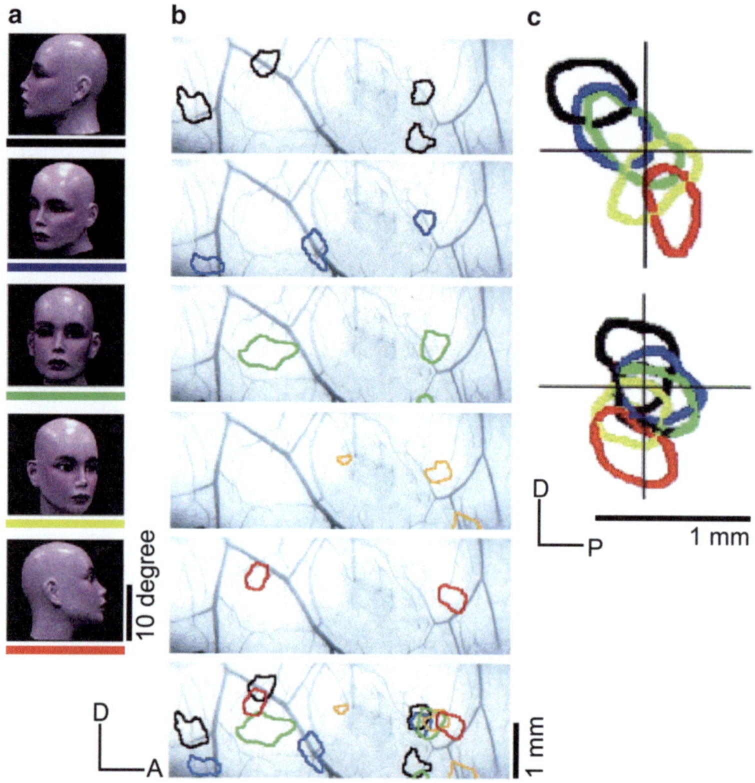

Fig. 5 Systematic shift of activation spot with the rotation of the face. (**a**) Face stimuli for mapping fine functional structures. We used doll face images with five different views; −90° (left profile), −45°, 0° (front face), +45°, and +90° (right profile). (**b**) Activation spots of five different views of the face. The contours circumscribing the pixels with *t*-values giving $p < 0.05$, compared with the cocktail reference averaged over the maps obtained for the five different views of the face, are shown. Colors of each contour are consistent with bars below the images in (**a**). The contours of spots for the different views are superimposed at the *bottom*. (**c**) Enlarged *t*-value contours for two cases obtained in reference to the face cocktail reference (modified from [32])

cortex where we recorded columnar activation, the question is whether some parts of IT cortex consist of patches and some others consist of feature columns, or whether the patches observed are composed of feature columns.

To investigate functional structures larger than columns, we investigated spatial patterns of raw intrinsic signals. As mentioned earlier, intrinsic signals consist of local and global signals. The local signal represents stimulus-specific columnar activation (Fig. 3). In contrast, the global signal was once thought to be the stimulus nonspecific component evoked by visual stimulation. However, careful analysis of global signal revealed that this was not the case (Fig. 6). For example, the raw intrinsic signals evoked by the silhouette of fire extinguisher revealed inhomogeneous darkening of

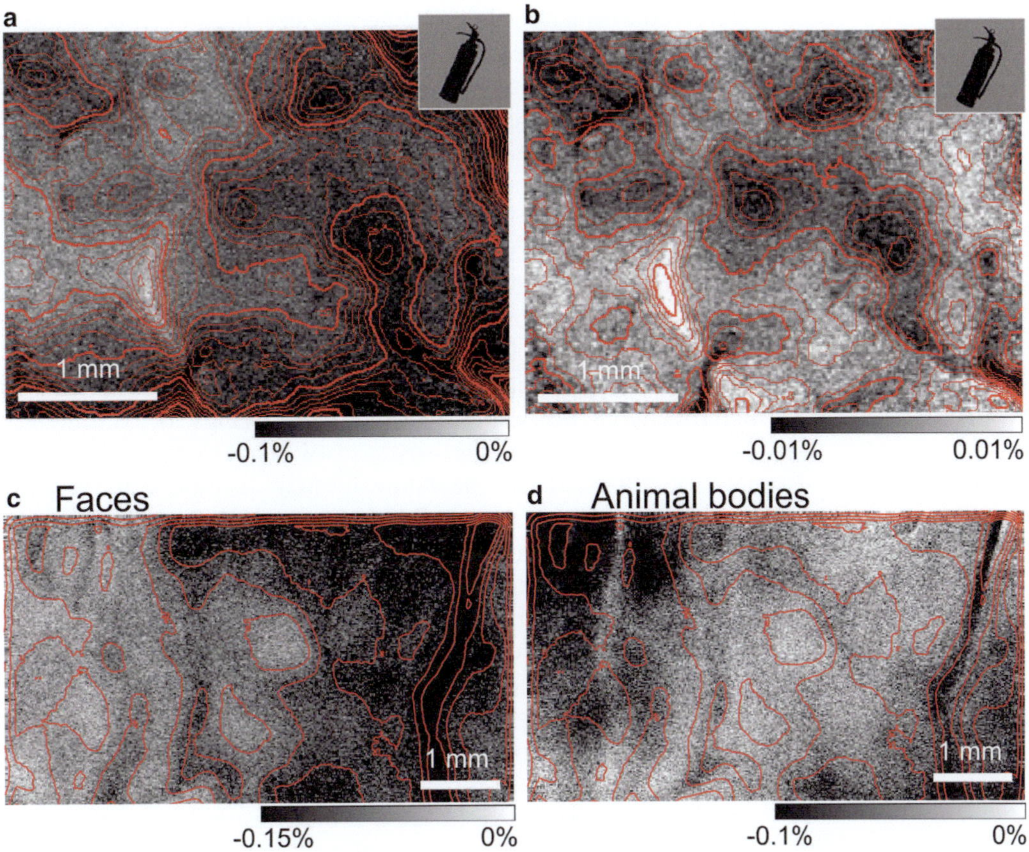

Fig. 6 Local and global signals of intrinsic signal in IT cortex. Activation maps represent global (**a**) and local (**b**) signals evoked by the stimulus in upper right corner. Activation maps for global signals evoked by monkey faces (**c**) and animal bodies (**d**). The intensity of the evoked intrinsic signals is represented by the gray scale. Intensity of signals is also indicated by red contours for clarity

cortex; the darkening was stronger in right side than in left side of the image (Fig. 6a). Columnar activation was seen on top of the global but inhomogeneous darkening. The columnar activation was extracted by high-pass spatial filtering (Fig. 6b). The signal sources of fMRI and OISI are not the same: OISI detects the early phase of deoxyhemoglobin increase while fMRI detects signals related to late phase of blood flow excess. However, both fMRI and intrinsic signals reflect blood oxygenation-related metabolic changes, and thus the inhomogeneity in raw intrinsic signals may reflect category-specific regional activation revealed by fMRI. We then used different category objects for OISI and investigated category-specific patterns of raw intrinsic signals in different IT cortex (Fig. 6c, d) [12]. We found face activated right half of the imaged region and animal body activated left upper and lower right regions, suggesting the existence of category-specific regions potentially corresponding to category-specific regional activation revealed by fMRI.

To address a question whether the size and spread of signals in fMRI and OISI reflect the veridical extent of neural activities, we conducted dense untargeted electrophysiological recordings in IT cortex [18]. We presented 104 object images to anesthetized monkeys and recorded neural activities from 39 recording sites in a wide region of IT cortex. We then classified recording sites based on similarity in object selectivity by hierarchical clustering analysis (Fig. 7a). We found that recording sites were clustered into seven groups based on similarity in object selectivity, and the sites belonging to the same groups formed domains on the cortical surface (Fig. 7b). Some of the domains responded to certain object categories such as faces and monkey bodies, and they overlapped with heterogeneous darkening evoked by OISI (Figs. 6c, d, and 7b). These results suggest the existence of category-specific regional activation in IT cortex; these domains may correspond to category-specific patches revealed by fMRI. Interestingly, the sites within a domain are different in sub-category selectivity (Fig. 7d, e). For example, site cc preferred monkey faces more than human faces, and site Q preferred human faces more than monkey faces while faces activated significantly better than non-face object in both sites (Fig 7e, 1 and 2). There were sites with no significant difference for monkey and human faces (site ff) (Fig. 7e, 3). The sites with differential selectivity for the face sub-category were rather randomly distributed in the face preferred domain (Fig. 7d). Finally, OISI revealed activation of distributed spots consistent with the above electrophysiological results. Altogether, the results revealed that regions with common functional properties (domains) consist of a finer functional structure (columns) in anterior IT cortex such as a mosaic. In a mosaic, locally dissimilar tiles form a coherent larger picture. Feature columns in IT cortex can be thought of as mosaic tiles, making up a larger picture of category selectivity.

8 Summary

In this chapter, we described that OISI in higher visual area (IT cortex) is a powerful technique for demonstrating object representation and functional organization in IT cortex. Though intrinsic signals do not directly reflect neural activities, the results can provide useful information that is consistent with neuronal responses. Two components of intrinsic signals were distinguished by their spatial extent. Local signals correspond to columnar activity representing visual features, and global signals correspond to domain activity, and some of the domains represent object categories in IT cortex.

Recently, fMRI and two-photon laser-scanning microscopy are widely used in many physiological studies. Yet there are still some ranges of spatial resolution or measurement area that neither of the techniques can achieve. Resolution of fMRI is around 1 mm that is not enough high to visualize individual columnar activity.

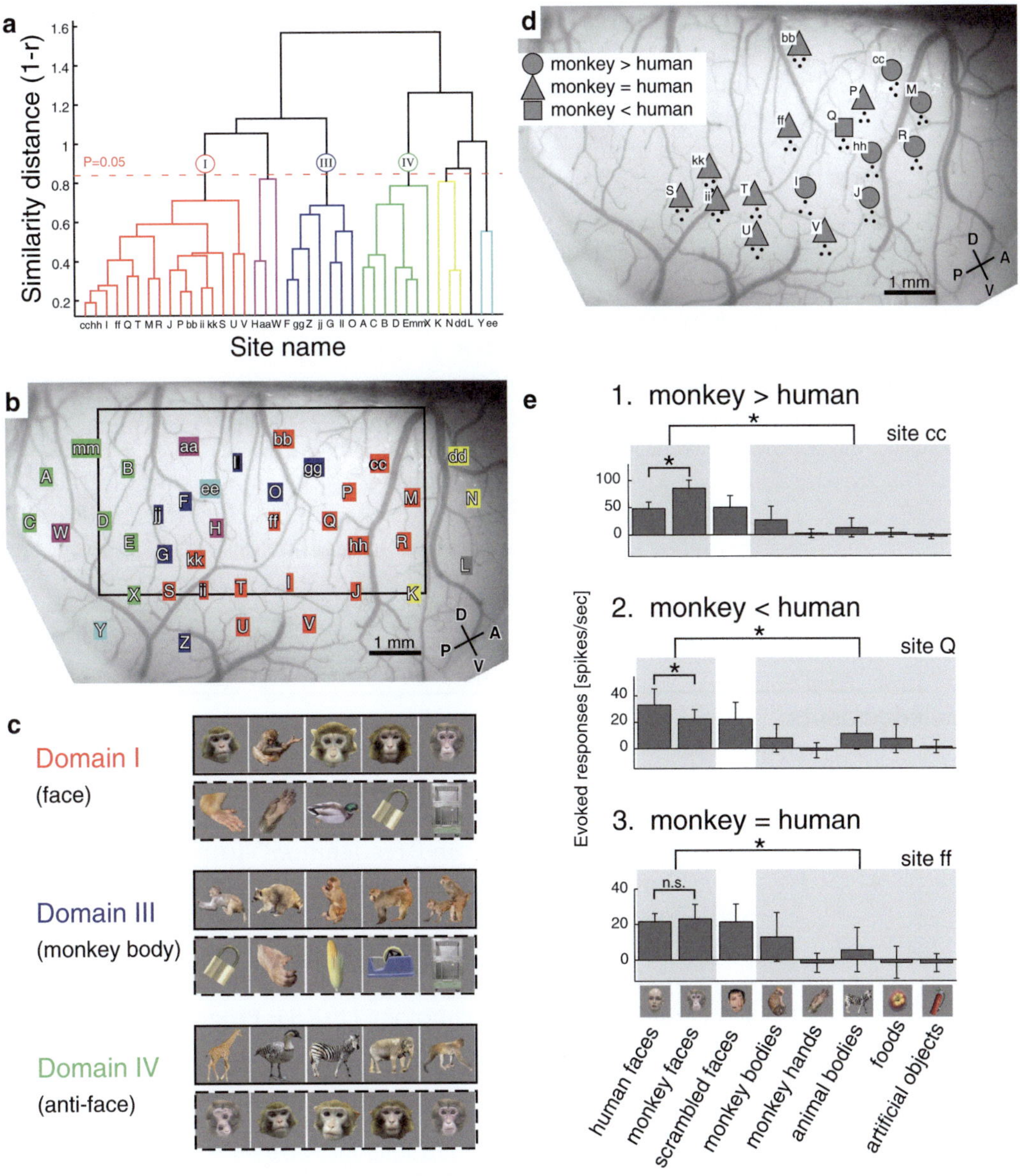

Fig. 7 Hierarchical clustering and cortical mapping by electrophysiology. (**a**) Dendrogram obtained by the hierarchical clustering of recording sites based on similarity in object selectivity across 104 object images. The vertical axis represents similarity distance defined by 1—correlation coefficient. The *horizontal broken red line* represents the threshold for statistically significant correlations between two sites ($p = 0.05$). The similarity distance lower than the threshold gives seven clusters. (**b**) Sites belonging to the same clusters are also clustered on the cortex. The sites and clusters are indicated with the same names and colors as in (**a**). The *solid rectangular* region corresponds to Fig. 6c, d. (**c**) Object selectivity of representative domains. Object selectivity was calculated by averaging evoked responses of sites within each domain. The *rectangular boxes* give the best 5 (the box with *solid line*) and the worst 5 (the box with *broken line*) stimuli. The color of the name of each domain is matched with the color in (**a**, **b**). (**d**, **e**) Sites in Domain I showing differential responses to monkey faces and human faces. In (**d**) sites in Domain I are differentiated by symbols indicating differential responses to monkey faces and human faces. (**e**) Representative site selectivity for different object categories and subcategories of faces. Vertical axis gives evoked responses of columnar responses recorded by electrodes. *Asterisks* give $p < 0.05$ (*t*-test) (modified from [18])

On the other hand, measurement range of two-photon laser-scanning microscopy is limited to a few hundred microns and the range is too small to visualize columnar organizations. The spatial resolution and measurement area of OISI (resolution, ~0.1 mm; measurement range, ~6 mm × 6 mm) is situated between these two methods.

One limitation of OISI is that OISI is a 2D imaging technique and signals along depth axis is integrated. Recently, functional optical coherence tomography (fOCT) is invented as a novel 3D functional imaging technique of intrinsic signals that has a higher spatial resolution than fMRI and a wider measurement area than two-photon imaging [34, 35]. fOCT may be a useful tool to overcome the limitation in OISI. In fact, the technique successfully visualized 3D structures of pinwheels in cat primary visual cortex [35].

We conducted OISI in IT cortex with anesthetized animals to avoid movement artifacts. However, some studies have succeeded in using OISI in V4 of awake monkeys [11]. One future direction would be to improve apparatus and data analysis and to make recording intrinsic signals in IT cortex from conscious monkeys possible. Then, we can address more cognitive aspects of object recognition.

Acknowledgements

This work is supported by Grant-in-AID for Scientific Research 22300137, Grant-in-Aid for Scientific Research on Innovative Areas, "Face perception and recognition" from the Ministry of Education, Culture, Sports, Science, and Technology (MEXT), and Grant-in-Aid for Scientific Research on Innovative Areas, "Sparse Modeling," (25120004), JSPS, Japan. This work is also supported by the "Funding Program for World-leading Innovative R&D on Science and Technology (FIRST Program)."

References

1. Perrett DI, Rolls ET, Caan W (1982) Visual neurones responsive to faces in the monkey temporal cortex. Exp Brain Res 47:329–342
2. Desimone R, Albright TD, Gross CG, Bruce C (1984) Stimulus-selective properties of inferior temporal neurons in the macaque. J Neurosci 4:2051–2062
3. Freiwald WA, Tsao DY (2010) Functional compartmentalization and viewpoint generalization within the macaque face-processing system. Science 330:845–851
4. Tanaka K, Saito H, Fukada Y, Moriya M (1991) Coding visual images of objects in the inferotemporal cortex of the macaque monkey. J Neurophysiol 66:170–189
5. Tsunoda K, Yamane Y, Nishizaki M, Tanifuji M (2001) Complex objects are represented in macaque inferotemporal cortex by the combination of feature columns. Nat Neurosci 4:832–838
6. Yamane Y, Tsunoda K, Matsumoto M et al (2006) Representation of the spatial relationship among object parts by neurons in macaque inferotemporal cortex. J Neurophysiol 96: 3147–3156
7. Fujita I, Tanaka K, Ito M, Cheng K (1992) Columns for visual features of objects in monkey inferotemporal cortex. Nature 360: 343–346
8. Blasdel GG, Salama G (1986) Voltage-sensitive dyes reveal a modular organization in monkey striate cortex. Nature 321:579–585
9. Grinvald A, Lieke E, Frostig RD et al (1986) Functional architecture of cortex revealed by

optical imaging of intrinsic signals. Nature 324:361–364

10. Malonek D, Tootell RB, Grinvald A (1994) Optical imaging reveals the functional architecture of neurons processing shape and motion in owl monkey area MT. Proc Biol Sci 258: 109–119
11. Tanigawa H, Lu HD, Roe AW (2010) Functional organization for color and orientation in macaque V4. Nat Neurosci 13:1542–1548
12. Sato T, Tanifuji M (2011) Optical intrinsic signal imaging from macaque inferotemporal cortex reveals functional structures in different spatial scales. Soc Neurosci Abstr 484.08
13. Tsao DY, Freiwald WA, Knutsen TA et al (2003) Faces and objects in macaque cerebral cortex. Nat Neurosci 6:989–995
14. Kourtzi Z, Tolias AS, Altmann CF et al (2003) Integration of local features into global shapes: monkey and human FMRI studies. Neuron 37:333–346
15. Pinsk MA, DeSimone K, Moore T et al (2005) Representations of faces and body parts in macaque temporal cortex: a functional MRI study. Proc Natl Acad Sci U S A 102:6996–7001
16. Tsao DY, Freiwald WA, Tootell RBH, Livingstone MS (2006) A cortical region consisting entirely of face-selective cells. Science 311:670–674
17. Bell AH, Hadj-Bouziane F, Frihauf JB et al (2009) Object representations in the temporal cortex of monkeys and humans as revealed by functional magnetic resonance imaging. J Neurophysiol 101:688–700
18. Sato T, Uchida G, Lescroart MD et al (2013) Object representation in inferior temporal cortex is organized hierarchically in a mosaic-like structure. J Neurosci 33:16642–16656
19. Vanzetta I, Grinvald A (1999) Increased cortical oxidative metabolism due to sensory stimulation: implications for functional brain imaging. Science 286:1555–1558
20. Vanzetta I, Slovin H, Omer DB, Grinvald A (2004) Columnar resolution of blood volume and oximetry functional maps in the behaving monkey; implications for FMRI. Neuron 42: 843–854
21. Fukuda M, Rajagopalan UM, Homma R et al (2005) Localization of activity-dependent changes in blood volume to submillimeter-scale functional domains in cat visual cortex. Cereb Cortex 15:823–833
22. MacVicar BA, Hochman D (1991) Imaging of synaptically evoked intrinsic optical signals in hippocampal slices. J Neurosci 11: 1458–1469
23. Tsunoda K, Oguchi Y, Hanazono G, Tanifuji M (2004) Mapping cone- and rod-induced retinal responsiveness in macaque retina by optical imaging. Invest Ophthalmol Vis Sci 45:3820–3826
24. Sato T, Uchida G, Tanifuji M (2009) Cortical columnar organization is reconsidered in inferior temporal cortex. Cereb Cortex 19: 1870–1888
25. Saleem KS, Tanaka K, Rockland KS (1993) Specific and columnar projection from area TEO to TE in the macaque inferotemporal cortex. Cereb Cortex 3:454–464
26. Bonhoeffer T, Grinvald A (1996) Optical imaging based on intrinsic signals: the methodology. In: Toga AW, Mazziotta JC (eds) Brain mapp. Methods. Academic, San Diego, pp 55–97
27. Arieli A, Grinvald A, Slovin H (2002) Dural substitute for long-term imaging of cortical activity in behaving monkeys and its clinical implications. J Neurosci Methods 114:119–133
28. Przybyszewski AW, Sato T, Fukuda M (2008) Optical filtering removes non-homogenous illumination artifacts in optical imaging. J Neurosci Methods 168:140–145
29. Wang G, Tanaka K, Tanifuji M (1996) Optical imaging of functional organization in the monkey inferotemporal cortex. Science 272:1665–1668
30. Ts'o DY, Frostig RD, Lieke EE, Grinvald A (1990) Functional organization of primate visual cortex revealed by high resolution optical imaging. Science 249:417–420
31. Bonhoeffer T, Grinvald A (1991) Iso-orientation domains in cat visual cortex are arranged in pinwheel-like patterns. Nature 353:429–431
32. Wang G, Tanifuji M, Tanaka K (1998) Functional architecture in monkey inferotemporal cortex revealed by in vivo optical imaging. Neurosci Res 32:33–46
33. Lin C, Sato T, Hung CP, Tanifuji M (2012) Continuous mapping of mirror face rotation in macaque inferior temporal cortex. Soc Neurosci Abstr 465.01
34. Maheswari RU, Takaoka H, Kadono H et al (2003) Novel functional imaging technique from brain surface with optical coherence tomography enabling visualization of depth resolved functional structure in vivo. J Neurosci Methods 124:83–92
35. Nakamichi Y, Kalatsky VA, Watanabe H et al (2012) The singularity string: a revised model of pinwheels in cat orientation columnar structures revealed by functional optical coherence tomography. Soc Neurosci Abstr 623.05

Part III

Neurovascular Coupling Techniques in Epilepsy

Chapter 9

Closed Cranial Window Applications in the Rat: Studies on Neurovascular Coupling Involving Pial Arterioles and the Glia Limitans

Haoliang Xu, Chanannait Paisansathan, and Dale A. Pelligrino

Abstract

Because of their intimate anatomic relationship with cerebral arterioles and neurons, astrocytes have been postulated to function as signal transducers, transferring information from activated neurons to the cerebral microcirculation. The above is labelled as "neurovascular coupling" and its components (i.e., neuron, astrocyte, and vascular cells) as the "neurovascular unit (NVU)." In the brain, NVUs can take a variety of forms. In the chapter, we highlight a well-represented cortical NVU consisting of pial arterioles, the glia limitans, and cortical neurons. This particular NVU can be readily examined in vivo using a closed cranial window system that permits one to monitor pial arteriolar diameter changes and manipulate astrocytic influences during periods of enhanced synaptic activity. In addition, this model can be used to evaluate mechanisms of upstream arteriolar relaxation. This dilation is necessary during periods of increased metabolic demand, in order to permit more blood to reach dilated downstream vessels, thereby improving nutrient supply to the activated neurons. Without pial arteriole dilation, downstream dilation in the vicinity of active synapses may be ineffective, placing neurons at risk, especially during episodes of intense neuronal activity, such as seizure.

Key words Neurovascular coupling, Neurovascular unit, Pial arteriolar, Cerebral blood flow, Astrocyte, Seizure

1 Background

The tight coupling between neuronal activity and blood flow is fundamental to brain function. When synaptic activity in a specific brain region goes up, cerebral blood flow (CBF) increases in a temporally and spatially coordinated manner. Many possibilities have been proposed to explain such neurovascular coupling, but uncertainty remains. It is widely recognized that the coupling process involves cellular elements (neurons, astrocytes, and vascular cells) that are collectively embodied in the neurovascular unit (NVU). It should be emphasized that the experimental findings arising from the approaches described herein, represent an expansion from the

Mingrui Zhao et al. (eds.), *Neurovascular Coupling Methods*, Neuromethods, vol. 88, DOI 10.1007/978-1-4939-0724-3_9, "traditional" applications of closed cranial window/intravital microscopy systems, which is to gain an understanding of mechanisms involved in the regulation of cerebral arteriolar tone in an in vivo setting. Thus, this model was modified to include generally overlooked, but nevertheless important, aspects of the neurovascular coupling process within the NVU. Namely, it was applied in the examination of the role of astrocytes in the vitally important process of pial arteriolar dilation during increased sensorimotor activity. During the past decade, we have conducted a series of studies, the results of which support the hypothesis that downstream parenchymal NVU elements sense an increase in local synaptic activity and transmit an astrocyte-dependent vasodilating signal upstream to pial arterioles. This is necessary during periods of increased metabolic demand, in order to permit more blood to reach dilated downstream vessels, thereby improving nutrient supply to activated neurons. Without it, that downstream dilation may be ineffective, placing neurons at risk, especially during episodes of intense neuronal activity (e.g., seizure [1]).

In contrast to neurons, astrocytes have multiple direct contacts with cerebral microvessels [2]. Astrocytes also display extensive contacts with neurons, to the extent that a single forebrain astrocyte may directly interact with 10^5 neuronal synapses in rodents and more than ten times that amount in humans [3]. That arrangement, coupled with the extensive level of interactions between astrocytes and microvessels, supports the postulate that astrocytes play a vital role in neurovascular coupling. Studies using astrocyte cell culture and brain slice models have provided valuable clues as to the possible role of astrocytes in CBF regulation. However, one cannot ignore questions related to the relevance of results obtained using in vitro models [4].

Over the course of multiple published reports, our laboratory has obtained evidence that the glia limitans, the layer of astrocytic processes that blankets the cortical surface, plays an important role in regulating pial arteriolar dilation in vivo [5, 6]. Due to the accessibility of the glia limitans, and its close contact with pial arterioles, as well as the availability of a highly effective and selective astrocyte toxin (L-α-aminoadipic acid (L-αAAA)), we have been able to examine the role of astrocytes (represented by the glia limitans) in modulating pial arteriolar dilating function in vivo under a variety of conditions (details below). It is of relevance to note that the glia limitans/pial arteriole relationship does not precisely recapitulate the astrocyte/arteriole arrangement present within the brain parenchyma. Thus, astrocytes do not contact pial vessels over the entire vessel circumference. Nevertheless, pial microvessels do reside in close apposition to the glia limitans, to the extent that factors arising from the glia limitans can influence the phenotype of pial microvessels (e.g., expression of endothelial barrier antigen [7]).

2 Equipment, Materials, and Procedures

2.1 Closed Cranial Window/Intravital Microscopy Methodology: Basic Preparation

The closed cranial window model has been employed in our laboratory for many years. Its inclusion in studies of cerebral vasoregulatory function [8, 9] and leukocyte trafficking [10–13] predates its use in experiments addressing neuro-glio-vascular coupling (e.g., [1, 14–16]). Despite the fact that the closed cranial window technology has been applied in multiple studies to address a wide variety of experimental objectives, the basic protocol for in vivo cranial window preparation among studies have not varied extensively. Sprague–Dawley rats (250–300 g) were used in most of our studies. Following endotracheal intubation, the rat is paralyzed (curare) and connected to a rodent ventilator. The surgical preparation is performed under 2 % isoflurane/70 % N_2O (balance O_2) anesthesia. Body temperature is maintained at 38 °C by a servo-controlled heating pad. The rat is then stabilized in the prone position, using a head-holding apparatus. Following placement of femoral arterial and venous catheters, a midline skin incision is made to expose the skull. A 9 mm diameter craniotomy is created over the skull midline using a fine-tip dental drill to cut through the bone. During this process, a continuous stream of compressed air is used to cool the drilling area and remove bone debris. A 26 gauge needle, with bent tip, is then employed as a tool for cutting and removing the dura. Great care is taken to preserve the integrity of the sagittal sinus. Following removal of the dura within the skull opening, a borosilicate glass window (10 mm in diameter and 2 mm in thickness) outfitted with three predrilled holes is prepared for subsequent attachment to the skull surrounding the craniotomy. Flame-tapered PE60 tubing is inserted into the holes in the window and trimmed so that they do not protrude from the underside of the window. The ports are used for inflow and outflow of artificial cerebrospinal fluid (aCSF) and intracranial pressure (ICP) monitoring. The window with the attached cannulae is then glued to the skull using cyanoacrylate gel (*see* Fig. 1). In some limited cases (e.g., when antisense and missense oligodeoxynucleotides (ODNs) or the gliotoxin, L-α-aminoadipic acid (L-αAAA), are administered into the space under the window prior to study [5, 8, 11, 17–19]), the rats are permitted to recover for ≥24 h. In that case, following injections, the ports are plugged and the scalp incision is sutured closed until the day of study. However, in most cases, the animals are used for immediate experimentation.

2.2 Pial Arteriolar Image Recording and Data Analysis

Intravital video microscopy is used to monitor pial arteriole diameter changes in response to a wide variety of stimuli, including topically applied agents and increased neuronal activity. Images are acquired using a CCD camera (Photometrics, Tucson, AZ) mounted on an upright stereomicroscope (Nikon, Melville, NY). Recorded images are saved for subsequent analysis. Arteriolar

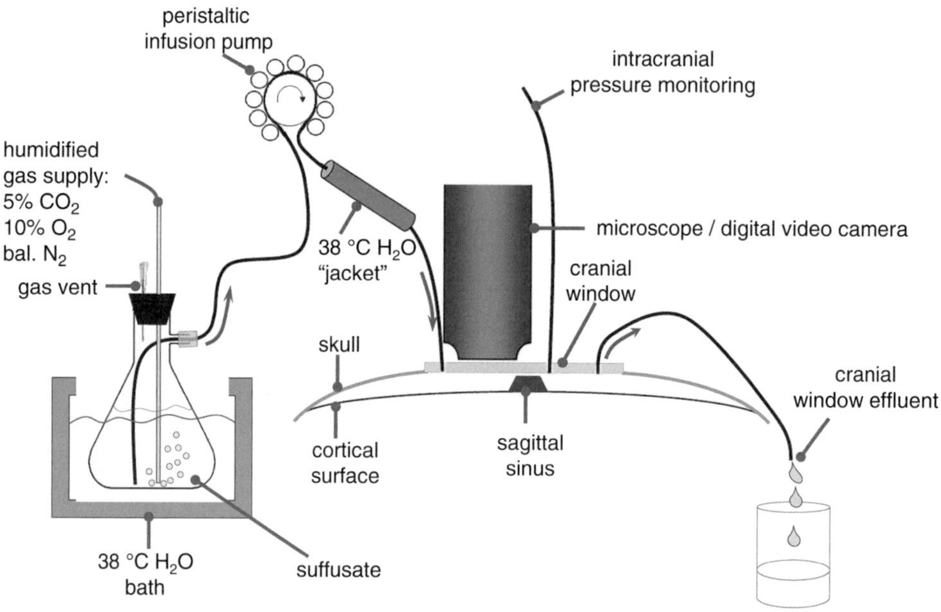

Fig. 1 Basic components of the cranial window artificial cerebrospinal fluid (aCSF) suffusion setup. The system permits a controlled suffusion of aCSF (typically at a rate of 0.2–1.0 mL/min) at a constant temperature (37–38 °C) and physiologically normal CSF PO_2, PCO_2, and pH, maintained via bubbling humidified gas into the delivery flask

diameter changes are measured off line using Metamorph software (MDS Analytical Technologies, Sunnyvale, CA). Irrespective of the stimulus, we typically measure the diameter values of three different pial arterioles and report the response as the average percent change in diameter relative to the baseline diameter.

2.3 Troubleshooting Cranial Window-Related Problems

In Table 1, we consider some of the key problems we have encountered when preparing rats with cranial windows, along with preventive and corrective measures. Factors such as bleeding during and after surgery, abnormal ICP levels, poor image quality (includes bubble removal measures), and abnormal arteriolar responses are addressed.

2.4 Post-surgical Experimental Preparation: Anesthesia/Analgesia Considerations

The actual experiments are performed in the presence of 70 % N_2O/30 % O_2 plus fentanyl, muscle relaxation (curare), and mechanical ventilation. Thus, upon completion of surgical procedures, the isoflurane anesthesia is discontinued and intravenous (iv) delivery of the opioid analgesic, fentanyl, is initiated. Specifically, the rats are given a 10 μg/kg iv fentanyl bolus, followed by infusion at 25 μg/kg/h. While provision of adequate anesthesia/analgesia is a necessity, one cannot ignore the possibility that agents employed in this endeavor might interfere with neurovascular coupling mechanisms and confound experimental interpretations.

Table 1
Troubleshooting: quick reference guide

Problem	Causes	Preventive measures or possible solutions
Bleeding during craniotomy	1. Sagittal sinus tear 2. Bleeding from bone vasculature	*Prevention*: 1. (1) Bluntly dissect the connecting membrane between sagittal sinus and skull; (2) Discontinue procedure until bleeding stops completely 2. Preventive measure not devised *Solution*: 1. Topical application of thrombin 2. Bone wax topical application may help
Bleeding under the installed cranial window	1. Mild bleeding is not fully controlled before cranial window installation 2. Later occurrence of bleeding from unseen vessel damage	*Prevention*: 1. Install cranial window only after all the bleeding sites are cleared 2. After dura removal, immerse brain surface under warm aCSF, check possible bleeding sites carefully *Solution*: Two methods to remove blood underneath the window: 1. Increase aCSF infusion rate. Mild bleeding may stop spontaneously. Blood can be removed by outflow stream 2. Uninstall cranial window as soon as bleeding occurs. Gently flush the bleeding area to remove blood. Reinstall cranial window when bleeding ceases
Low *intracranial pressure*	1. Cranial window leakage 2. Low infusion rate 3. Low outflow outlet	*Solution*: 1. Reseal the leaking sites using cyanoacrylate gel 2. Adjust to appropriate infusion rate 3. Increase the height of the outlet of outflow tube
High *intracranial pressure*	1. Infusion rate is too high 2. Outflow outlet is too high 3. Blockade of outflow stream 4. Abnormal arterial blood gas	*Solution*: 1. Reduce infusion rate 2. Reduce the height of outflow tube 3. Gently flush the outflow tube to remove the blockade 4. Check artificial ventilation system, and correct the problem accordingly

(continued)

Table 1 (continued)

Problem	Causes	Preventive measures or possible solutions
Poor *image quality*	1. Air bubble appears under cranial window space 2. Reflection of microscope light 3. Uneven brain surface in observation field	*Prevention*: 1. (1) Air bubble that is flushed into cranial window can be prevented by prefilling aCSF from the proximal end of inflow tube using a syringe equipped with a long needle; (2) check the entire aCSF pumping system, and replace the broken connecting segment where the air is sucked into irrigating solution; (3) the possibility of air comes in from the leaking site on cranial window circumference should be considered 2. Craniotomy should be performed at the midline of the skull to avoid uneven window installation due to the natural curve of the rat skull 3. Preventive measure not devised *Solution*: 1. Three measures to remove bubbles: (a) Tilt the rat's head up and gently tap on cranial window surface using a forceps. Air bubble may move to the top edge of the window, and then escape with outflow stream (b) Adjust the outflow tube to extremely low level. This may generate negative pressure and suck out the bubbles (c) Temporarily increasing inflow suffusion rate can remove bubbles in some instances 2. Adjust the cranial window to full horizontal position 3. Consider using an alternative observation area
Unexpected abnormal *vascular response*	1. Low vascular dilating response 2. Significant vasomotion "breathing vessel"	*Prevention*: 1. Can be prevented by: (1) keeping brain surface moist using warm aCSF during craniotomy, (2) initiate aCSF suffusion as soon as possible; (3) limit surgical bleeding 2. Check arterial blood gases, check suffusate pH, sufficiency of anesthesia, and correct problems accordingly

Certainly, fentanyl, which is a μ-opioid receptor agonist, warrants some consideration, in this regard. Mu-opioid receptors are well expressed in superficial neocortical GABAergic interneurons, as well as in a vasoactive intestinal peptide (VIP)-rich subpopulation of those interneurons. In view of the μ-opioid receptor abundance in neocortex interneurons, which may have far-reaching influences on both excitatory pyramidal neurons and other inhibitory interneurons within the forebrain, one cannot easily predict whether or to what degree excitatory/inhibitory balance within the brain will be affected. Moreover, evidence indicates expression of VIPergic receptors (e.g., $VPAC_{1-2}$) in pial arterioles [20, 21]. Since topical application of VIP, via a closed cranial window, elicits pial arteriolar dilation [22], one might envisage an arrangement where μ-opioid-linked mechanisms not only modulate interneuronal signalling, but also neurovascular signalling as well. Although the inclusion of inhaled 70 % N_2O in our studies permits us to achieve adequate analgesia with lower fentanyl exposure, confounding influences of fentanyl on neurovascular coupling cannot be dismissed.

As an alternative to fentanyl plus N_2O, one could consider using the hypnotic drug, α-chloralose. In fact, this is often the agent of choice for in vivo neurovascular coupling experiments; although there are some drawbacks to its use. First, it is a much poorer analgesic compared to fentanyl [23]. Second, with α-chloralose, stability of anesthesia is difficult to maintain, and its use is recommended only in terminal experiments [24]. Third, both agents have been shown to affect brain metabolic activity. Thus, while fentanyl may be accompanied by as much as a 40 % increase in brain glucose metabolism in normal cortical tissue [25], relative to awake or lightly anesthetized rats, α-chloralose has been reported to be associated with a ~50 % reduction [26]. Such baseline effects may influence vascular reactivity to brain activation when performing, for example, functional magnetic resonance imaging (fMRI) analyses, to the extent that greater vascular responses could occur when starting from a lower state of synaptic activity [27].

2.5 Synaptic Activation Induces Pial Arteriolar Dilation

We imposed either "physiological" or "supraphysiological" neuronal activation, by sciatic nerve stimulation (SNS) or topical application of bicuculline, respectively. Both activation paradigms elicit significant increases in pial arteriolar diameters. Thus, a 20 min suffusion of an aCSF solution containing the GABA-A antagonist, bicuculline (100 μM), is accompanied by a rapid and sustained pial arteriolar dilation (Fig. 2b). SNS is associated with increases in pial arteriolar diameters, in the hindlimb projection area of the contralateral somatosensory cortex, beginning 1–2 s after initiation of the stimulus, and peaking at ~30 s, followed by a gradual return to baseline, over the next 1–2 min (e.g., *see* Fig. 3a). Quantitative comparisons of the evoked pial arteriolar responses during SNS utilize the peak response values. During seizure, the enhanced neuronal activity can be tracked through monitoring cortical EEG patterns and applying a power analysis algorithm (Fig. 2a, c). For SNS, the contralateral sciatic nerve is exposed in the region of the sciatic notch. The proximal end is

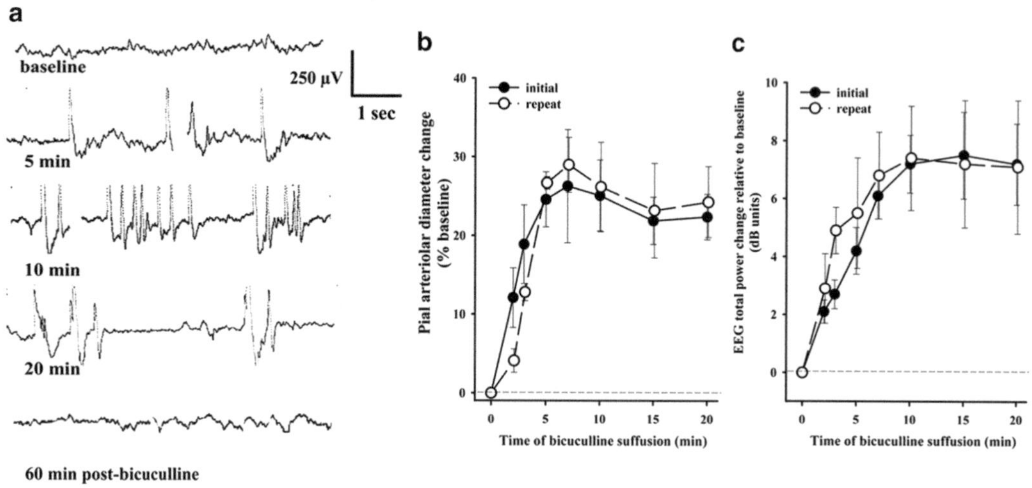

Fig. 2 A 20-min topical suffusion of bicuculline methiodide (100 μM) is accompanied by changes in the cortical electroencephalogram (EEG) and pial arteriolar dilations. (**a**) Representative EEG recordings showing the burst and suppression patterns typical of seizure. Brain electrical activity returns to normal within 1 h of cessation of bicuculline suffusion. (**b**) Pial arteriolar diameter increases during seizure (expressed as a percent increase from resting diameter). The data at each time point are presented as means ± SE ($n = 4$ rats). Bicuculline suffusion elicits a rapid and sustained pial arteriolar dilation that is reproducible upon repeat bicuculline suffusion, initiated 60 min after cessation of initial exposure. (**c**) EEG total power is expressed in decibel unit increases [32] relative to baseline levels (means ± SE at each time point; $n = 4$ rats). A nearly identical pattern of EEG power increase is observed upon repeating the bicuculline suffusion 1 h after the initial exposure. The data in B and C represent time controls for subsequent experiments. (Modified from H Xu et al. (2008) Am J Physiol Heart Circ Physiol 294, H622-32)

placed in a silver ring electrode, which is insulated with silicone rubber. The nerve is electrically stimulated for 20 s with square-wave pulses (current: 0.7 mA, frequency: 2 Hz, and pulse duration: 0.1 ms) using a stimulator (Stimulator HC, ML155, AD Instruments, Sydney, Australia). Somatosensory evoked potentials generated during SNS are recorded from the somatosensory cortex. Bipolar recordings are made from stainless steel screws fastened to the skull, with a reference placed in the occipital bone. Peak-to-peak (N1P1) amplitude differences were acquired and analyzed using Powerlab 4/30 and Labchart 7 (AD Instruments). Electrophysiological signals were digitized at 2 kb/s and bandpass filtered (0.1–1 kHz). Additionally, repeated stimulations (applied following 60 min of drug-free aCSF suffusion after the previous exposure to bicuculline or 30 min after previous SNS) evoked nearly identical patterns of pial arteriolar dilations and EEG power changes (Fig. 2b, c). Such a demonstration of the reversibility and reproducibility of the response (i.e., time control) is important in establishing the stability and validity of the model. Topical application of the nerve action potential blocker, tetrodotoxin (TTX,

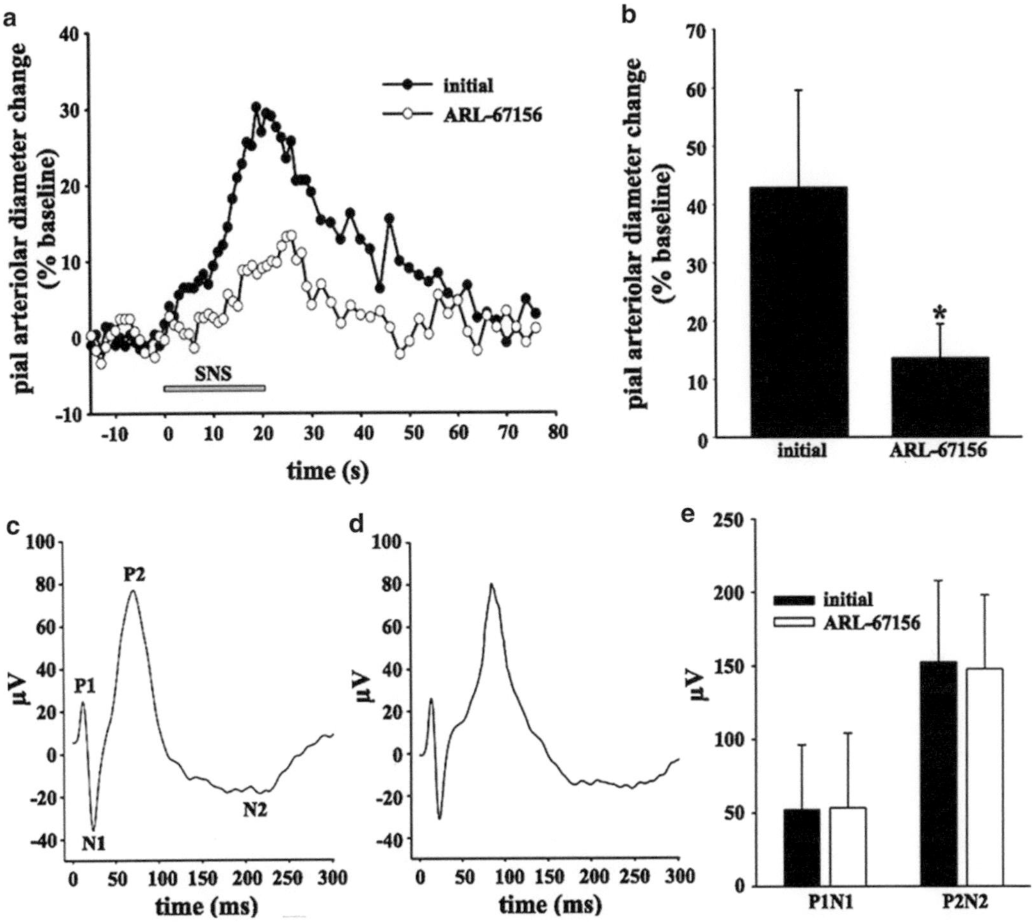

Fig. 3 Sciatic nerve stimulation (SNS)-related data. (**a**) Representative time course of changes in pial arteriolar diameters associated with a 20-s SNS (*gray bar*). Each curve represents the average response measured in two animals, first in the absence (initial; *filled circle*) and, subsequently, in the presence of the ecto-diphosphohydrolase blocker, ARL-67156 (100 μM; *open circle*). (**b**) Effect of ARL-67156 (100 μM) on SNS-evoked pial arteriolar dilations. Values are means ± SD. **P* < 0.05 vs. initial; *n* = 5. Representative SNS-generated somatosensory-evoked potentials (SEPs) recorded from the contralateral cortical surface before (**c**) and after (**d**) ARL-67156 application is also shown. The discussion regarding ARL-67156 is revisited in Sect. 3.3. (**e**) Note that no differences in the SEP peak-to-peak amplitudes were observed, irrespective of whether the P1N1 or the P2N2 amplitude was measured. Results in (**b**) and (**e**) are means ± SD; *n* = 5. (Modified from F Vetri et al. (2011) Am J Physiol Heart Circ Physiol 301, H1369-77)

1 μM), eliminated bicuculline-induced EEG changes, and blocked the concomitantly evoked pial arteriolar responses as well. TTX showed the same effect on SNS-induced dilations. These findings demonstrate that, despite the scarcity of direct intrinsic connections to pial arterioles, signals generated from activated neurons trigger pial arteriolar dilations.

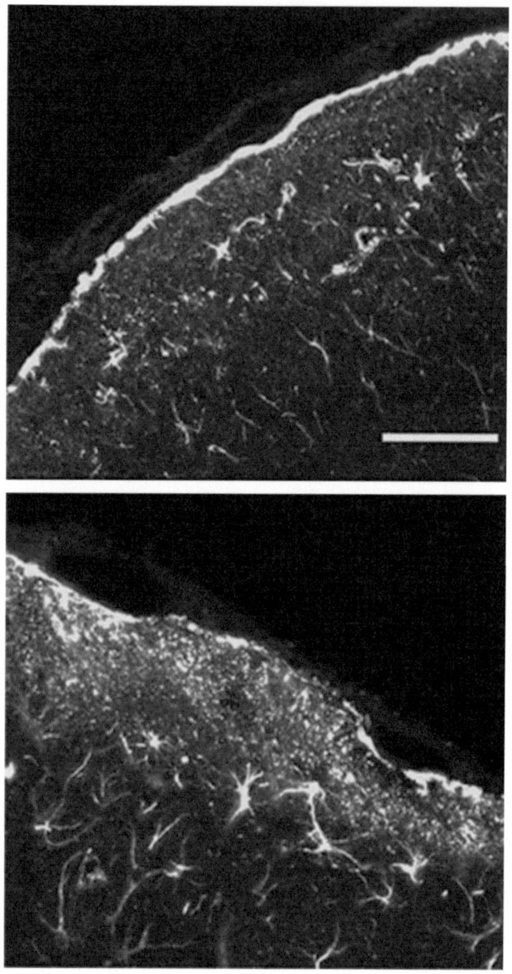

Fig. 4 Representative glial fibrillary acidic protein (GFAP) expression patterns in vehicle-treated (*upper panel*) and L-α-aminoadipic acid (L-αAAA)-treated (*lower panel*) rats. Each panel corresponds to a different rat. Note that, in the normal (vehicle treated) brain, GFAP is highly concentrated in the glia limitans. However, in rats exposed to topical L-αAAA for 24 h, GFAP expression in the glia limitans takes on a patchy appearance, exhibiting a loss of continuity. Additionally, the L-αAAA-treated brains often exhibit a fragmented expression of GFAP near the brain surface. Scale bar = 50 μm. (Modified from H Xu et al. (2004) Am J Physiol Heart Circ Physiol 287, H331-9)

2.6 The Role of Astrocytes (Glia Limitans) in Neurovascular Coupling

Twenty-four hours exposure to the highly selective gliotoxin, L-αAAA (2 mM), applied to the cortical surface produces an astrocytic injury that is characterized by a histological and a physiological "footprint." The former is represented by a fragmented as opposed to a continuous pattern of GFAP expression at the normal cortical surface (Fig. 4). Furthermore, this injury was confined to the superficial glia limitans, since no overt signs of damage to astrocytes lying >100 μm below the cortical surface were observed (*see* Fig. 4). The "physiological footprint" expressed itself in a number of ways. First, in rats treated with topically applied L-αAAA, only neural activation-induced pial arteriolar dilation was completely diminished. There were no signs of any loss of pial arteriolar reactivity to the endothelium-dependent dilator, acetylcholine, or to the direct smooth muscle-acting NO donor, *S*-nitroso-acetyl-penicillamine (SNAP). These findings provided confirmation that L-αAAA treatment lacked vascular cell toxicity. Second, to rule out L-αAAA effects on neurons, either in the case of SNS or during bicuculline-induced seizure, we monitored somatosensory-evoked responses (SSERs) or performed EEG power analyses, respectively. Because neither indicator of the strength of synaptic activation was affected in the presence of L-αAAA, we were able to conclude an absence of L-αAAA influence on neurons. Moreover, since astrocytes modulate synaptic function, the fact that the above indicators of synaptic activity were not affected suggests that the injury produced by topically applied L-αAAA does not extend into the neuropil. These findings also represent an important contrast to results obtained using TTX, where a primary blockade of neuronal activity prevented the pial arteriolar vasodilating response.

2.7 Role of Endothelium and Smooth Muscle Cells in Neurally Evoked Pial Arteriolar Dilations

Evidence does not support endothelium participation in pial arteriolar dilations accompanying neural activation. Light plus fluorescein dye (L/D) endothelium injury can selectively damage pial arteriolar endothelium, without affecting smooth muscle cells [17]. This is also demonstrated in experiments where L/D injury completely prevented vasodilations elicited by the endothelial nitric oxide synthase (eNOS) activator, acetylcholine, without altering the response to the NO donor, SNAP. We observed that neither bicuculline- nor SNS-associated pial arteriolar dilations were altered by L/D, suggesting the lack of involvement of the endothelium in the transmission of a neuronally evoked vasodilating signal to the pial arterioles. The issue of whether vascular smooth muscle may provide a signalling conduit for the coupling of neuronal activation to smooth muscle dilation cannot be completely disregarded, since there is no strategy we can envisage that would restrict vascular smooth muscle intercellular signalling while leaving the contractile apparatus intact. However, the fact that L-αAAA (*see* above) completely prevented SNS- and seizure-induced dilations leaves little room for contributions from a vascular smooth muscle signalling pathway.

3 Experimental Results, Validations, and Endogenous Mediators

3.1 Connexin-43-Related Communication Contributes to Neuronal Activation-Related Dilation

In a study published by us a decade ago [17], immunohistochemical analyses indicated the presence of several connexin (Cx) subtypes in cortical surface tissue in rats—namely, Cx43, Cx40, and Cx37. However, of these, only Cx43 showed any meaningful expression in astrocytes, being especially concentrated in the glia limitans. In fact, in a subsequent study, the expression of Cx43 was generally found to track that of GFAP [1]. Moreover, Cx43 displayed limited expression in pial vascular endothelium, and no detectable expression in smooth muscle cells [17]. Published data demonstrated that Cx43-containing gap junctions participate in a variety of pial arteriolar dilating responses. This was revealed through topical applications of the Cx-selective blocking peptides, Gap 27 (an inhibitory peptide towards Cx43 and Cx37) and Gap 26 (selective inhibitory peptide towards Cx40 and Cx37). Experiments were then performed to ascertain whether there were any indications of Cx43 involvement in bicuculline- or SNS-induced pial arteriolar dilations. Thus, following the initial measurement of pial arteriolar diameter changes during bicuculline exposure or SNS, baseline conditions were reestablished. A suffusion of Gap 27 (SRPTEKTIFII, 300 μM) or Gap 26 (VCYDQAFPISHIR, 300 μM) was then initiated. Forty-five minutes later, the neural activation was repeated. Gap 27 attenuated bicuculline-induced pial arteriolar dilation (by ~50 %), without altering EEG power. A similar result was obtained with the SNS-evoked pial arteriolar response, where no changes in SSERs were observed (unpublished observation) despite an 80 % reduction in the vasodilating response [1]. No effects of Gap 26 on pial arteriolar reactivity to neural activation were seen. These findings hint at the participation of gap junctional (or hemichannel-related) communication in pial arteriolar responses to both "supraphysiological" and "physiological" neuronal activation, with some suggestion that contributions from Cx43-containing structures may play a greater role during physiological activation. Furthermore, the aforementioned findings that glial injury completely prevented the pial arteriolar response, but endothelial injury had no effect, coupled with the apparent lack of Cx43 expression in pial vascular smooth muscle, suggests that the sites of the Cx43-dependence of the pial arteriolar response are more likely to be on astrocytes than blood vessels.

3.2 SNS-Induced Pial Arteriolar Dilation: Cross Talk Among Endogenous Elements via A_2 Receptor-Linked, PKA-Mediated Phosphorylations

The discussion above considered the important role played by astrocytes in cell-to-cell communication within an NVU composed of activated somatosensory neurons and pial arterioles, with an astrocytic component represented by the superficial glia limitans. Data derived from both in vivo (cranial window/intravital microscopy (e.g., [15, 28]) and in vitro (e.g., brain slice [29]) approaches have provided clues as to the likely factors contributing to that signalling—namely, adenosine; large-conductance, Ca^{2+}-operated K^+ (BK_{Ca}) channels in astrocytes/glia limitans; and inward-rectifier K^+ (K_{ir}) channels concentrated in pial arteriolar smooth muscle. Thus, in a recent study, we examined the effects of adenosine receptor, BK_{Ca}, and K_{ir} blockers, either singly or in combination, on SNS-evoked pial arteriolar dilations [14]. For the BK_{Ca}- and K_{ir}-channel blockers, paxilline and Ba^{2+}, the 10 and 100 μM doses were chosen for maximum efficacy in repressing pial arteriolar dilations to their respective openers, NS-1619 and KCl, without compromising selectivity [16]. For the A_2 blocker, ZM 241385, a 10 μM dose was used. As shown in a recent report [16], that dose completely blocks pial arteriolar dilations to topically applied adenosine. However, at 10 μM, ZM 241385 likely exceeds its selectivity towards the A_{2A} over the A_{2B} adenosine receptor subtype. Nevertheless, that dose should permit retention of ZM 241385 selectivity towards blocking adenylyl cyclase (AC)-activating A_2 receptors (A_2Rs) in relation to AC-inhibitory A_1 and A_3 adenosine receptors [30].

Table 2 provides a summary of the findings regarding A_2R blockade, BK_{Ca} blockade and K_{ir} blockade on SNS-evoked pial arteriolar dilations. It derives largely from data obtained in a recent publication from our laboratory [14]. Individually, the three

Table 2
Purinergic/K^+-channel cross talk in SNS-evoked pial arteriolar dilation

Target and blocker	% Reduction SNS-evoked dilation (single inhibitor)	% Reduction SNS-evoked dilation (combined inhibitors)	Indication
adenosine A_2R, ZM 241385 (10 μM)	64	+Paxilline (10 μM) 73 %	Nonadditive influence of A_2R and BK_{Ca} channel activation suggests one pathway potentiates function of the other. Speculate this relates to A_2R-mediated, PKA-linked phosphorylation of BK channel (*see* Table 3)
BK_{Ca}, paxilline (10 μM)	55	+$BaCl_2$ (100 μM) 46 %	Nonadditive nature of BK_{Ca} and K_{ir} channel activation suggests that both channels represent key elements in the same signalling pathway. Likely related to K^+release from glia limitans BK_{Ca} channels driving activity of pial arteriolar K_{ir} channels
K_{ir}, $BaCl_2$ (100 μM)	57	+ZM 241385 (10 μM) 58 %	Nonadditive influence of A_2R and K_{ir} channel activation suggests one pathway potentiates function of the other. Speculate this relates to A_2R-mediated, PKA-linked phosphorylation of K_{ir} channel (*see* Table 3)

Data taken from [14]
A_2R adenosine A_2 receptor, *PKA* cAMP-dependent protein kinase

blockers reduced SNS-evoked dilations by 55–64 %. The apparently nonadditive contributions of A_2Rs, BK_{Ca} channels, and K_{ir} channels were supported by findings of no further significant reductions in the presence of combined blocker applications. There are two key indications arising from the finding of overlapping influences among the three blocking regimens. First, the overlap between the effects of the A_2R antagonist (ZM-241385) and either of the K^+ channel blockers (paxilline or Ba^{2+}) led us to speculate that A_2R activation potentiates BK_{Ca} and K_{ir} function. Second, the overlap in the effects of paxilline and Ba^{2+} implies that both channels are key components of the same signalling pathway. For the first, we hypothesized that the common thread was A_2R activation of AC and subsequent phosphorylation of BK_{Ca} and K_{ir} channels via cAMP-dependent protein kinase (PKA). For the second, supported by findings published by Nelson and co-workers (*see*, for example, [29]), we postulated that increased glia limitans BK_{Ca} activity potentiates pial arteriolar smooth muscle K_{ir} channel function. Those two issues are addressed in Table 3.

3.3 Modelling Purinergic Contributions to SNS-Evoked Pial Arteriolar Dilations

Our postulated model for purine-driven signal propagation involving astrocytes and pial arterioles begins with a cellular ATP efflux component linked to increased synaptic activity [15]. As illustrated in Fig. 5, there are a number of fates and functions for extracellular ATP. This includes interactions with G-protein-coupled metabotropic P_2Y receptors on neighboring astrocytes, which may contribute to the generation and propagation of a Ca^{2+}/ATP "wave" among astrocytes, eventually reaching pial arterioles. A second fate is the rapid hydrolysis of ATP via the action of ecto-nucleotidases, leading to elevations in extracellular fluid (ECF) AMP and adenosine levels and, perhaps under certain conditions, elevations in intermediate hydrolytic products, like ADP. It should also be noted that the combination of increases in ECF ATP and adenosine may also participate in regulating synaptic function (for additional details, *see* [15]).

For the remainder of this section, we will give consideration to experimental findings addressing the role of ATP ecto-hydrolysis in gliovascular signalling, using closed cranial window/intravital microscopy technology. The general approach used in these studies was the monitoring of pial arteriolar responses to topical applications of ATP and its hydrolytic products (i.e., ADP, AMP, and adenosine). Topically applied ATP, at concentrations of 1, 10, and 100 μM, was associated with a dose-dependent dilation of pial arterioles (Fig. 6b). In the presence of the A_2R blocker, ZM 241385, the responses to 1 and 10 μM ATP were completely blocked, while the attenuation in the response to 100 μM ATP was incomplete (reduced by 67 %). However, upon addition of the selective P_2Y_1 antagonist, MRS-2179 (10 μM), to the suffusate, the response to 100 μM ATP was totally prevented (Fig. 6d).

Table 3
cAMP and PKA dependence of pial arteriolar dilations accompanying SNS and suffusions of adenosine, BK_{Ca} channel opener, and K_{ir} channel opener

Suffusate composition	% Reduction in dilation in the presence of AC blocker, MDL-12330A (5 μM)	% Reduction in dilation in the presence of PKA blocker, KT-5720 (1 μM)	Implications
Adenosine 10 μM; 100 μM	 53 55	 42 59	Consistent with AC and PKA inhibitors restricting the activation of the AC-cAMP-PKA pathway that normally results from adenosine engagement of its A_2R. Implies that, under normal conditions, the likely effect of this phosphorylation is vasodilation, arising from: ↑opening of vsm K_{ir} and GL BK_{Ca} channels; and ↓Ca^{2+}-sensitivity of vsm contractile proteins [33] (*see* below and Fig. 5)
BK_{Ca} opener, NS-1619 10 μM; 50 μM	 44 83	 57 71	Implies that a "basal" level of phosphorylation needs to be present for BK_{Ca} channel opening, via NS-1619, to elicit pial arteriolar relaxation. In a previous report [14], we found that GL ablation, via L-AAA, only restricts NS-1619-, but not KCl- or adenosine-induced dilations. This indicates that BK_{Ca} channels are primarily expressed on the glia limitans, as opposed to pial vascular cells
K_{ir} opener, KCl 6 mM; 12 mM	 40 65	 58 69	Blocking PKA-linked phosphorylations nonadditively restricts pial arteriolar dilations linked to both K_{ir} and BK_{Ca} channel opening. This finding also implies that a "basal" level of phosphorylation needs to be present for K_{ir} channel opening, via 6–12 mM K^+, to elicit pial arteriolar dilation. In addition, in a recent study [14], we reported that the selective Kir blocker, Ba^{2+} (100 μM), restricts dilation to BK_{Ca} channel opening (via NS-1619) and adenosine. This implies an interdependence, whereby GL BK_{Ca} channel-linked pial arteriolar dilations involve opening of vsm K_{ir} channels
AC activator, forskolin 1 μM; 10 μM	 65 66	–	Confirms efficacy of AC blockade via MDL-12330As

(continued)

Table 3 (continued)

Suffusate composition	% Reduction in dilation in the presence of AC blocker, MDL-12330A (5 μM)	% Reduction in dilation in the presence of PKA blocker, KT-5720 (1 μM)	Implications
SNS	85	65	SNS-evoked pial arteriolar dilation involves adenosine A_2R-dependent activation of AC, promoting ↑cAMP and ↑PKA-linked phosphorylation of targets in *pial arteriolar smooth muscle*, leading to potentiation of relaxation via two main pathways [1] K_{ir} channel opening → hyperpolarization → VDCC closure → ↓ $[Ca^{2+}]_I$; and [2] Ca^{2+} desensitization of contractile proteins [34, 38] SNS-evoked dilation also is speculated to involve pathways within the *glia limitans* that act to modulate the smooth muscle pathways. This includes inter-astrocytic movement of ATP and Ca^{2+} (ATP/Ca^{2+} "wave") from synaptic sites to perivascular sites, where released ATP is hydrolyzed, via ecto-nucleotidases, to AMP and adenosine, either of which can activate A_2R's and promote PKA-linked phosphorylations at selected GL sites (e.g., intracellular Ca^{2+} storage pools [35] and BK_{Ca} channels [39]), releasing Ca^{2+} while enhancing release of K^+ via BK_{Ca} channels and potentiating smooth muscle K_{ir} function

Data taken from [14]

AC adenylyl cyclase, *PKA* cAMP-dependent protein kinase, *GL* glia limitans, *vsm* vascular smooth muscle, *A_2R* adenosine A_2 receptor, *VDCC* voltage-dependent Ca^{2+} channel

It should be noted that, in the presence of MRS-2179, by itself, a near-complete blockade of ADP-induced dilations was observed (Fig. 6c; *see* also [6]). These findings suggest that the remaining ZM 241385-resistant vasodilation during suffusion of 100 μM ATP (Fig. 6b) reflects some ATP conversion to ADP at high ATP levels (100 μM). In the presence of ARL-67156 (100 μM), which selectively inhibits the ecto-diphosphohydrolases, ecto-nucleoside triphosphate diphosphohydrolase-1 (e-NTPDase-1) and ecto-pyrophosphatase/phosphodiesterase (e-NPP)) (Fig. 6e), most (but not all) of the dilating response to ATP was prevented. It is of some relevance to note that, in the presence of ARL-67156, SNS-evoked dilations are reduced by ~70 % (Fig. 3b). That decrease is quite close to the 65–85 % level of reduction in pial arteriolar dilating responses to SNS associated with inhibition of

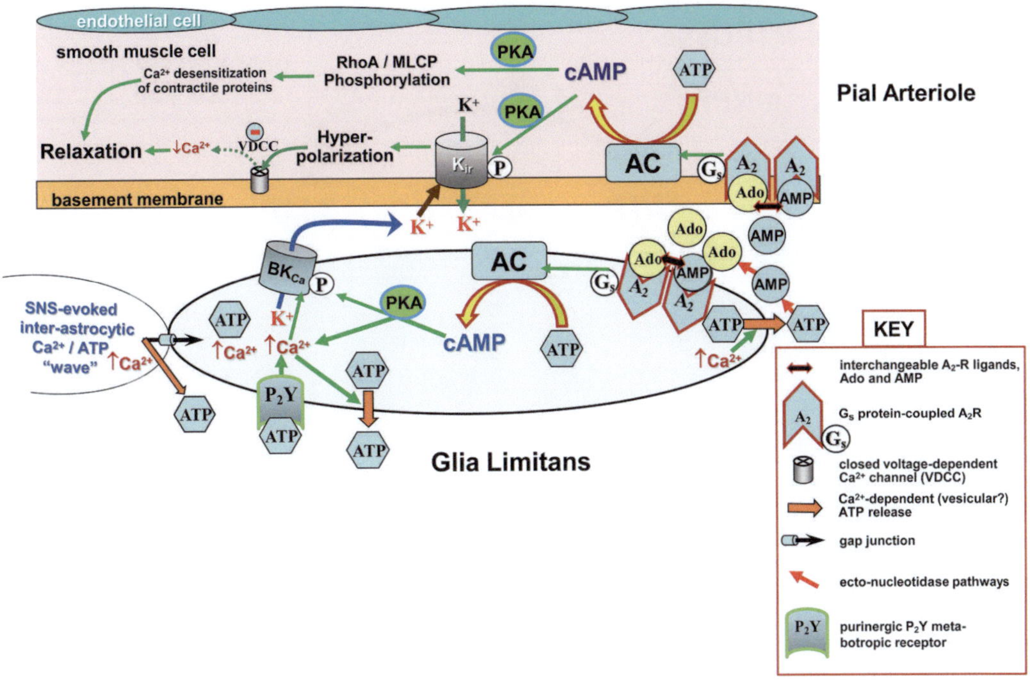

Fig. 5 Postulated scheme describing purinergic/K^+-channel cross talk in SNS-evoked pial arteriolar dilation. Key components include adenosine (Ado) A_2 receptor A_2R) activation (by either Ado or AMP), stimulation of adenylyl cyclase (AC), and potentiation of cyclic AMP-dependent protein kinase (PKA)-mediated phosphorylation (and enhanced function) of BK_{Ca} and K_{ir} channels. PKA also been linked to repression of smooth muscle contractility (e.g., via phosphorylation of RhoA and myosin light chain phosphatase (MLCP)) [33, 34], and enhanced mobilization of Ca^{2+} from storage sites in the cytosol (e.g., in astrocytes [35, 36])

PKA or adenylyl cyclase (*see* Table 3 and [14]). The partial attenuation of the pial arteriolar response to 100 μM ATP may be explained by an activation of the ecto-ATPase, e-NTPDase-2 (Fig. 6a), only under circumstances of *excessive* ATP release [16].

We next explored the ecto-5′-nucleotidase (e-5′-N)-sensitive portion of the ecto-phosphohydrolysis pathway. Thus, we addressed whether the A_2R-dependence of the pial arteriolar dilating response to topically applied ATP required total ATP dephosphorylation to adenosine. As an initial step, we established that topical application of AMP (similar to adenosine [16]]) was associated with a robust dose-dependent pial arteriolar dilation (Fig. 7b). We then determined that the AMP response was completely prevented in the presence of A_2R blockade (Fig. 7b). To establish whether that reflected total AMP conversion to adenosine, via e-5′-N, or included a direct influence of AMP on the A_2R [31], we evaluated AMP responses in the absence and presence of the selective e-5′-N blocker, AOPCP. The selectivity and efficacy of the 300 μM AOPCP dose used was established in recent publications from our laboratory [14, 16]. Unlike the total blockade of the pial arteriolar

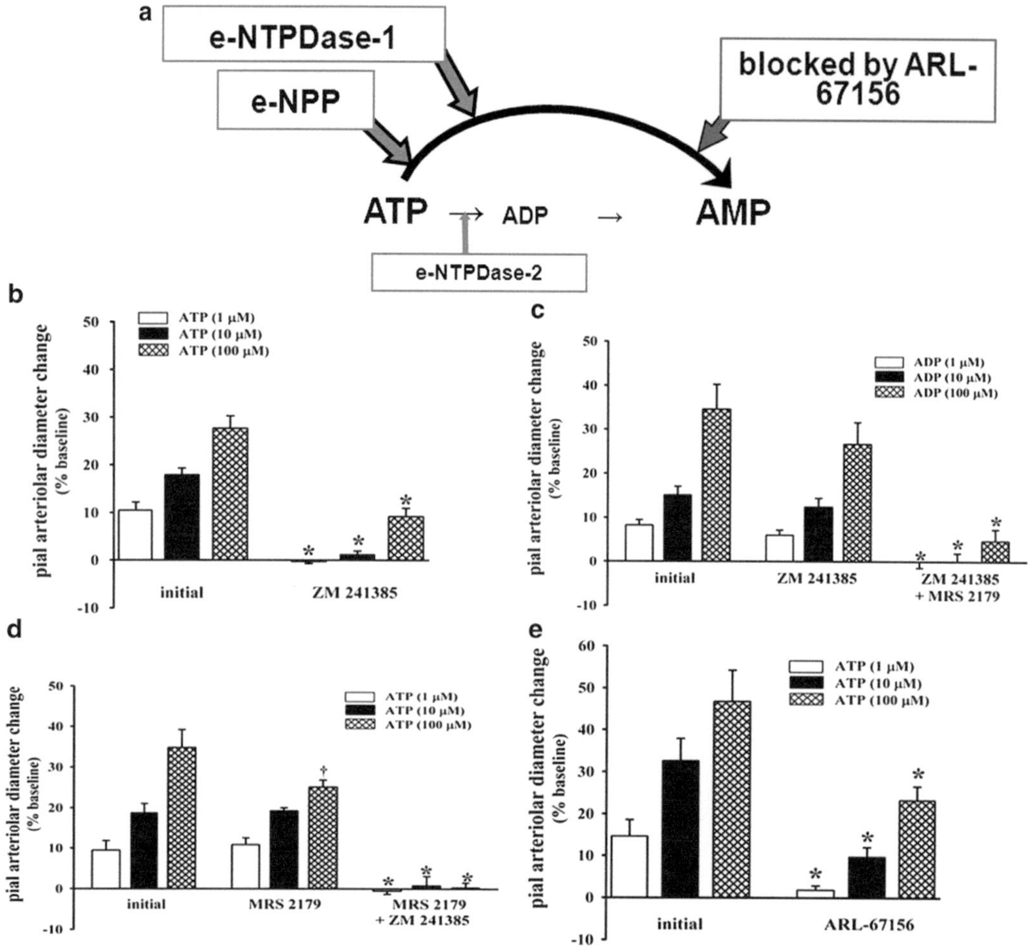

Fig. 6 (**a**) Fate of topically applied ATP and ADP, with respect to pial arteriolar dilating responses: Primary role of direct ATP to AMP conversions (via the actions of ecto-nucleoside triphosphate diphosphohydrolase-1 (e-NTPDase-1) and/or ecto-pyrophosphatase/phosphodiesterase (e-NPP)) in topical ATP-mediated dilations; but a minimal contribution from ecto-ATP (or ecto-ADP?) mono-hydrolase actions, mediated by NTPDase-2. (**b**) The finding of near-complete loss of ATP-evoked responses in the presence of A_2-receptor blockade, with ZM 241385, suggests that ATP release into the extracellular compartment promotes pial arteriolar dilation primarily through its conversion to adenosine. However, as shown in Fig. 7, both AMP and adenosine appear capable of activating A_2-receptors. (**c**) On the other hand, ADP-evoked dilations are unaffected by A_2-receptor blockade, but show complete repression in the presence of selective P_2Y1-receptor blockade with MRS 2179. (**d**) In support of the findings in (**c**), MRS 2179 has virtually no effect on ATP-induced dilations, except for a ~30 % reduction at 100 μM ATP (may reflect a limited ATP to ADP conversion under conditions of excessive ATP release). (**e**) Inhibition of ecto-diphosphohydrolase activity almost completely prevents dilating responses at 1 and 10 μM ATP. The partial attenuation in the response to 100 μM ATP may be a reflection of NTPDase-2-mediated conversion of ATP to ADP (the latter of which is insensitive to ARL-67156 influences). (Modified from F Vetri et al. (2011) Am J Physiol Heart Circ Physiol 301, H1369-77)

dilations evoked by suffusions of 1–100 μM AMP in the presence of the A_2R blocker, ZM 241385, exposure to AOPCP was accompanied by 100 %, 61 %, and 46 % reductions in AMP responses at 1, 10, and 100 μM, respectively (Fig. 7c). The finding of a

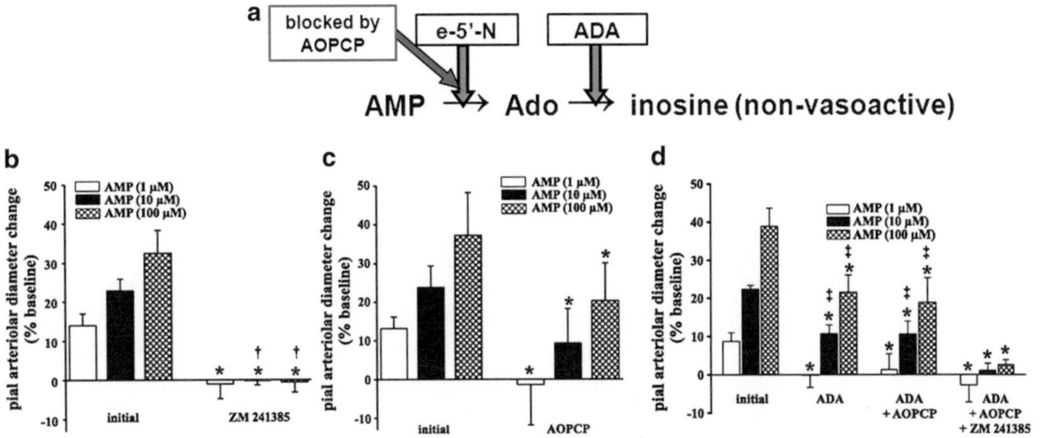

Fig. 7 (**a**) Fate of topically applied AMP: Hydrolytic conversion to adenosine (Ado), via ecto-5′-nucleotidase (e-5′-N), and subsequent deamination to non-vasoactive inosine in the presence of adenosine deaminase (ADA). (**b**) Adenosine (Ado) A_2-receptor blockade with 10 μM ZM 241385, given topically, is associated with complete loss of pial arteriolar dilating response to topically applied AMP (1–100 μM)—a finding seemingly consistent with complete in vivo AMP hydrolysis to adenosine. (**c**) However, in the presence of the selective ecto-5′-nucleotidase blocker, AOPCP (300 μM), a significant response remained at 10 and 100 μM AMP. This could either be attributable to insufficient inhibition of ecto-5′-nucleotidase allowing some AMP to Ado conversion OR that AMP itself activates the A_2 receptor. (**d**) To test the latter, AMP was suffused in the presence of Ado depletion via topically applied ADA, either alone or in combination with AOPCP. ADA converts Ado to inosine, which is not vasoactive [37]. Despite substantially greater removal of Ado, the pial arterioles remained responsive to AMP. Only when the A_2-receptor blocker was added, was complete loss of AMP reactivity observed. This is likely explained by A_2-receptor reactivity to AMP (e.g., [31]). (Modified from F Vetri et al. (2011) Am J Physiol Heart Circ Physiol 301, H1369-77)

significant residual dilation to AMP in the presence of AOPCP is consistent with AMP acting as an A_2R agonist. To obtain additional confirmation for that conclusion, we monitored reactivity to 1–100 μM AMP in the presence of adenosine deaminase (ADA; 2 units/mL) to deplete adenosine levels relative to levels of AMP. This strategy was associated with a ~50 % reduction in the dilating response to 10 and 100 μM AMP (Fig. 4d). To ensure that all adenosine was essentially removed, AOPCP was added to the suffusate, without incurring further losses in AMP reactivity (Fig. 4d). Subsequent addition of ZM 241385 resulted in a complete loss of the pial arteriolar response to AMP (Fig. 4d). These findings support a role for AMP as an A_2R agonist.

4 Conclusions

It is well documented that localized elevations in brain synaptic activity are accompanied by increases in CBF within such neuronally active regions. This coupled response is designed to ensure adequate delivery of O_2 and glucose to meet increased local energy demands. However, those demands may be largely unmet without the concomitant occurrence of dilation in upstream arteriolar segments. In the cerebral cortex, the processes that permit vasodilating signals, originating at sites of enhanced synaptic firing, to reach the pial arterioles, which are largely isolated from parenchymal neurons by the glia limitans, has been a focus of our laboratory for many years. In this endeavor, closed cranial window/intravital microscopy technology has proven to be indispensable, in that it permits the direct measurement of pial arteriolar diameters during conditions of high (seizure) and more normal (SNS) enhancements in evoked synaptic activity. The cranial window model also allows us to apply agents to the cortical surface, which are capable of preventing the influences of specific cells or cellular structures. This includes disruption of the glia limitans (using the selective gliotoxin, L-αAAA); and disruption of interastrocytic communication through connexin-43-containing gap junctions (using gap junction mimetic peptides). In addition, the easy accessibility to the cortical surface allows us to introduce a variety of pharmacological and molecular [5, 8, 9, 18, 19] interventions that address the specific and interactive (cross talk) roles of K^+ channels (K_{ir} and BK_{Ca}), as well as ATP and its extracellularly formed hydrolytic products, ADP, AMP, and adenosine.

References

1. Xu HL, Mao L, Ye S, Paisansathan C et al (2008) Astrocytes are a key conduit for upstream signaling of vasodilation during cerebral cortical neuronal activation in vivo. Am J Physiol Heart Circ Physiol 294:H622–H632
2. Iadecola C, Nedergaard M (2007) Glial regulation of the cerebral microvasculature. Nat Neurosci 10:1369–1376
3. Oberheim NA, Tian GF, Han X et al (2008) Loss of astrocytic domain organization in the epileptic brain. J Neurosci 28:3264–3276
4. Filosa JA (2010) Vascular tone and neurovascular coupling: considerations toward an improved in vitro model. Front Neuroenergetics. doi:16. 10.3389/fnene.2010.00016
5. Xu HL, Koenig HM, Ye S et al (2004) Influence of the glia limitans on pial arteriolar relaxation in the rat. Am J Physiol Heart Circ Physiol 287:H331–H339
6. Xu HL, Ye S, Baughman VL et al (2005) The role of the glia limitans in ADP-induced pial arteriolar relaxation in intact and ovariectomized female rats. Am J Physiol Heart Circ Physiol 288:H382–H388
7. Allt G, Lawrenson JG (1997) Is the pial microvessel a good model for blood–brain barrier studies? Brain Res Brain Res Rev 24: 67–76
8. Xu HL, Gavrilyuk V, Wolde HM et al (2004) Regulation of rat pial arteriolar smooth muscle relaxation in vivo through multidrug resistance protein 5-mediated cGMP efflux. Am J Physiol Heart Circ Physiol 286:H2020–H2027
9. Xu HL, Galea E, Santizo RA et al (2001) The key role of caveolin-1 in estrogen-mediated regulation of endothelial nitric oxide synthase function in cerebral arterioles *in vivo*. J Cereb Blood Flow Metab 21:907–913
10. Santizo R, Pelligrino DA (1999) Estrogen reduces leukocyte adhesion in the cerebral circulation of female rats. J Cereb Blood Flow Metab 19:1061–1065
11. Santizo RA, Xu HL, Galea E et al (2002) Combined endothelial nitric oxide synthase upregulation and caveolin-1 downregulation decrease leukocyte adhesion in pial venules of ovariectomized female rats. Stroke 33:613–616
12. Santizo RA, Anderson S, Ye S et al (2000) Effects of estrogen on leukocyte adhesion after transient forebrain ischemia. Stroke 31:2231–2235
13. Xu HL, Salter-Cid L, Linnik MD et al (2006) Vascular adhesion protein-1 plays an important role in postischemic inflammation and neuropathology in diabetic, estrogen-treated ovariectomized female rats subjected to

transient forebrain ischemia. J Pharmacol Exp Ther 317:19–29

14. Paisansathan C, Xu HL, Vetri F et al (2010) Interactions between adenosine and potassium channel-related pathways in the coupling of somatosensory activation and pial arteriolar dilation. Am J Physiol Heart Circ Physiol 299:H2009–H2017
15. Pelligrino DA, Vetri F, Xu HL (2011) Purinergic mechanisms in gliovascular coupling. Semin Cell Dev Biol 22:229–236
16. Vetri F, Xu H, Mao L et al (2011) ATP hydrolysis pathways and their contributions to pial arteriolar dilation in rats. Am J Physiol Heart Circ Physiol 301:H1369–H1377
17. Xu HL, Santizo RA, Baughman VL et al (2002) ADP-induced pial arteriolar dilation in ovariectomized rats involves gap junctional communication. Am J Physiol Heart Circ Physiol 283:H1082–H1091
18. Xu HL, Wolde HM, Gavrilyuk V et al (2004) cAMP modulates cGMP-mediated cerebral arteriolar relaxation in vivo. Am J Physiol Heart Circ Physiol 287:H2501–H2509
19. Xu HL, Vetri F, Lee HK et al (2009) Estrogen replacement therapy in diabetic ovariectomized female rats potentiates postischemic leukocyte adhesion in cerebral venules via a RAGE-related process. Am J Physiol Heart Circ Physiol 297:H2059–H2067
20. Taki K, Kaneko T, Mizuno N (2000) A group of cortical interneurons expressing mu-opioid receptor-like immunoreactivity: a double immunofluorescence study in the rat cerebral cortex. Neuroscience 98:221–231
21. Fahrenkrug J, Hannibal J, Tams J et al (2000) Immunohistochemical localization of the VIP1 receptor (VPAC1R) in rat cerebral blood vessels: relation to PACAP and VIP containing nerves. J Cereb Blood Flow Metab 20:1205–1214
22. Lenti L, Zimmermann A, Kis D et al (2009) PACAP and VIP differentially preserve neurovascular reactivity after global cerebral ischemia in newborn pigs. Brain Res 1283:50–57
23. Tremoleda JL, Kerton A, Gsell W (2012) Anaesthesia and physiological monitoring during in vivo imaging of laboratory rodents: considerations on experimental outcomes and animal welfare. EJNMMI Res 2:44
24. Tsurugizawa T, Uematsu A, Uneyama H et al (2010) Effects of isoflurane and alpha-chloralose anesthesia on BOLD fMRI responses to ingested L-glutamate in rats. Neuroscience 165:244–251
25. Statler KD, Janesko KL, Melick JA et al (2003) Hyperglycolysis is exacerbated after traumatic brain injury with fentanyl vs. isoflurane anesthesia in rats. Brain Res 994:37–43
26. Austin VC, Blamire AM, Allers KA et al (2005) Confounding effects of anesthesia on functional activation in rodent brain: a study of halothane and alpha-chloralose anesthesia. Neuroimage 24:92–100
27. Hyder F, Rothman DL, Shulman RG (2002) Total neuroenergetics support localized brain activity: implications for the interpretation of fMRI. Proc Natl Acad Sci U S A 99: 10771–10776
28. Meno JR, Nguyen TS, Jensen EM et al (2005) Effect of caffeine on cerebral blood flow response to somatosensory stimulation. J Cereb Blood Flow Metab 25:775–784
29. Dunn KM, Nelson MT (2010) Potassium channels and neurovascular coupling. Circ J 74:608–616
30. Fredholm BB, IJzerman AP, Jacobson KA et al (2011) International Union of Basic and Clinical Pharmacology. LXXXI. Nomenclature and classification of adenosine receptors—an update. Pharmacol Rev 63:1–34
31. Ross FM, Brodie MJ, Stone TW (1998) Adenosine monophosphate as a mediator of ATP effects at P1 purinoceptors. Br J Pharmacol 124:818–824
32. Pelligrino DA, Gay RL, Baughman VL et al (1996) Nitric oxide synthase inhibition modulates N-methyl-D-aspartate-induced changes in cerebral blood flow and EEG activity. Am J Physiol Heart Circ Physiol 271:H990–H995
33. Pfitzer G (2001) Invited review: regulation of myosin phosphorylation in smooth muscle. J Appl Physiol 91:497–503
34. Somlyo AP, Somlyo AV (2003) Ca^{2+} sensitivity of smooth muscle and nonmuscle myosin II: modulated by G proteins, kinases, and myosin phosphatase. Physiol Rev 83:1325–1358
35. Nishizaki T (2004) ATP- and adenosine-mediated signaling in the central nervous system: adenosine stimulates glutamate release from astrocytes via A2a adenosine receptors. J Pharmacol Sci 94:100–102
36. Pelligrino DA, Xu HL, Vetri F (2010) Caffeine and the control of cerebral hemodynamics. J Alzheimers Dis 20(suppl 1):S51–S62
37. Ngai AC, Winn HR (1993) Effects of adenosine and its analogues on isolated intracerebral arterioles—extraluminal and intraluminal application. Circ Res 73:448–457
38. Ko EA, Han J, Jung ID et al (2008) Physiological roles of K^+ channels in vascular smooth muscle cells. J Smooth Muscle Res 44:65–81
39. Price DL, Ludwig JW, Mi H et al (2002) Distribution of rSlo Ca^{2+}-activated K^+ channels in rat astrocyte perivascular endfeet. Brain Res 956:183–193

Chapter 10

Multi-Spectral Imaging of Blood Volume, Metabolism, Oximetry, and Light Scattering

Mingrui Zhao, Hongtao Ma, Samuel Harris, and Theodore H. Schwartz

Abstract

Advances in functional imaging techniques including fMRI, PET, and SPECT have improved our understanding of the relationship between brain activity and brain energy supply. Neurovascular and neurometabolic coupling are critical to supply the energy demands of brain tissue during both normal physiological function and pathological conditions. With the use of multi-spectral imaging techniques, one can simultaneously measure changes in cerebral blood volume, oxyhemoglobin, deoxyhemoglobin, light scattering, and local metabolism during epilepsy.

Key words Neurovascular coupling, Epilepsy, Intrinsic optical imaging, Autofluorescence, Brain activity map

1 Background and Historical Overview

Advances in functional imaging techniques including functional magnetic resonance imaging (fMRI), positron emission tomography (PET), and single-photon emission computed tomography (SPECT) have improved our understanding of the relationship between brain activity and brain energy supply [1, 2]. Neurovascular coupling concerns the relationship between neuronal activity, blood flow, and tissue oxygenation. Neurovascular and neurometabolic coupling are critical to supply the energy demands of brain tissue during both normal physiological function and pathological conditions [3]. Optical imaging of intrinsic signals (OIS) is a powerful technique for measuring intrinsic activity-related changes in brain tissue reflectance related to hemodynamic events. Functional physiological changes, such as blood volume changes, hemoglobin oximetry changes, light scattering changes, and neurometabolic changes, can result in intrinsic tissue reflectance changes [4–9]. Mapping these changes can be used to generate a brain activity map. The brain activity map can then be used as a functional

Mingrui Zhao et al. (eds.), *Neurovascular Coupling Methods*, Neuromethods, vol. 88,
DOI 10.1007/978-1-4939-0724-3_10, © Springer Science+Business Media New York 2014

connectome, which can provide invaluable information to understand physiological and pathological brain processes [10].

Unlike extrinsic signal from applied dyes such as voltage or calcium-sensitive dyes, which may cause phototoxicity, OIS offers a distinct advantage since extrinsic dyes can alter the normal physiology of the sample, especially in in vivo preparations. Furthermore, OIS, unlike penetrating electrodes, does not require any contact with the tissue of interest and is ideally suited to studying chronic preparations, in which an investigator may wish to image a sample over a period of days, weeks, or months, and for intraoperative mapping of the human cortex during neurosurgery.

Activity-related intrinsic changes in tissue reflectance associated with electrical activity or metabolism were first demonstrated by the pioneering work of Hill, Keynes, Chance, Jobsis, and their colleagues more than 60 years ago [11–13], the intrinsic optical signal is usually very small or noisy and is very hard to record. It was not until the 1980s that these intrinsic optical changes were used to map cortical activity in vivo developed by Grinvald and coworkers [14, 15]. OIS offers both high-spatial and high-temporal resolution simultaneously and has been used to study both acutely and chronically functional cortical maps and their architecture in different species including humans [16–18]. OIS is a minimally invasive optical mapping technique, which can be used for both normal brain function and pathological brain conditions. Improvement in computer speed, digital camera technology, and light sources has made this the OIS technique grow rapidly [19].

The sources of the intrinsic signal include reflectance changes from several optically active processes [20]. Those processes indirectly correlate with neuronal firing. There are at least three physiological components which originate from different sources: (1) changes in the blood volume; (2) chromophore redox, including the oxy/deoxyhemoglobin ratio (oxymetry); intracellular cytochrome oxidase and electron carriers; and (3) light scattering. At wavelengths such as 570 nm, an isosbestic wavelength of hemoglobin, OIS provides a direct measure of total hemoglobin (Hbt). Hbt is equivalent to cerebral blood volume (CBV) if the hematocrit remains constant, and CBV is proportional to cerebral blood flow (CBF) [21]. Under more “red” light, such as 610 nm, deoxygenated hemoglobin (Hbr) absorbs light more strongly than oxygenated hemoglobin (HbO_2) and it is possible to directly quantify both Hbr and Hbt concentration with the appropriate calculations [4, 6]. At 800 nm, or near-infrared wavelengths, the optical signal is largely derived from light scattering related to cell swelling as well as intra- and extracellular fluid shifts, which provide an indirect representation of neuronal activity, less influenced by the changes in CBV and hemoglobin oxygenation that dominate the intrinsic signal at lower wavelengths [7–9]. In addition, endogenous fluorescence from important biomolecules such as nicotinamide adenine dinucleotide (NADH) and flavin adenine dinucleotide (FAD) can be used as endogenous contrast to assess the brain metabolic state [22]. NADH and FAD play a critical role in cellular respiration. Flavoprotein autofluorescence imaging (AFI) measures changes in the redox state of mitochondrial flavoproteins [23–27], primarily FAD since the oxidized form (FAD+) is more fluorescent than the reduced form of FADH [28, 29]. Increases in neuronal activity result in increases in intracellular Ca^{2+} and depletion of ATP and production of ADP, which leads to a reduction in the proton gradient across the inner mitochondrial membrane and an increase in flavoprotein fluorescence [23–26, 30–32]. The excitation wavelength of flavoprotein autofluorescence is 430–500 nm, and the emission is 520–590 nm [25, 33, 34]. Furthermore, NADH also plays an important role in oxidative phosphorylation. Local energy metabolism in the brain is reflected as changes in the autofluorescence of endogenous NAD(P)H. NADH imaging has been used to monitor the metabolic state of the brain. The peak excitation of NADH autofluorescence is 340–360 nm and the peak emission is 450–480 nm [29, 35–41].

Because the intrinsic optical signals come from different sources and they are sensitive to different wavelengths, simultaneous multichannels are possible to directly compare each of those physiological/pathological components. Thus, multi-spectral imaging can help us to understand neurovascular and neurometabolic coupling of a variety of brain functions. Recently, a number of multichannel imaging methods have been developed. Multi-spectral imaging can be performed by acquiring light at multiple wavelengths or manipulating the light sources to multiple wavelengths.

1. One approach for multichannel imaging is to record imaging data using two or more camera systems. The exposed cortical surface is illuminated with white light using a halogen lamp and a DC regulated power supply through two fiberoptic light guides. Reflected light from the cortical surface can be collected by a 50-mm camera lens and divided into two separate paths by a beamsplitter. Each light beam passes through a second 50-mm camera lens and an interference filter tuned to one of two (or more) different wavelengths (e.g., 570 ± 10 and 610 ± 10 nm) (Fig. 1a) [42]. There is, for example, a commercial two-channel full-field simultaneous-imaging system from Photometrics. This commercial example is the DC2, an emission splitting system, enabling a user to acquire two spatially identical but spectrally distinct images simultaneously. Spectral comparisons can be performed with colocalization analysis, emission ratiometric analysis, or polarization studies of a single wavelength channel. The advantage of this method is to record full field of view imaging with high-spatial resolution. However, the use of multi-image acquisition systems significantly increases the cost of the whole system and the complexity of image registration.

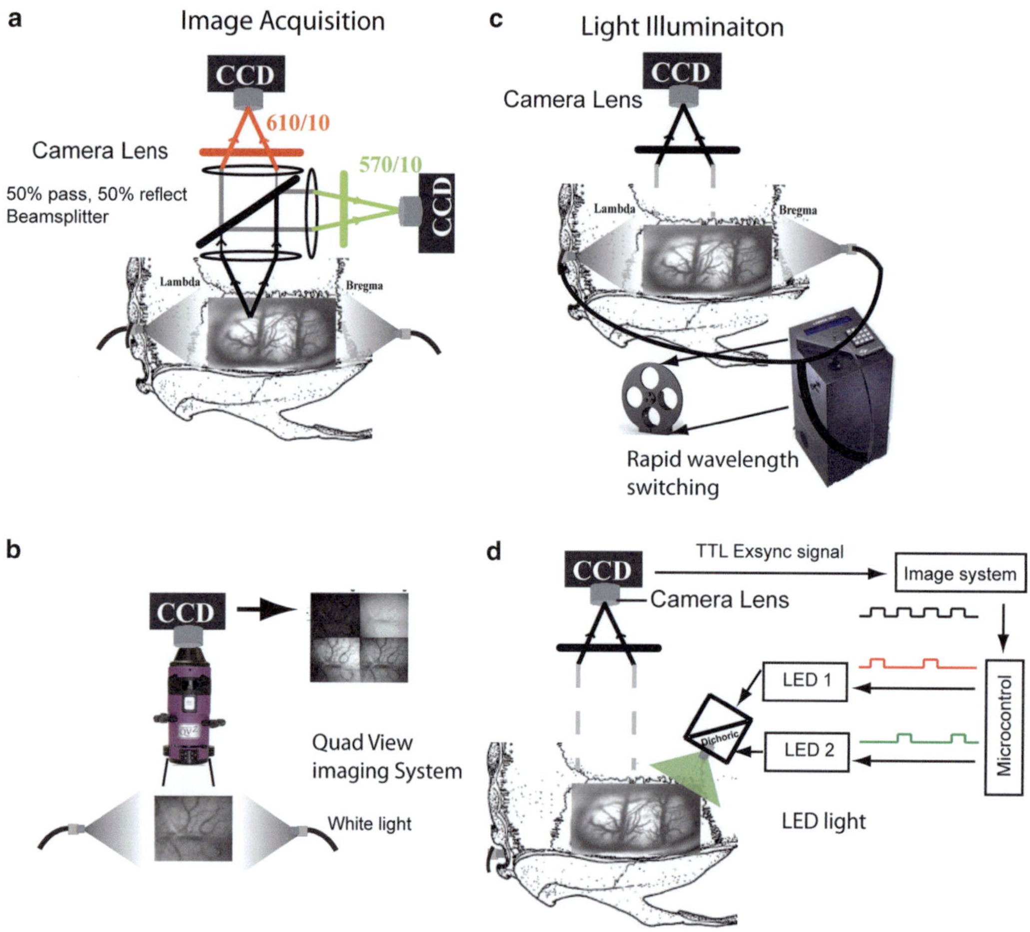

Fig. 1 The strategy for multi-spectral imaging. (**a**) A dual-camera system. Reflected light from the cortical surface is divided into two separate paths by a beamsplitter. Each light beam passed through an interference filter tuned to one of two different wavelengths. (**b**) A Quad-view system. The camera is attached with a Quad-view system. The Quad-view splits the light into four sub-images by different filters. (**c**) A filter-wheel method. (**d**) The fast LED switching method. A TTL signal from the camera frame-grabber is used to synchronize multi-wavelength LEDs with a microcontroller

2. Another approach is to use an image splitter and a single camera system. The Dual-view or Quad-view imaging system produces two or four non-overlapping images on a single image detector array. Using a set of dichroic and spectral filters or polarizers, all the images are produced simultaneously and no moving parts are used in image acquisition. The Photometrics QV2™ allows simultaneous acquisition of up to four emission channels in a single exposure. It uses a series of beamsplitters to split the emission light from a lens into four separate channels. All four channels are projected onto the CCD at the same time (Fig. 1b) [43]. The advantage of this method is that it is a true simultaneous imaging at multiple wavelengths. There is no time delay between different channels. Only one camera

is needed. The disadvantage that it requires a powerful white light source and the field of view is decreased by one fourth.

3. The most common multi-wavelength optical spectroscopy method uses a filter-wheel (Fig. 1c). The high-speed wavelength wheel such as Lambda DG-4 Plus can switch between adjacent filters at 0.5 ms (1.0 ms in the standard DG-4). But this technique still has limited spectral and temporal resolution [25–27].
4. The more contemporary technique for multi-imaging is to use a fast switching light-emitting diode (LED) system. A TTL signal from the camera is used to synchronize multi-wavelength LEDs with a microcontroller (Fig. 1d). The LEDs reach their full illumination intensity in <25 μs, and can be modulated at rates up to 10 kHz. The cost is low and the speed is fast with full field of image view [44–46]. However, based on the sampling rate of the camera, there may be a slight delay between the different channels.

We have used both the dual camera method [47, 48] and the fast LED switch method to record multi-spectral imaging. In this chapter we will demonstrate how to use a commercially-available imager system and a multichannel LED light system with custom control to measure OIS signals to study changes of blood volume, metabolism, and light scattering during acute seizure activity.

Several studies by our laboratory and others have shown that OIS can be used to map the onset and spread of epileptic events via their hemodynamic sequellae with very high-spatial and -temporal resolution, as well as high-spatial sampling [24, 33–39]. Epileptic events initiate a large focal increase in metabolism and a resulting increase in CBF at the seizure focus. Early studies to epileptic neurovascular coupling, studies using techniques with limited spatial and temporal resolution such as fMRI, PET, SPECT, and autoradiography proposed that the relative increase in CBF more than meets the increase in metabolic demand leading to an increase in blood oxygenation. However, more recent studies using higher temporal resolution techniques such as near-infrared spectroscopy (NIRS) and OIS showed that, for a variable period of time starting at the onset of the seizure, CBF is in fact inadequate to meet the metabolic demands of the epileptic tissue leading to a decrease in both tissue and hemoglobin oxygenation [22]. Herein we describe a multi-spectral imaging set-up used to investigate neurovascular epileptic coupling.

2 Equipment, Materials, and Setup

2.1 *In Vivo Surgery*

The surgical setup includes gas mixer (Smith Equipment, Watertown, SD), isoflurane Vaporizer (VetEquip, Inc. Pleasanton, CA), homeothermic Heating system (Harvard Apparatus, Holliston, MA), stereotaxic Frame (David Kopf Instruments,

Tujunga, CA), Micromotors drill system (Foredoom, Bethel, CT), MouseOX™ Plus Oximeter (Starr Life Science Corp.), and a floating table (Newport, Irvine, CA). Materials for surgery are 0.9 % Sodium Chloride Solution (Baxter International Inc., Deerfield, IL) and Bupivacaine (0.25 %, Penn Veterinary Supply, Inc.).

2.2 Electrophysiology Setup

Electrophysiological equipment includes a Kopf 730 Needle Pipette Puller (David Kopf Instruments, Tujunga, CA), Impedance check (FHC, Bowdoinham, ME), Power 1401 high-performance data acquisition interface, and ADC16 expansion, Spike 2 software (Cambridge Electronic Design, Cambridge, UK), and AC/DC differential amplifier (Model 3000, A-M system, Carlsborg, WA).

2.3 Imaging Setup

The main imaging system is the Imager 3001 system with Adimec-1000M/D camera (Optical Imaging Ltd., Rehovot, Israel). The camera positioning is controlled by a Robotic Camera Mount frame (Optical Imaging Ltd., Rehovot, Israel).

The Optical Imaging Tandem Lens assembly provides a high numerical, high-optical quality, and narrow depth of field, all optimized for optical imaging of the brain over relatively large areas, such as several millimeters square. The magnification of the Tandem Lens is the ratio of the top lens' focal length divided by the bottom lens' focal length. A two 50-mm Nikon Nikkor lens (NA = 1.2, Nikon, Japan) gives a 10× final magnification [49].

The light was provided through a liquid light guide (0.3in Core, Newport Corporation, Irvine, CA) by four high-power LEDs (M470L2: 470 nm, 1,600 mA; M530L2; 530 nm, 1,600 mA; M617L2: 617 nm, 1,600 mA; and M780L2: 780 nm, 1,000 mA. Throlabs, Newark, NJ) passed through a 455 ± 70-nm interference filter for blue (Chroma D455_70X), 528 ± 10 nm for green (Omega Optics 528BP10), 615 ± 10 nm for red (FF01-615/20 Semrock), and 785 ± 31 (FF01-785/61, Semrock) for NIR. Light was collected off the cortical surface after passing through a long-pass interference filter (Chroma HQ520LP; cutoff = 520 nm). Thus, using a single filter, we simultaneously measured changes in AFI, 530, 617, and 780 nm. AFI is a non-hemodynamic measure of cellular metabolism derived from the fluorescence of flavoproteins associated with the electron transport chain in mitochondria. Imaging flavoprotein autofluorescence used a band-pass excitation filter (455 ± 35 nm) and a >515 nm long-pass emission filter [25] (Fig. 2a).

Frame toggle is the most important signal generated by Vdaq of imager 3001 system that helps other instruments sense or record vdaq's timing. The intent of frame toggle is to provide a signal that reverses its sense at the end of every data frame. In intrinsic imaging experiments, several camera frames are typically summed into each data frame. This TTL signal is routed to an Arduino microcontroller

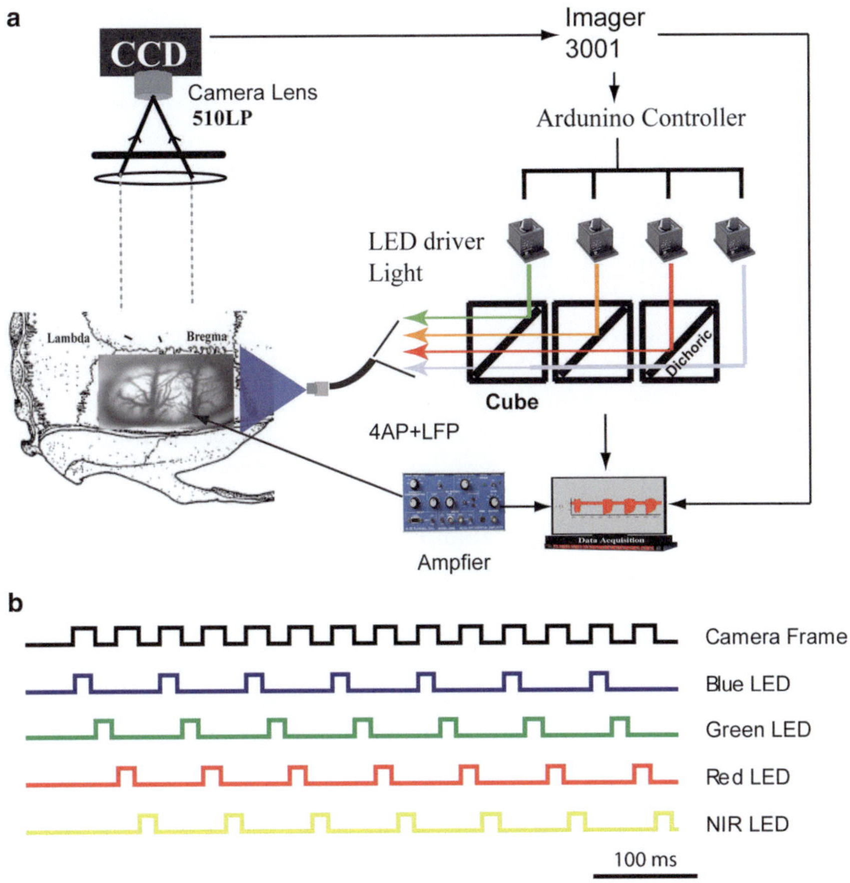

Fig. 2 Multispectral imaging diagram (**a**) When the camera acquires frames, the Imager 3001 outputs an EXSYNC TTL signal in real-time, which is sent to the interrupt ports of the Arduino Diecimila microcontroller. Using a custom strobe function, downsampled strobe signals are generated by the Arduino and sent to the LED drivers which strobe the LEDs (*blue*, *green*, *red*, and *gray arrows*). The camera frames and strobe signals are recorded by the computer, along with electrophysiological signals. (**b**) An example of camera frame and LED signals

(Arduino Diecimila, Sparkfun Electronics). A custom software uploaded to the microcontroller will serially strobe the camera TTL signal four LEDs. The camera is allowed to free-run and multi-wavelength illumination is locked precisely to each frame grab (Fig. 2b).

2.4 Focal Drug Injection

4-Aminopyridine (4-AP, Sigma, St. Louis, MO) was injected by a Nanoject II injector (Drummond Scientific, Broomall, PA).

2.5 Data Analysis

The offline analysis was performed using custom analysis software written by Matlab (MathWorks, Natick MA) (*see* below for more details).

3 Procedures

3.1 Surgery

All experimental procedures were approved either by the Weill Cornell Medical College of Cornell University Animal Care and Use Committee following NIH guidelines. Adult male mouse were initially anesthetized with 1.5–3 % isoflurane in 70 % N_2:30 % O_2 by facemask and maintained on 1–3 % isoflurane in 70 % N_2: 30 % O_2. Body temperature was maintained at 37 °C with a regulated heating blanket. The heart rate, breath rate, pulse distention, breath distention, and arterial blood oxygen saturation (SpO_2) were carefully monitored with a MouseOX™ Plus Oximeter and were maintained stable at normal values throughout the experiment. Animals were placed in a stereotaxic frame. Rodents were supplemented with hourly subcutaneous injections of 5 % glucose (wt/vol) in saline (0.1 mL/100 g per animal) to maintain hydration and physiological parameters were recorded every 20 min to ensure the animal was stable.

A midline scalp incision was made to expose the bone over the parietal cortex of one or both hemispheres. The bone over the hemisphere to be imaged was thinned with a dental drill and removed. For the majority of the experiments, a ~4 × 4 mm cranial window was opened by a dental drill over one hemisphere between Lambda and Bregma to expose somatosensory cortex and the surrounding brain. The dura was carefully removed. For intrinsic signal and flavoprotien imaging, a custom chamber was adhered to the skull and filled with silicone oil after positioning all electrodes and/or probes to increase the translucency.

3.2 Light Illumination Setting

Before doing the real experiment using an animal, it is better to perform the artificial cortex experiment, which mimics the conditions of a real experiment as closely as possible. It troubleshoots vany of the imaging problems and assigns the problem to either the hardware or to the biology. It is not necessary to perform before each experiment, but the first step is to perform or evaluate an artificial cortex test whenever changing the imaging setup or when in doubt. Attach the artificial cortex LED (a part of Imager 3001 system) the port on the Imager's laboratory interface panel. Illuminate the artificial cortex as the actual experiment. Run the experiment and focus the LED display. Control the level of the LED illumination itself with the potentiometer on the Laboratory Interface. Select the default color map to show the live image with a 16-color pseudo-color display (Fig. 4a). Then do the measurement of the shot noise level and the dark noise level.

Shot noise level: Do a normal artificial cortex test when the LED of artificial cortex is disconnected or its power level is down to 0. Make sure the background illumination is on. Shot noise is the high-spatial frequency noise in an image that is due to the statistical nature of light hitting the camera's detector. It increases with the square root of the light collected.

Dark noise level: Run the experiment when the illumination light is turn off. Examine the clipping values of the online maps. Dark noise is the noise inherent in the camera's electronics. Measuring this value provides another baseline value of the imaging system's performance.

Camera dual chip adjust: The Adimec 1000M/D camera, which is used on the Imager 3001/S, is a high-quality, fully digital CCD camera. It can operate at speeds from 25 to 50 Hz at full resolution (1,004 × 1,004), or at 75 Hz at half resolution. The Adimec 1000M/D uses two sensor channels including one for the left half of the image and one for the right half image. Because these two halves of the image are independently converted from analog to digital, it is possible to have small but noticeable differences in intensity. The AdimecTune camera adjustment utility balances the two halves of the Adimec 1000M/D camera, so that they have the same intensity. It is recommended to do an experiment in a camera whose halves are balanced, if for no other reason, so that a saved green reference image won't have any obvious imbalance.

The camera set-up utility can view the live imaging. It is very important to illuminate the animal's cortex properly in order to get high-quality data. Three considerations are important in setting the illumination: first, the camera should be near saturation, to get the full dynamic range of the camera into play. Second the camera must not actually be saturated: every pixel must be below its maximum, or else the image will be clipped, and the data acquired will be invalid. Third, the illumination must be as even as possible. If the illumination is uneven, the signal-to-noise ratio in different parts of the image will differ greatly, making analysis difficult. Roughly adjust the position of the light guides and the overall illumination level using the voltage control knob on the LED light driver. The intensity and uniformity of light illumination should be maximized. The final pattern of ideal illumination should look like Fig. 4b. Adjust each wavelength of LED light to get the best images.

Adjust the camera's position so that it is pointed at the preparation, and at the proper distance to achieve adequate focus. To get a maximum-sized region in focus, remember to choose the camera's angle to be perpendicular to the preparation. The definition of "good focus" depends on what you are doing at the moment. For differential imaging, a 200–400-μm focus under the surface of the cortex will reduce artifacts due to blood vessels in images by blurring. A floating bench is often used to minimize motion artifacts. In order to get the best optical signals, the camera lens is focused to infinity and the maximum opening in aperture (the lowest f number for 50-mm Nikon lens, f/1.2).

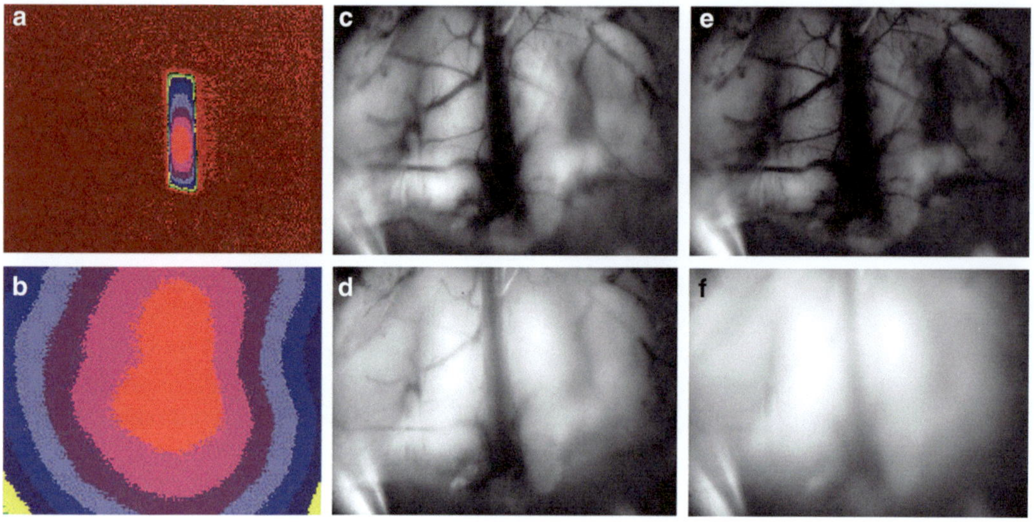

Fig. 3 Light illuminations and gray raw image of mouse neocortex exposed at multi-wavelength illumination. (**a**) The artificial cortex LED. (**b**) A typical illumination pattern. (**c**) 470-nm wavelength. The 510-nm long-pass filter was used to view to get the AFI imaging (**d**) 530-nm image of cortical surface. The 610-nm image (**e**) and 780-nm (**f**) of cortical surface. The blood vasculature are less clear under these two wavelengths

Before beginning an experiment, it is recommended to take a static reference image of the area to be functionally imaged using each wavelength illumination (Fig. 3c–f).

3.3 Baseline Recording

For most of our experiments, pixels were binned 2×2 (21 μm/pixel) and collected by one CCD camera at a frame-rate of 50 Hz. The camera was focused ~500 μm below the cortical surface. A floating bench was used to minimize motion artifacts. The LongDaq continuous recording software can simultaneously grab and process images and save image data in a series of block files. Grabbing only from the ROI will lower the data rate to dish, which can help with reliability, save disk space, and reduce the size of the data file.

Extracellular local field potential (LFP) was recorded with the same single-barreled glass microelectrode (impedance, 2–4 MΩ) filled with 0.9 % saline, positioned <1 mm from the 4-AP electrode and lowered to a depth of 300–500 μm into the neocortex. The signal was amplified and filtered between 0.1 and 500 Hz using an AC/DC differential amplifier, and digitized at 1,000 Hz by a CED Power 1401. The imaging frame signal was recorded by the CED Power 1401 and Spike2 software.

3.4 Acute Epileptic Model

Ictal discharges were induced by injecting the potassium channel blocker 4-AP (15 mM, 0.5 μL) through a single-barreled glass microelectrode using a Nanoject II injector. Spontaneous ictal discharges occurred periodically for up to 2 h. Interictal spikes also

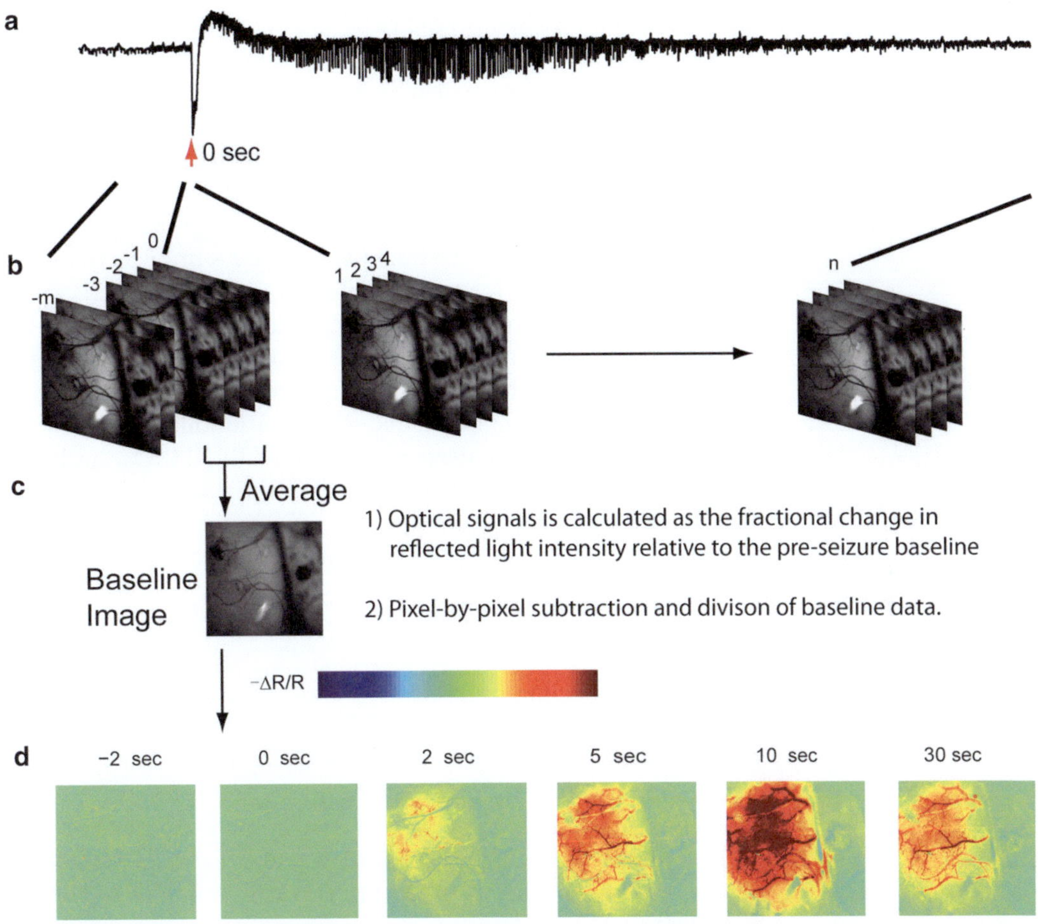

Fig. 4 Data analysis. (**a**) A typical LFP recording of 4-AP seizure. The initial spike of the epileptic event was determined as the ictal onset (*red arrow*: 0 s). (**b**) The image block data includes a series of images in which the camera frame timing marker was matched with the LFP timing. The 0 s image is based on the seizure onset. (**c**) Averaged baseline image. Optical signal is calculated as the fractional change in reflected intensity related to the baseline as pixel-by-pixel intensity. (**d**) A functional map is generated

occasionally occurred with varying frequency during the interseizure intervals. 4-AP ictal discharges typically began with a large biphasic spike. The initial spike was typically followed by a low-amplitude fast activity with progressive increase in amplitude (recruiting rhythm), then a period of rapid spike and wave activity and a gradual decrease in spike frequency and amplitude at the offset of the seizure (Fig. 4).

3.5 Data Analysis

The offline analysis was performed using custom analysis software written by Matlab (MathWorks, Natick, MA). Seizure onset was determined by visual analysis of the EEG from the initial negative deflection of the large spike that preceded all seizures. Since ictal events can have a varying electrographic morphology, we only analyzed those that had no interictal spikes during the 10s prior to the ictal onset and that began with a large population spike followed by a recruiting rhythm to eliminate any confusion regarding the onset of the seizure. Termination was determined at the cessation of spike and wave events and the return to baseline, pre-ictal activity.

Imaging data were converted to percent change (Δ) from baseline by subtracting, then dividing the average baseline value obtained over a 2 s block of time beginning 2 s prior to the onset of the ictal discharge. It was calculated as the fractional change in reflected light intensity relative to the pre-ictal baseline by pixel-by-pixel division and subtraction (Fig. 4). A low-pass filter was used to remove noise induced by heartbeat and mechanical vibration.

For the optical hemodynamic signals, a modified form of the Beer-Lambert Law was used to calculate Hbr, HbO_2, and Hbt changes from the 530 and 617-nm data with pathlength correction [6, 50] using Eq. 1.

$$\log\frac{I_0^\lambda}{I(t)} = \left(\varepsilon_{\text{Hbr}}^\lambda \times \Delta[\text{Hbr}](t) + \varepsilon_{\text{HbO}_2}^\lambda \times \Delta[\text{HbO}_2](t)\right) \times l^\lambda + \Delta S \quad (1)$$

I_0 is the pre-stimulus intensity, $I(t)$ the measured intensity time course, ε extinction coefficients, l the pathlength through the tissue, S scattering, $\Delta[\text{Hbr}](t)$ and $\Delta[\text{HbO}_2](t)$ is the time course of change in absorption, and superscript λ indicates wavelength dependency. By measuring I_0 and $I(t)$ at two wavelengths and using the known wavelength dependency of the optical pathlength ($\alpha = \varepsilon \times l$), we determine a pathlength corrected value of $\Delta[\text{Hbr}](t)$ and $\Delta[\text{HbO}_2](t)$ [51].

Functional spectral time courses were reported as relative changes from baseline. Each frame following the onset of each seizure was divided by an average of the 20 frames prior to the onset. Spatial low-pass filtering was achieved by convolution with a Gaussian kernel ($\sigma = 3$ pixels, about 63 μm). A region of interest (ROI, 5×5 pixels) on the ictal focus was selected to avoid large blood vessels. An average of percent fractional changes of optical signals at each time point in the ROI was used for further analysis.

The fluorescence of FAD was imaged simultaneously with changes in total tissue hemoglobin concentration ("cerebral blood volume") by either an excitation wavelength for FAD (blue light at 470-nm LED) or an isosbestic wavelength for hemoglobin (green light at 530-nm LED). To correct the FAD fluorescence signals for distortions due to the stimulus-evoked blood volume changes we divided the observed fractional changes in fluorescence by the simultaneously measured blood volume signal scaled by an appropriate correction factor [27]. Reflectance changes at intrinsic signals were expressed in percentage of a negative change from the baseline ($-\Delta R/R$), while flavoprotein autofluorescence is expressed in percentage of fluorescence change from the baseline ($\Delta F/F$).

4 Experimental Variables

The present protocol uses a 2 s block of time beginning 2 s prior to the onset of the ictal discharge as the baseline. If one is interested in analyzing pre-ictal activity, it is possible to select the other periods as the baseline that occur more than 10 s before ictal onset. For investigating the long-term effects of 4-AP and the slow development of seizure activity, the time before 4-AP injection or another period between ictal discharges can be used as the baseline. For analysis the small epileptic events such as the interictal spike or polyspike, a spike-triggered average (STA) method can be used in order to increase the signal-to noise ratio [52, 53]. The STA is given by using Eq. 2.

$$\mathrm{STA} = \frac{1}{n_{\mathrm{sp}}} \sum_{i=1}^{T} y_i \mathbf{x}_{\mathrm{i}} \tag{2}$$

where $\mathbf{x}_{\mathrm{i}}$ is the spatio-temporal stimulus vector preceding the ith time bin, and y_i the spike count in that bin., $n_{\mathrm{sp}} = \sum y_i$, is the total number of spikes.

The LED light is an expensive, stable, and easy-to-use lighting source [54, 55]. It instantly illuminates at full intensity as soon as electrical current is applied. Unlike arc lamps that are turned on continuously, LEDs can be switched on or off instantly when needed with no deleterious effects to their lifespan. Additionally, with no moving parts, the all-electronic system is vibration free. It is ideally suited to the fast image acquisition. High-power LED can be obtained at most of the wavelengths today. It is easy for the researcher to modify illumination wavelengths by changing the LED light and beamsplitter to assess other intrinsic optical signals or NADH autofluorescence. New neuroimaging and neuromodulatory techniques such as "optogenetics," which combine optical and genetic techniques, have emerged as a popular tool to probe and control neuronal function with light [56–60]. One can modify the optical recording mechanisms to be used for photostimulation system with the combination of optogenetic techniques but that is beyond the scope of this chapter [61].

5 Typical/Anticipated Results

A typical multi-spectral mapping experiment of blood volume, light scatting, oximetry, and metabolism is presented here.

5.1 CBV and Light Scatting

Optical recording at 530 nm can provide a map of CBV during seizure activity (Fig. 5a). In this experiment, each seizure was accompanied by a dramatic, focal change in CBV. The 780-nm signal, on the other hand, indirectly measures changes in neuronal

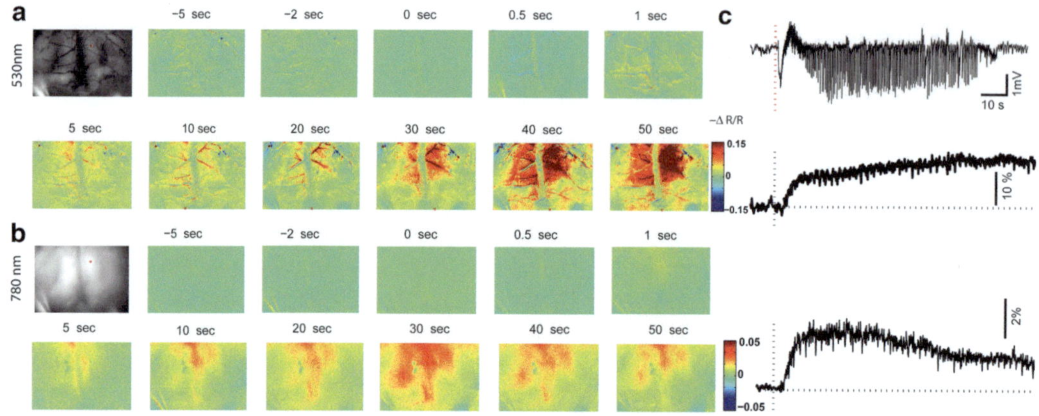

Fig. 5 Blood volume and light scattering maps. (**a**) The 530-nm (blood volume) intrinsic changes of the ictal discharge. The *red boxes* inside the 530 and 780-nm image demonstrate the location of the superpixel in (**c**). (**b**) The 780-nm NIR (light scattering) map of the same seizure. (**c**) The timecourse of LFP recording and light reflectance changes in the ictal focus. *Top*: LFP; *middle*: 530 nm, and *bottom*: 780 nm. There are clear increases in both wavelengths after seizure onset

activity through light scatter. The light scattering signal increased after ictal onset (Fig. 5b). Unlike the blood volume map, the signal is not maximal over the blood vessel.

5.2 Oximetry

Using the signals from 530 to 610-nm wavelengths, the concentration of Hbr, HbO_2, and Hbt were calculated. The Hbr signal also showed a transient dip in hemoglobin oxygenation (increase in Hbr) at the onset of the seizures, and there was a delayed increase in hemoglobin oxygenation (decrease in Hbr) in the focus (Fig. 6a). Similar to the 530-nm signal, Hbt dramatically increased right after the seizure onset (Fig. 6b).

5.3 Flavoprotein Imaging Metabolism Map

The AFI of an ictal discharge is typically biphasic, with an early increase (oxidation phase, “light phase”) followed by a prolonged decrease (reduction phase, “dark phase”) (Fig. 7). It is apparent in these experiments that the AFI signal was very different with the blood volume signal recorded as 530 nm. The early fluorescence was very short and the spatial extent was much smaller than the late negative fluorescence phase. The oxidation phase is due to fast Ca^{2+} and ADP evoked increases in the rate of FADH2 utilization in the electron transport chain. The reduction phase is thought to be due to a somewhat more delayed increase in tricarboxylic acid (TCA) cycle activity that replenishes FADH2 at a faster rate than it can be utilized by the electron transport chain. The early increase in metabolism in the epileptic focus is the likely etiology of the dip in hemoglobin and tissue oxygenation, or “initial dip” previously demonstrated in this model [42, 51].

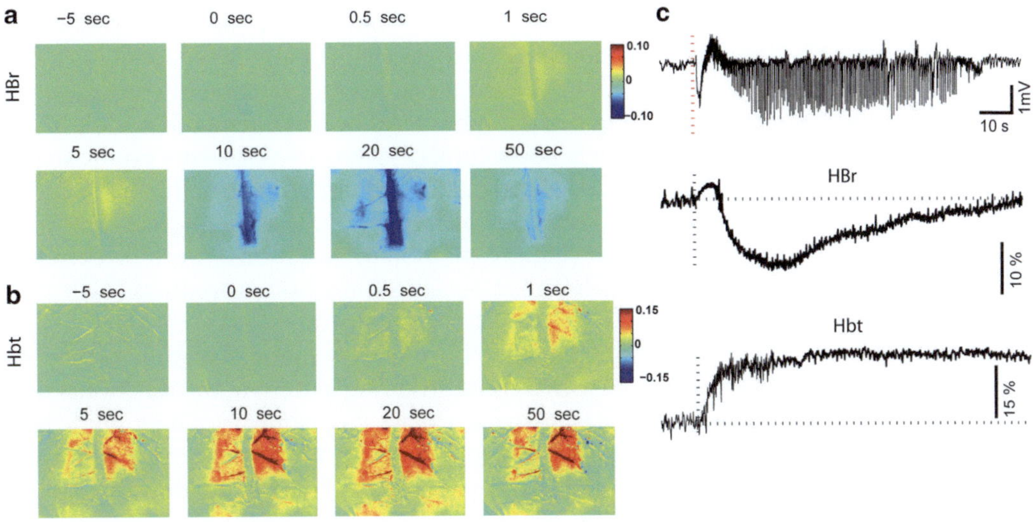

Fig. 6 Hemodynamic map. (**a**) The Hbr changes over the time during a 4-AP seizure. The Hbr shows biphasic changes including an initial dip in oxygenation and later hyperoxygenation signal. (**b**) The Hbt map (blood volume) of the same seizure. There is a dramatic increase in the blood volume after ictal onset. (**c**) Timecourses showing the change in Hbr (*middle*) and Hbt (*bottom*). The *top panel* shows the LFP recording. The *dotted line* shows the moment of seizure onset

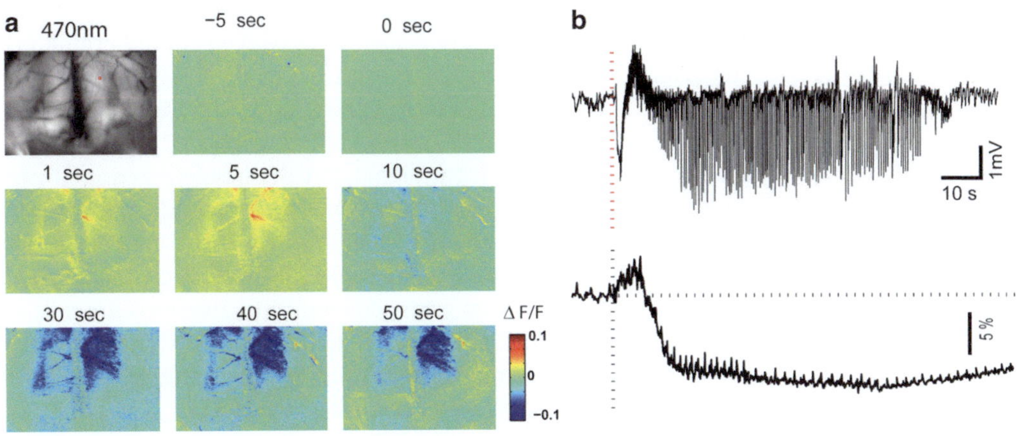

Fig. 7 Flavoprotein autofluorescence-based metabolic functional map. (**a**) The autofluorescence signal over the ictal discharge. The *red boxes* inside the 470-nm image demonstrate the location of the superpixel in (**b**). (**b**) The timecourse of LFP recording and fluorescence changes in the ictal focus. *Top*: LFP; *bottom*: Autofluorescence. The metabolic changes in seizure focus are biphasic

6 Troubleshooting

Here we describe several practical troubleshooting tips that may help researchers to obtain more reliable and reproducible optical imaging data.

6.1 Stability

Vibration noise can originate from relative movements of any of the components along the optical light path. Mechanical instability may be due to: (1) a floating table is not floating due to lack of air, or not floating due to tight cables between the table and floor; (2) Camera holder is not secure; (3) animal is not secure on the stereotaxic frames; (4) light guides are not secure; or (5) large movements of the cortex due to respiration. If constant periodical noise is predominant (e.g., respiration or heartbeat), it can be corrected by proper synchronization and subtraction procedures [62].

6.2 Ring-Shape Artifact

The curvature of the cortex creates inhomogeneity in reflected light intensity with a concentric-ring pattern in the functional map. This is called a ring-shape artifact. Adjustments of light guide position are necessary in order to make the cortex image more uniform. A Gaussian band-pass spatial filter can remove this kind of artifact [63].

6.3 Spatial Biological Artifacts

Biological sources of noise are very common during optical mapping. These can result from changes in cortical vasculature and parenchyma owing to heart pulsations and respiration as well as 0.1 Hz "vasomotion" noise [64] and other types of noise such as that caused by cortical motion. If the amplitude of the measured signal is low, sufficient averaging should remove many of these sources of noise. Accordingly, the averaged data should contain the characteristic biological artifacts captured by the imaging setup, such as vascular artifacts related to heartbeat and respiration. Differential imaging may then achieve near-complete elimination of biological artifacts, because if both the signal and the baseline averaged maps contain similar artifact residues they can be removed by simple subtraction. If the signal is adequately high, such as with epileptic events, then averaging is not necessary since the noise signal is relatively small. However, in some cases, conventional methods are still not always satisfactory. Other options include principal component analysis (PCA) or more simply using local similarity minimization maybe of help to remove spatial biological artifacts [65].

7 Conclusion

Optical imaging has emerged as a potent tool to map the brain activity over large areas of cortex with very high-spatial resolution. We have tried to outline the technical aspects of a simultaneous multi-wavelength experiment for imaging CBV, oxymetry, light scatting, and local metabolism in the case of an epileptic event. Multi-wavelength imaging is a powerful tool, if employed correctly and carefully, to investigate neurovascular and neurometabolic coupling in both normal and pathological brain states.

Acknowledgements

This work was supported by the NINDS RO1 NS49482 (T.H.S), CURE Taking Flight Award (H.M.), the Clinical and Translational Science Center (CTSC) Grant UL1 RR 024996 Pilot Grant (M.Z), and the Cornell University Ithaca-WCMC seed grant (M.Z.). We thank Dr. Yevgeniy B. Sirotin for help with LED setup and AFI analysis.

References

1. Logothetis NK, Wandell BA (2004) Interpreting the BOLD signal. Annu Rev Physiol 66:735–769
2. Raichle ME, Mintun MA (2006) Brain work and brain imaging. Annu Rev Neurosci 29:449–476. doi:10.1146/annurev.neuro.29.051605.112819
3. Girouard H, Iadecola C (2006) Neurovascular coupling in the normal brain and in hypertension, stroke, and Alzheimer disease. J Appl Physiol 100(1):328–335. doi:10.1152/japplphysiol.00966.2005
4. Frostig RD, Lieke EE, Ts'o DY et al (1990) Cortical functional architecture and local coupling between neuronal activity and the microcirculation revealed by in vivo high-resolution optical imaging of intrinsic signals. Proc Natl Acad Sci 87:6082–6086
5. Malonek D, Grinvald A (1996) Interactions between electrical activity and cortical microcirculation revealed by imaging spectroscopy: implications for functional brain mapping. Science 272:551–554
6. Sheth SA, Nemoto M, Guiou M et al (2004) Columnar specificity of microvascular oxygenation and volume responses: implications for functional brain mapping. J Neurosci 24(3):634–641
7. Salzberg BM, Obaid AL, Gainer H (1985) Large and rapid changes in light scattering accompany secretion by nerve terminals in the mammalian neurohypophysis. J Gen Physiol 86(3):395–411
8. Obaid AL, Flores R, Salzberg BM (1989) Calcium channels that are required for secretion from intact nerve terminals of vertebrates are sensitive to omega-conotoxin and relatively insensitive to dihydropyridines. Optical studies with and without voltage-sensitive dyes. J Gen Physiol 93(4):715–729
9. MacVicar BA, Hochman D (1991) Imaging of synaptically evoked intrinsic optical signals in hippocampal slices. J Neurosci 11(5):1458–1469
10. Alivisatos AP, Chun M, Church George M et al (2012) The brain activity map project and the challenge of functional connectomics. Neuron 74(6):970–974
11. Hill DK, Keynes RD (1949) Opacity changes in stimulated nerve. J Physiol 108(3):278–281
12. Jobsis F (1977) Noninvasive, infrared monitoring of cerebral and myocardial oxygen sufficiency and circulatory parameters. Science 198(4323):1264–1267. doi:10.1126/science.929199
13. Chance B, Cohen P, Jobsis F et al (1962) Intracellular oxidation-reduction states in vivo: the microfluorometry of pyridine nucleotide gives a continuous measurement of the oxidation state. Science 137(3529):499–508. doi:10.1126/science.137.3529.499
14. Grinvald A (1985) Real-time optical mapping of neuronal activity: from single growth cones to the intact mammalian brain. Annu Rev Neurosci 8(1):263–305. doi:10.1146/annurev.ne.08.030185.001403
15. Grinvald A, Lieke E, Frostig RD et al (1986) Functional architecture of cortex revealed by optical imaging of intrinsic signals. Nature 324(6095):361–364
16. Bonhoeffer T, Grinvald A (1991) Iso-orientation domains in cat visual cortex are arranged in pinwheel-like patterns. Nature 353(6343):429–431
17. Schwartz TH, Bonhoeffer T (2001) In vivo optical mapping of epileptic foci and surround inhibition in ferret cerebral cortex. Nat Med 7(9):1063–1067
18. Schwartz TH, Chen LM, Friedman RM et al (2004) Intraoperative optical imaging of human face cortical topography: a case study. Neuroreport 15(9):1527–1531
19. Zepeda A, Arias C, Sengpiel F (2004) Optical imaging of intrinsic signals: recent developments in the methodology and its applications. J Neurosci Methods 136(1):1–21

20. Cohen LB (1973) Changes in neuron structure during action potential propagation and synaptic transmission. Physiol Rev 53(2):373–418
21. Sheth SA, Nemoto M, Guiou G et al (2004) Linear and nonlinear relationships between neuronal activity, oxygen metabolism, and hemodynamic response. Neuron 42:347–355
22. Georgakoudi I, Quinn KP (2012) Optical imaging using endogenous contrast to assess metabolic state. Annu Rev Biomed Eng 14(1):351–367.doi:10.1146/annurev-bioeng-071811-150108
23. Chance B, Ernster L, Garland PB et al (1967) Flavoproteins of the mitochondrial respiratory chain. Proc Natl Acad Sci U S A 57(5): 1498–1505
24. Reinert KC, Dunbar RL, Gao W et al (2004) Flavoprotein autofluorescence imaging of neuronal activation in the cerebellar cortex in vivo. J Neurophysiol 92(1):199–211. doi:10.1152/jn.01275.2003, pii: 01275.2003
25. Reinert KC, Gao W, Chen G et al (2007) Flavoprotein autofluorescence imaging in the cerebellar cortex in vivo. J Neurosci Res 85(15):3221–3232. doi:10.1002/jnr.21348
26. Llano DA, Theyel BB, Mallik AK et al (2009) Rapid and sensitive mapping of long-range connections in vitro using flavoprotein autofluorescence imaging combined with laser photostimulation. J Neurophysiol 101(6):3325–3340. doi:10.1152/jn.91291.2008, pii: 91291.2008
27. Sirotin YB, Das A (2010) Spatial relationship between flavoprotein fluorescence and the hemodynamic response in the primary visual cortex of alert macaque monkeys. Front Neuroenergetics 2:6. doi:10.3389/fnene.2010.00006
28. Chance B, Sager R (1957) Oxygen and light induced oxidations of cytochrome, flavoprotein, and pyridine nucleotide in a chlamydomonas mutant. Plant Physiol 32(6):548–561
29. Chance B, Schoener B, Oshino R et al (1979) Oxidation-reduction ratio studies of mitochondria in freeze-trapped samples. NADH and flavoprotein fluorescence signals. J Biol Chem 254(11):4764–4771
30. Gao W, Chen G, Reinert KC et al (2006) Cerebellar cortical molecular layer inhibition is organized in parasagittal zones. J Neurosci 26(32):8377–8387. doi:10.1523/JNEUROSCI.2434-06.2006, pii: 26/32/8377
31. Theyel BB, Llano DA, Sherman SM (2010) The corticothalamocortical circuit drives higher-order cortex in the mouse. Nat Neurosci 13(1):84–88. doi:10.1038/nn.2449, pii: nn.2449
32. Benson RC, Meyer RA, Zaruba ME et al (1979) Cellular autofluorescence—is it due to flavins? J Histochem Cytochem 27(1):44–48
33. Lee IY, Chance B (1968) Activation of malate-linked reductions of NAD and flavoproteins in Ascaris muscle mitochondria by phosphate. Biochem Biophys Res Commun 32(3):547–553
34. Hassinen I, Chance B (1968) Oxidation-reduction properties of the mitochondrial flavoprotein chain. Biochem Biophys Res Commun 31(6):895–900
35. Chance B, Graham N, Mayer D (1971) A time sharing fluorometer for the readout of intracellular oxidation-reduction states of NADH and flavoprotein. Rev Sci Instrum 42(7):951–957
36. Lee IY, Chance B (1977) Regulatory factors of acetaldehyde metabolism in isolated rat liver mitochondria. Adv Exp Med Biol 85A:203–224
37. Barlow CH, Harden WR III, Harken AH et al (1979) Fluorescence mapping of mitochondrial redox changes in heart and brain. Crit Care Med 7(9):402–406
38. Rehncrona S, Mela L, Chance B (1979) Cerebral energy state, mitochondrial function, and redox state measurements in transient ischemia. Fed Proc 38(11):2489–2492
39. Duchen MR, Biscoe TJ (1992) Mitochondrial function in type I cells isolated from rabbit arterial chemoreceptors. J Physiol 450: 13–31
40. Kosterin P, Kim GH, Muschol M et al (2005) Changes in FAD and NADH fluorescence in neurosecretory terminals are triggered by calcium entry and by ADP production. J Membr Biol 208(2):113–124. doi:10.1007/s00232-005-0824-x
41. Takano T, Tian G-F, Peng W et al (2007) Cortical spreading depression causes and coincides with tissue hypoxia. Nat Neurosci 10(6):754–762
42. Zhao M, Nguyen J, Ma H et al (2011) Preictal and ictal neurovascular and metabolic coupling surrounding a seizure focus. J Neurosci 31(37):13292–13300. doi:10.1523/jneurosci.2597-11.2011
43. Prakash N, Biag JD, Sheth SA et al (2007) Temporal profiles and 2-dimensional oxy-, deoxy-, and total-hemoglobin somatosensory maps in rat versus mouse cortex. Neuroimage 37(suppl 1):S27–S36, http://dx.doi.org/10.1016/j.neuroimage.2007.04.063
44. Kozberg MG, Chen BR, DeLeo SE et al (2013) Resolving the transition from negative to positive blood oxygen level-dependent responses in the developing brain. Proc Natl Acad Sci. doi:10.1073/pnas.1212785110

45. Sun R, Bouchard MB, Hillman EM (2010) SPLASSH: open source software for camera-based high-speed, multispectral in-vivo optical image acquisition. Biomed Opt Express 1(2):385–397. doi:10.1364/boe.1.000385
46. Bouchard MB, Chen BR, Burgess SA et al (2009) Ultra-fast multispectral optical imaging of cortical oxygenation, blood flow, and intracellular calcium dynamics. Opt Express 17(18):15670–15678
47. Zhao M, Suh M, Ma H et al (2007) Focal increases in perfusion and decreases in hemoglobin oxygenation precede seizure onset in spontaneous human epilepsy. Epilepsia 48(11):2059–2067. doi:10.1111/j.1528-1167.2007.01229.x, pii: EPI1229
48. Ma H, Zhao M, Suh M et al (2009) Hemodynamic surrogates for excitatory membrane potential change during interictal epileptiform events in rat neocortex. J Neurophysiol 101(5):2550–2562. doi:10.1152/jn.90694.2008, pii: 90694.2008
49. Ratzlaff EH, Grinvald A (1991) A tandem-lens epifluorescence macroscope: hundred-fold brightness advantage for wide-field imaging. J Neurosci Methods 36(2–3):127–137
50. Sato C, Nemoto M, Tamura M (2002) Reassessment of activity-related optical signals in somatosensory cortex by an algorithm with wavelength-dependent path length. Jpn J Physiol 52(3):301–312
51. Zhao M, Ma H, Suh M et al (2009) Spatiotemporal dynamics of perfusion and oximetry during ictal discharges in the rat neocortex. J Neurosci 29(9):2814–2823. doi:10.1523/JNEUROSCI.4667-08.2009, pii: 29/9/2814
52. Suh M, Bahar S, Mehta AD et al (2005) Temporal dependence in uncoupling of blood volume and oxygenation during interictal epileptiform events in rat neocortex. J Neurosci 25(1):68–77
53. Arieli A, Shoham D, Hildesheim R et al (1995) Coherent spatiotemporal patterns of ongoing activity revealed by real-time optical imaging coupled with single-unit recording in the cat visual cortex. J Neurophysiol 73(5):2072–2093
54. Hohman B (2007) LED light source: major advance in fluorescence microscopy. Biomed Instrum Technol 41(6):461–464
55. Albeanu DF, Soucy E, Sato TF et al (2008) LED arrays as cost effective and efficient light sources for widefield microscopy. PLoS One 3(5):e2146. doi:10.1371/journal.pone.0002146
56. Boison D (2005) Adenosine and epilepsy: from therapeutic rationale to new therapeutic strategies. Neuroscientist 11(1):25–36. doi:10.1177/1073858404269112
57. Adamantidis AR, Zhang F, Aravanis AM et al (2007) Neural substrates of awakening probed with optogenetic control of hypocretin neurons. Nature 450(7168):420–424
58. Boison D (2006) Adenosine kinase, epilepsy and stroke: mechanisms and therapies. Trends Pharmacol Sci 27(12):652–658
59. Zhang F, Wang L-P, Brauner M et al (2007) Multimodal fast optical interrogation of neural circuitry. Nature 446(7136):633–639
60. Zayat L, Noval MG, Campi J et al (2007) A new inorganic photolabile protecting group for highly efficient visible light GABA uncaging. Chembiochem 8(17):2035–2038. doi:10.1002/cbic.200700354
61. Deisseroth K, Feng G, Majewska AK et al (2006) Next-generation optical technologies for illuminating genetically targeted brain circuits. J Neurosci 26(41):10380–10386. doi:10.1523/jneurosci.3863-06.2006
62. Grinvald A, Shoham D, Shmuel A et al (1999) In-vivo optical imaging of cortical architecture and dynamics. In: Windhorst U, Johansson H (eds) Modern techniques in neuroscience research. Springer, Berlin, pp 893–968
63. Przybyszewski AW, Sato T, Fukuda M (2008) Optical filtering removes non-homogenous illumination artifacts in optical imaging. J Neurosci Methods 168(1):140–145, http://dx.doi.org/10.1016/j.jneumeth.2007.09.006
64. Mayhew JEW, Askew S, Zheng Y et al (1996) Cerebral vasomotion: a 0.1-Hz oscillation in reflected light imaging of neural activity. Neuroimage 4(3):183–193, http://dx.doi.org/10.1006/nimg.1996.0069
65. Fekete T, Omer DB, Naaman S et al (2009) Removal of spatial biological artifacts in functional maps by local similarity minimization. J Neurosci Methods 178(1):31–39, http://dx.doi.org/10.1016/j.jneumeth.2008.11.020

Chapter 11

Physiological Basis of BOLD fMRI Decreases

Robert Kim, Fahmeed Hyder, and Hal Blumenfeld

Abstract

Recent studies suggest that positive functional magnetic resonance imaging (fMRI) signal changes in the cortex under normal physiological conditions are closely related to increased neuronal activity. However, the physiological basis of negative blood oxygen level-dependent (BOLD) fMRI signal changes in the cortex as well as in subcortical structures are less well understood. Animal models that employ combined neuroimaging and direct electrophysiological measurements provide a powerful avenue for studying the complex relationships between negative fMRI changes and neuronal activity under normal and abnormal conditions (such as epilepsy). Several animal studies have shown that fMRI decreases do not necessarily translate to decreased electrophysiological signals. Here, we review three such scenarios: (1) increased local field potentials (LFP) occur during cortical slow oscillations despite reduced mean neuronal firing, resulting in decreased BOLD, cerebral blood flow (CBF) and cerebral blood volume (CBV); (2) massive increases in neuronal activity during hippocampal seizures exceed the ability of CBF to meet metabolic demands leading to decreased local BOLD fMRI signals; (3) reverse coupling of neuronal activity and CBF in the basal ganglia leads to increased neuronal firing and LFP during normal somatosensory stimulation and spike-wave seizures, along with a paradoxical decrease in BOLD fMRI, CBF, and CBV. A better understanding of the physiological basis underlying BOLD fMRI decreases can provide improved interpretation of fMRI data in health and disease.

Key words fMRI, Epilepsy, BOLD decreases, Neuronal activity, Cerebral blood flow, Neurovascular coupling, Negative BOLD

1 Introduction

Dynamic functional neuroimaging based on the blood oxygen level-dependent (BOLD) signal has become indispensable to mapping brain activity and understanding the neuronal and physiological basis of brain activity changes under normal and abnormal states. However, the BOLD signal is a result of complex interactions between neuronal activity and hemodynamic components such as cerebral blood volume (CBV) and cerebral blood flow (CBF), as well as cerebral metabolic rate of oxygen ($CMRO_2$) [1]. Since these hemodynamic and neuroenergetic determinants of the BOLD functional magnetic resonance imaging (fMRI) signal are

Mingrui Zhao et al. (eds.), *Neurovascular Coupling Methods*, Neuromethods, vol. 88,
DOI 10.1007/978-1-4939-0724-3_11, © Springer Science+Business Media New York 2014

related to neuronal activity, it is of great importance to understand the coupling between these parameters under normal conditions as well as how these relationships are altered during abnormal brain activity such as epilepsy.

Increases in fMRI BOLD signals have been extensively studied and shown to be linked to increased neuronal electrical activity followed by increased CBV and CBF [2–7]. On the other hand, the negative BOLD signal has not been investigated as much, and its origin is less well understood [8]. Under normal conditions cortical BOLD decreases may be related to decreases in neuronal activity [9]. The potential relevance of BOLD decreases has been highlighted by several neuroimaging studies which observed sustained negative BOLD signals in default mode regions with task initiation under normal conditions [10–13]. Under abnormal conditions such as epilepsy, several studies have reported negative BOLD responses in specific cortical and subcortical structures during seizures both in humans and animal models [14–20]. Thus it is critical to investigate and understand the relationship between BOLD decreases and neuronal activity to better understand the physiological basis of these changes.

Epileptic seizures can be divided into two categories: generalized seizures that affect both hemispheres of the brain, and partial seizures that are limited to more localized changes mainly in one hemisphere [21, 22]. The uncertainty in this field can be highlighted by an illustrative example. Bilateral 3–4 Hz spike-wave discharges (SWD) with strong amplitude are often observed in absence epilepsy which is a form of generalized epilepsy [23]. During absence seizures, patients experience brief episodes (usually lasting less than 10 s) of impaired attention and unresponsiveness. Results from previous human and animal studies suggest that abnormal electrophysiological oscillations during SWD could originate from thalamocortical networks [23–28] and give rise to a mixture of BOLD fMRI increases and decreases. However, whether these oscillations lead to an overall decrease or increase in cortical and subcortical neuronal activity during SWD is inconclusive.

In this chapter, we will discuss possible relationships between fMRI decreases and neuroenergetics (neuronal activity, CBV, CBF, and $CMRO_2$) under normal conditions as well as during epileptic seizures. We will then review three scenarios where unexpected or paradoxical decreases in BOLD have been observed during increases in neuronal activity. These findings have important implications for the interpretation of cortical and subcortical fMRI data and suggest that negative BOLD signals could arise from multiple mechanisms.

2 Relationships Between BOLD fMRI Decreases and Neuronal Activity

Continuous metabolism of energy substrates such as glucose and oxygen is crucial for sustaining neuronal activity (v) which reflects membrane voltage changes in presynaptic and postsynaptic neurons [29–32]. Energy (ATP) consumption by neuronal activity is largely due to restoration of ion concentration gradients (i.e., Na^+ and K^+ concentration gradients) generated by postsynaptic potentials and action potentials [33].

CBF and $CMRO_2$ consumption are two main underlying components that determine oxygen levels in the brain [34, 35]. It is widely held that an increase in neuronal activity leads to increased metabolism, which, in turn, causes CBF to elevate. Blood and tissue oxygenation level (pO_2) increases when increased oxygen delivery resulting from elevated CBF exceeds metabolic demands. Since the BOLD signal intensity (S) in fMRI is inversely related to deoxygenated hemoglobin concentration, increased oxygenation level results in an increase in S.

$CMRO_2$ is considered the most direct estimate of neuronal activity, because oxidative metabolism is the main energy source for meeting high cerebral energy demands [36]. It is possible to measure $CMRO_2$ directly with magnetic resonance spectroscopy (MRS), but MRS techniques suffer from low spatiotemporal resolution. BOLD fMRI, which has relatively high spatiotemporal resolution, can be utilized to estimate $CMRO_2$ and neuronal activity indirectly using the following relationship [34, 37–39]:

$$\Delta S / S = A\left[\left(\Delta \mathrm{CBF} / \mathrm{CBF} - \Delta \mathrm{CMRO}_2 / \mathrm{CMRO}_2\right) / \left(1 + \Delta \mathrm{CBF} / \mathrm{CBF}\right) - \Delta \mathrm{CBV} / \mathrm{CBV}\right]$$

where A is a magnetic field-dependent constant, and CBV refers to cerebral blood volume. The term $\Delta S/S$ is determined by changes in oxygenation. Because deoxyhemoglobin is paramagnetic, it causes a decrease in the spin-echo (T_2) relaxation time or in the gradient-echo (T_2^*) relaxation time, resulting in decreased BOLD signal [35].

Overall, increases and decreases in neuronal activity are often associated, respectively, with increases and decreases in the BOLD signal. However, this relationship between neuronal activity and BOLD changes does not always hold, and it is important to use caution when interpreting fMRI BOLD signal changes, especially decreases, under abnormal conditions such as epilepsy.

3 BOLD fMRI Increases

Positive BOLD response changes have been linked to increased neuronal activity (v) [3, 4, 40]. This increase in v and $CMRO_2$ is then followed by a sufficiently large increase in hyperemic response

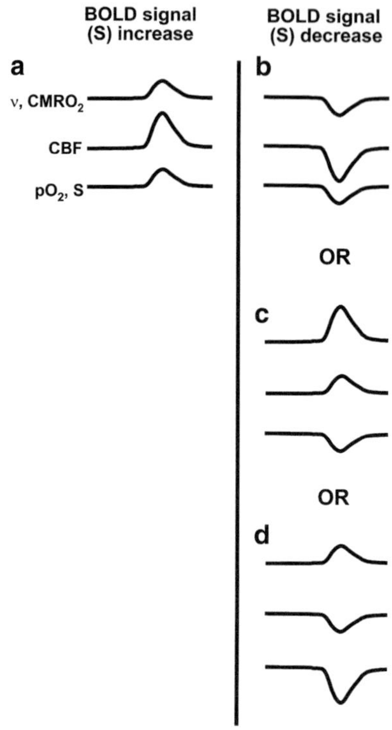

Fig. 1 Schematic illustration of increases and decreases in the BOLD signal (*S*). Increases and decreases in the BOLD signal are related to changes in neuronal activity (ν), cerebral metabolic rate of oxygen consumption ($CMRO_2$), and cerebral blood flow (CBF). (**a**) An example of changes resulting in increases in *S* and oxygenation levels (pO_2). (**b–d**) Examples of changes that can lead to decreases in *S*. *Note*: Figures not drawn to scale. Modified from Blumenfeld, H., *Functional MRI studies of animal models in epilepsy*. Epilepsia, 2007. **48 Suppl 4**: p. 18–26

(CBF) to exceed oxidative metabolic demands, thereby resulting in increased pO_2 and BOLD signal, *S* (Fig. 1a). However, this is not the only mechanism that could result in increased BOLD signal. For example, abnormally high increase in CBF could lead to BOLD increases even when there are no significant changes in neuronal activity [41].

4 BOLD fMRI Decreases

Sustained decreases in the BOLD signals have been observed in numerous functional neuroimaging studies [18, 42–44]. Negative BOLD changes are usually assumed to be correlated with decreases in neuronal activity. Shmuel et al. found that a large portion of BOLD decreases originated from decreased neuronal activity in monkey visual cerebral regions (V1) under normal physiological conditions, and observed decreased neuronal activity accompanied by large decreases in CBF [9]. The tight coupling between fMRI

BOLD and neuronal activity decreases does not necessarily hold in subcortical brain regions or under epileptic conditions, and the origin of sustained negative BOLD responses is poorly understood. Therefore, additional investigation of these relationships is warranted to further understand mechanisms behind fMRI decreases.

Recent imaging studies of animal seizure models revealed that negative changes in BOLD can result from three different mechanisms (Fig. 1b–d) [16–18]: (1) decreased neuronal activity (v) and $CMRO_2$, followed by a large decrease in CBF which, in turn, causes net oxygen levels (pO_2) and S to drop (Fig. 1b), (2) significantly elevated v and $CMRO_2$, accompanied by a smaller increase in CBF that is not sufficient to meet metabolic demands, leading to a decrease in pO_2 and S (Fig. 1c), or (3) increased v and $CMRO_2$ combined with a paradoxical decrease in CBF leading to a large, sustained depression of pO_2 and S (Fig. 1d).

In the subsequent sections, we will discuss each of the three mechanisms listed above in more detail, which provide a cautionary message about the interpretation of BOLD fMRI decreases.

5 fMRI Decreases with Increases in LFP Activity

Animal studies of epilepsy that employ neuroimaging (fMRI), electrophysiological recordings, and hemodynamic measurements (i.e., CBF and CBV) make it possible to investigate fMRI BOLD changes and its neural correlates. Englot et al. performed such studies using Sprague–Dawley rats to investigate changes in neuronal activity, CBF, CBV, and fMRI BOLD during partial limbic seizures [16].

Spontaneous partial limbic seizures in epileptic rats or seizures induced acutely by brief hippocampal stimuli revealed slow oscillations (1–2 Hz) with large amplitude in the frontal cortex [16]. The hippocampus displayed fast poly-spike activity (9–12 Hz) in both local field potentials (LFP) and MUA during partial limbic seizures, while LFP recordings from the orbitofrontal (OFC) cortex presented large amplitude ictal 1–2 Hz slow waves [16]. In addition, MUA recordings from the frontal cortex showed distinct neuronal firing patterns during slow-wave activity: each slow-wave trough corresponded to decreased neuronal firing (low amplitude MUA), while each slow-wave peak was aligned with increased neuronal firing (high amplitude MUA) [16].

Partial limbic seizures and neocortical slow activity were further investigated through fMRI experiments and CBV measurements [16]. Several subcortical structures including the hippocampus (HC), thalamus (Thal), and septal nuclei (Septum) showed fMRI BOLD increases during electrically-stimulated partial limbic seizures (Fig. 2a) [16]. On the contrary, pronounced BOLD decreases were observed in cortical regions including the orbitofrontal (OFC), retrosplenial (RSC), and anterior cingulate

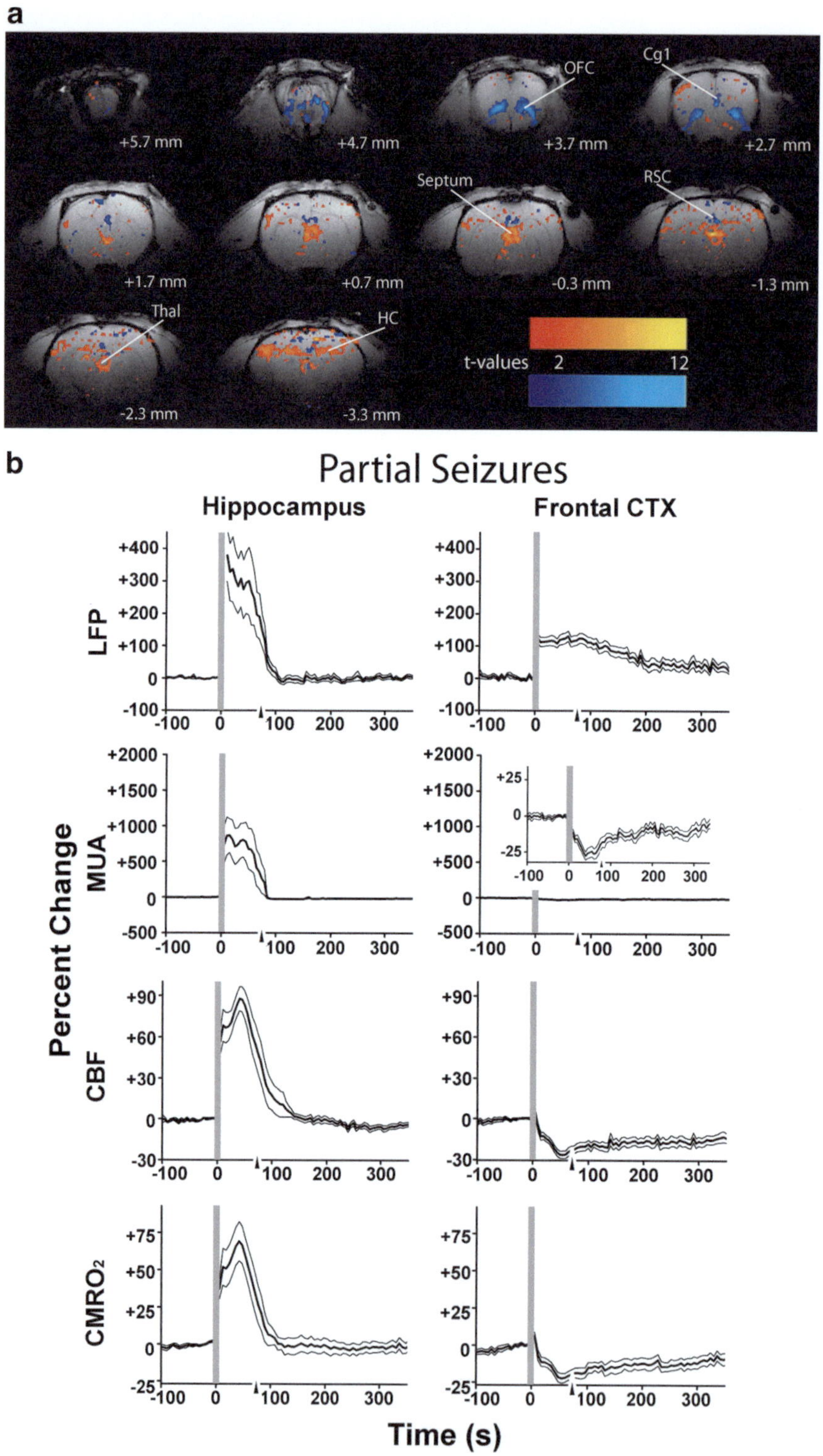

Fig. 2 Decreased BOLD with increased LFP amplitude in orbitofrontal cortex during partial limbic seizures. (**a**) During electrically-stimulated partial limbic seizures, BOLD fMRI signal increases were observed in the hippocampus (HC), thalamus (Thal), and septal nuclei (Septum). Pronounced BOLD decreases were observed in the orbitofrontal (OFC), anterior cingulate (Cg1), and retrosplenial/posterior cingulate (RSC) cortices.

(Cg1) cortices (Fig. 2a) [16]. Flow-related CBV recordings during partial seizures were measured to determine the origin of the BOLD decreases [16]. Interestingly, the timecourse of CBV signal changes closely matched that of BOLD changes: elevated CBV in the hippocampus, thalamus, and septum, and negative CBV changes in the frontal, anterior cingulate, and retrosplenial/posterior cingulate cortices [16]. These findings suggest that BOLD decreases observed during partial limbic seizures might be due to reduced hemodynamic responses which, in turn, lead to decreased blood and oxygen supply.

Since negative CBV changes do not necessarily indicate decreased neuronal activity, direct recordings of electrophysiological activity (LFP and MUA) and CBF signal changes (via laser Doppler flowmetry) were measured in the hippocampus and OFC cortex [16]. During hippocampal-stimulated partial seizures (fast poly-spike activity), the hippocampus revealed large increases in LFP, MUA, and CBF (Fig. 2b, left column) [16]. In the frontal cortex, the expected decreases in MUA and CBF were observed during ictal neocortical slow-wave activity (Fig. 2b, right column, middle, and bottom traces) [16]. Cerebral metabolism ($CMRO_2$) changes estimated from fMRI BOLD, CBV, and CBF data were consistent with CBF and BOLD changes during partial seizures: increased $CMRO_2$ in the hippocampus and decreased $CMRO_2$ in the frontal cortex (Fig. 2b [16]; see also Fig. 1b).

Of note, despite the decreases in BOLD in OFC there were prominent *increases* in LFP signals in this region (Fig. 2b, right column, top trace). This can be explained by the fact that the LFP signal in this situation is produced by intermittent cessation of neuronal activity (action potentials and synaptic activity) during slow oscillations [16]. Therefore, despite the large fluctuating field potentials the mean rate of neuronal firing and synaptic activity—and hence energy metabolism—averaged across the cortical slow waves is actually decreased (Fig. 2b, right column, middle trace).

In summary, Englot et al. showed that ictal slow waves during partial limbic seizures in lightly-anesthetized rats were associated with reduced cortical function [16]. Furthermore, the multimodal study demonstrates one of the mechanisms that could lead to decreased fMRI BOLD signal changes: decreased neuronal activity and $CMRO_2$, followed by a concomitant depression in CBF and CBV which causes net oxygen levels and fMRI BOLD intensity to drop despite an increase in LFP amplitude (Fig. 1b).

Fig. 2 (continued) (**b**) Neuronal activity as measured by LFP and MUA, CBF, and $CMRO_2$ changes during partial limbic seizures differed significantly during ictal neocortical slow activity in the frontal cortex (Frontal CTX) vs. fast seizure activity in the hippocampus. In the hippocampus, all signals increased during seizure, and returned to baseline postictally. In the frontal cortex, there was a moderate increase in LFP ictally, but MUA, CBF, and $CMRO_2$ displayed decreases during seizure. *Vertical grey bars* represent synchronized beginning and arrowheads signify mean end of all seizures in each dataset. Modified from Englot, D.J., et al., *Remote effects of focal hippocampal seizures on the rat neocortex.* J Neurosci, 2008. **28**(36): p. 9066–81

6 fMRI Decreases with Extreme Increases in Neuronal Activity

While Englot et al. showed fMRI decreases associated with decreases in neuronal activity and CBF (Fig. 1b), Schridde et al. found increased neuronal activity linked to depression in the BOLD signal (Fig. 1c) [18]. Using adult Wistar rats as an animal model of generalized epilepsy, Schridde et al. performed fMRI experiments as well as combined electrophysiological and laser Doppler flowmetry CBF recordings to investigate the interplay between neuronal activity, CBF, and fMRI BOLD signals during generalized tonic-clonic seizures [18].

During bicuculline-induced seizures, fMRI BOLD activation maps revealed not only widespread increases in the cortex and thalamus, but also pronounced decreases in the hippocampus (Fig. 3a) [18]. The activation maps from all the fMRI experiments showed significantly elevated BOLD activity in the cortex during seizures, but the ictal BOLD signal changes (from the baseline) in the hippocampus were not as consistent: most of the experiments showed prominent decreases, but some maps showed slight increases or no significant changes [18].

In order to confirm that the negative BOLD changes in the hippocampus during seizures were not caused by decreased blood flow and reduced oxygen delivery (by either vascular steal or vasospasm [45]) to the region, Schridde et al. carried out additional experiments using an exogenous paramagnetic agent to measure flow-related CBV during seizures [18]. Unlike the fMRI BOLD activation maps (Fig. 3a) which showed increases throughout the cortex and thalamus and decreases in the hippocampus, the CBV activation maps revealed consistently high CBV responses throughout the whole brain during seizures [18]. These results suggest that decreases in blood supply did not cause negative BOLD changes in the hippocampus during seizures.

The next set of experiments measured changes in CBF and neuronal activity in cortex and hippocampus during seizures to further investigate the origin of negative BOLD signals [18]. Both LFP and MUA recordings were used to measure neuronal activity, while laser Doppler flowmetry was used to record CBF signals [18]. Large increases in LFP and MUA were observed during seizures in both cortex (S1BF) and hippocampus (Fig. 3b*a–d*) [18]. These increases in neuronal activity were then followed by increased CBF in both areas (Fig. 3b*e*, *f*) [18]. The findings further confirmed that negative BOLD responses in the hippocampus were not due

Fig. 3 (continued) in scale between (**b**) part c and (**b**) part d). Average time courses of percent signal change compared with 30 s of baseline data. Time courses are shown as mean ± SEM (unidirectional) and straight vertical line at time = 0 marks seizure onset. Modified from Schridde, U., et al., *Negative BOLD with large increases in neuronal activity.* Cereb Cortex, 2008. **18**(8): p. 1814–27

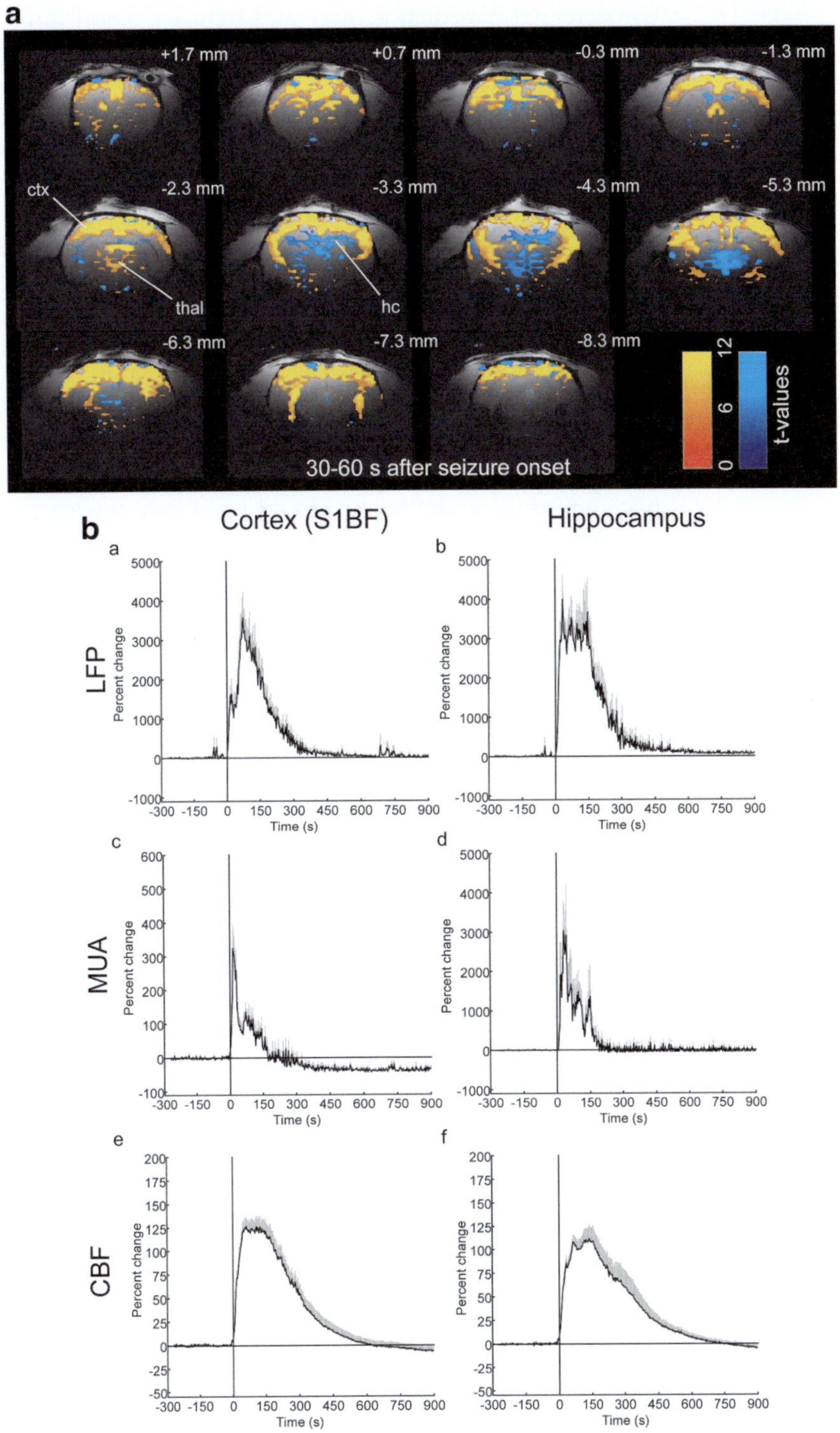

Fig. 3 Decreased BOLD with increased neuronal activity in the hippocampus during tonic-clonic seizures. (**a**) During bicuculline-induced generalized seizures, the BOLD signal showed widespread increases in the cortex (ctx), thalamus (thal), and prominent decreases in the hippocampus (hc). (**b**) Neuronal activity (LFP and MUA) and CBF in both cortex (S1BF) and hippocampus showed only increases during bicuculline-induced generalized seizures, with MUA increases in the hippocampus greatly exceeding those in the cortex (note difference

to decreased neuronal activity and reduced blood flow, and that decreases in fMRI BOLD can occur despite increases in neuronal activity and CBF.

Importantly, although increases in LFP and CBF were comparable between cortex and hippocampus, the MUA increase in hippocampus was much larger than the MUA increase in cortex (Fig. 3b*c*, *d*; note difference in scale) [18]. From this observation, Schridde et al. hypothesized that the oxidative metabolism rate ($CMRO_2$) in the hippocampus exceeded its blood supply during seizures [18]. Previous studies showed that increased metabolism during prolonged seizure activity could result in possible hypoxia [46–48] and even brief epileptic events can cause regional negative oxygenation signals [19, 20, 49–51]. The changes in $CMRO_2$ were then estimated from the BOLD, CBF, and CBV signals, and revealed a significantly larger oxygen consumption rate in the hippocampus than in the cortex [18]. In addition, the increase in CBF was far greater than the increase in $CMRO_2$ in the cortex during seizures, indicating that oxygenated blood supply greatly exceeded metabolic demands in the region [18]. This oversupply of oxygenated hemoglobin then resulted in positive BOLD signal changes in the cortex [18]. On the other hand, the increase in $CMRO_2$ in the hippocampus was nearly identical to the increase in CBF during seizures [18]. These findings suggest that oxygenated blood supply could have fallen short of metabolic energy demands in hippocampus at certain times, which could lead to elevated deoxygenated hemoglobin levels and negative BOLD changes [18].

Thus, the study demonstrates that negative BOLD responses do not always result from decreased neuronal activity, but can also occur from large increases in neuronal activity combined with smaller increases in CBF (Fig. 1c). Furthermore, depending on the brain region (i.e., cortical vs. subcortical structures) and state (i.e., normal vs. abnormal states), negative BOLD responses could result from either increased or decreased neuronal activity.

7 fMRI Decreases with Reverse Coupling of Neuronal Activity and CBF

The third mechanism of unexpected BOLD decreases involves reverse neurovascular coupling of neuronal activity and CBF: increases in neuronal activity are accompanied by decreases in CBF, resulting in reduced net oxygenation and negative BOLD signals (Fig. 1d). Mishra et al. demonstrated this altered relationship between hemodynamic and neuronal activity responses in the striatum during seizures (spontaneous spike-wave episodes) in the Wistar albino Glaxo rat of Rijswijk (WAG/Rij) as well as during normal somatosensory stimulation [17].

fMRI studies were used to measure BOLD and CBV, while LFP and MUA recordings were used for estimation of neural activity, and laser Doppler flowmetry experiments to measure CBF [17].

fMRI BOLD activation maps during spike-wave seizures (SWD) displayed mixed signal changes in cortical and subcortical structures (Fig. 4a) [17]. Prominent bilateral increases in somatosensory barrel cortex (S1BF), thalamus (Thal), anterior cingulate, and posterior cingulate regions were observed, while significant bilateral decreases originated from the striatum, or caudate-putamen (CPu) during SWD (Fig. 4a) [17]. It is also important to note that several brain regions including hippocampus (Hc) and primary visual cortex (V1M) showed no significant changes during SWD [17].

As previously shown by Schridde et al., it is feasible to observe negative BOLD changes despite increases in hemodynamic responses (CBV and CBF) and neuronal activity [18]. To test if there were hemodynamic increases or decreases during negative striatal BOLD signals, Mishra et al. measured CBV changes during spike-wave seizures [17]. CBV maps during SWD revealed similar patterns of changes as those shown by the fMRI activation maps: bilateral increases in the S1BF, thalamus, anterior cingulate, and posterior cingulate but symmetric decreases in the CPu [17]. One possible explanation for negative BOLD changes based on these results is decreased blood flow during SWD.

Even though decreased blood flow is often associated with decreased neuronal activity [40, 52, 53], further analyses were required to investigate CBF and neuronal activity changes and the origin of diminished blood supply to the CPu during SWD. Combined electrophysiological and laser Doppler CBF recordings acquired from S1BF and VPM displayed distinct increases in both LFP and MUA, followed by elevated CBF (from the baseline) during SWD (Fig. 4b–d) [17]. On the other hand, CPu presented rather unexpected signal changes: a large increase in LFP and a smaller increase in MUA were accompanied by a marked decrease in CBF (Fig. 4b–d) [17]. The discrepancy in amplitude between increases in LFP and MUA of CPu could be caused by intense activity of striatal inhibitory neurons during SWD leading to a sharp increase in synaptic activity (LFP) but only a slight increase in mean neuronal firing rate (MUA) [17, 54]. However, the reverse coupling between neuronal activity (both LFP and MUA) and CBF in the striatum was more paradoxical and difficult to explain. Of note, reverse coupling of neuronal activity and CBF was found in the rat caudate-putamen not only during SWD but also during *normal* somatosensory stimulation [17]. This suggests that seizures are not necessary for this reverse coupling to occur and that it may also be seen under normal conditions.

Human studies have demonstrated fMRI decreases in the basal ganglia including the caudate-putamen during somatosensory stimulation without seizures, as well as during SWD [14, 15, 55–57]. Although there are several mechanisms that could lead to decreased BOLD signals, the findings made by Mishra et al. suggest that a primary vascular decrease in CBF is the most reasonable explanation

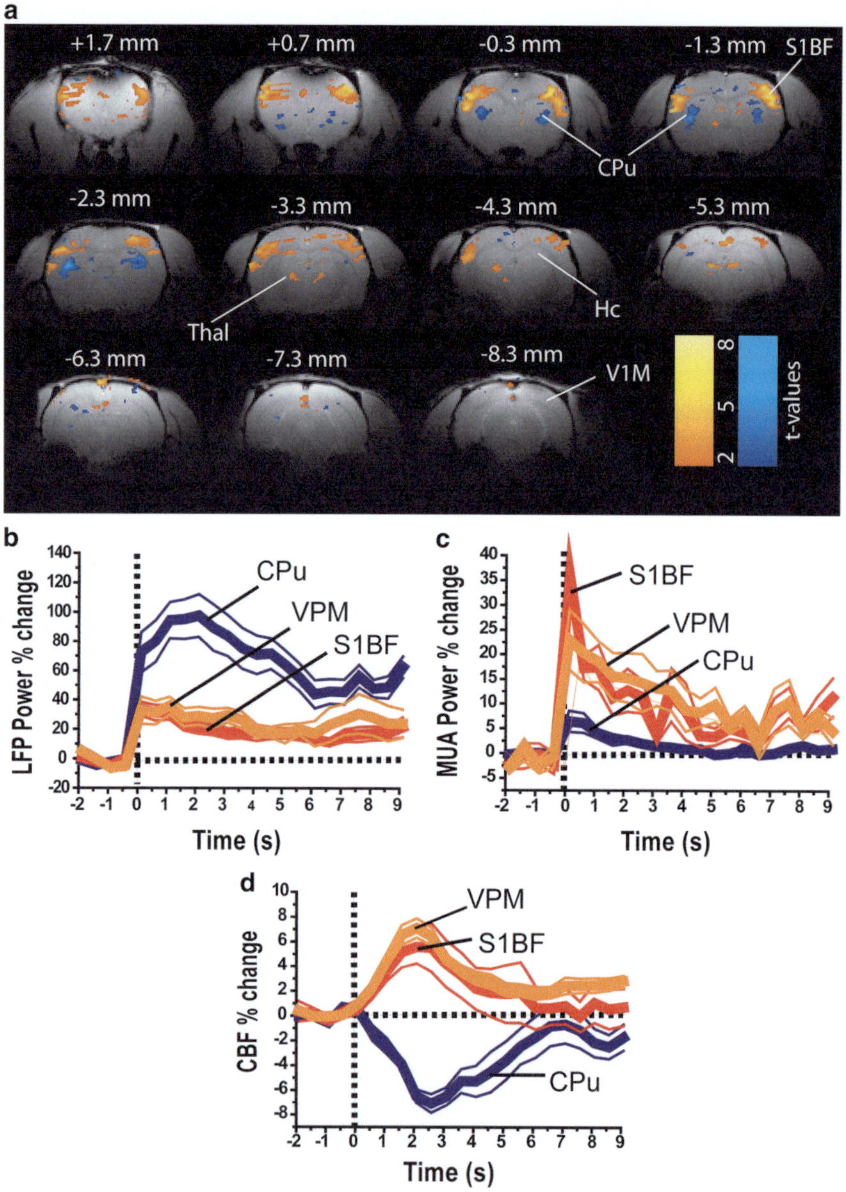

Fig. 4 Decreased BOLD and decreased CBF despite increased neuronal activity during SWD. (**a**) During SWD, somatosensory barrel cortex (S1BF) and thalamus (Thal) showed marked increases in BOLD signal. On the other hand, caudate-putamen (CPu) displayed prominent BOLD decreases. No changes were observed in primary visual cortex (V1M) or hippocampus (Hc). (**b–d**) Mean percent change (compared to 2 s baseline) timecourses of electrophysiology (LFP and MUA) and CBF measurements from the somatosensory cortex (S1BF), ventral posterior medial thalamic nucleus (VPM), and caudate-putamen (CPu) are shown. During SWD, both LFP and MUA from S1BF and VPM showed prominent increases, followed by large increases in CBF. In contrast, CPu measurements showed a large increase in LFP associated with a small increase in MUA, and a strong decrease in CBF during SWD. Timecourses are displayed as mean ± SEM, with 0.5 s time bins. Vertical line at time = 0 denotes SWD onset. Modified from Mishra, A.M., et al., *Where fMRI and electrophysiology agree to disagree: corticothalamic and striatal activity patterns in the WAG/Rij rat.* J Neurosci, 2011. **31**(42): p. 15053–64

for the negative BOLD signals observed in CPu [17]. Vascular steal or vasospasm could lead to a primary vascular decrease [17, 45]. Recent studies suggest that D2-receptor activation by dopamine release during SWDs and normal somatosensory stimulation may produce direct vasoconstriction in the caudate-putamen, leading to decreases in BOLD, CBF, and CBV despite local increases in neuronal activity [58–60].

In summary, Mishra et al. found hemodynamic changes (i.e., BOLD, CBV, and CBF) during spike-wave seizures in the WAG/Rij rat to be closely correlated with neural activity as measured by LFP and MUA in the somatosensory barrel cortex and thalamus [17]. However, this close coupling was found to be reversed in caudate-putamen where decreased CBF (possibly by vasoconstriction) resulted in decreased BOLD signals despite increases in neuronal activity [17]. This is another scenario where a negative BOLD response is not necessarily associated with a decreased neural activity signal (Fig. 1d), and as we have discussed in the previous sections, extra caution must be taken when interpreting fMRI BOLD decreases.

8 Summary/Conclusions/Future Directions

Based on the animal studies reviewed above, mechanisms of BOLD fMRI decreases can be classified into two categories: primarily neural mechanism and altered neurovascular coupling [7].

In a primarily neural mechanism, a decrease in neuronal activity leads to a large reduction in CBF, resulting in negative BOLD changes (Fig. 1b). Even though seizures often cause neuronal activity to increase in regions affected by severe epileptic activity, remote network effects could produce depressed neuronal activity in other brain regions that might be indirectly affected by seizures. Importantly, some regions may show large LFP signals despite an overall decrease in mean neuronal firing and synaptic activity. This mechanism was demonstrated by Englot et al. who showed that neocortical slow-wave activity during hippocampal-stimulated seizures corresponded to reduced neuronal firing which caused CBF and BOLD signals to decrease [16]. The mean rate of neuronal activity was reduced in this scenario, because cortical slow waves consist of periodic "Up" and "Down" states [16, 61–63] with neuronal activity greatly reduced during the Down states, leading to an overall decrease in mean energy metabolism.

Negative BOLD signal changes could also result from altered neurovascular coupling. This mechanism involves increased neuronal activity which is accompanied by either an insufficient increase in CBF to meet cerebral metabolic demands (Fig. 1c) or a paradoxical decrease in blood flow (Fig. 1d). Both cases lead to net decreases in oxygenation levels and BOLD signals. Schridde et al. observed extreme increases in neuronal activity in the hippocampus along with an inadequate increase in CBF during tonic-clonic seizures, while Mishra et al. showed reverse neurovascular coupling in the caudate-putamen that led to a decrease in CBF during SWD [17, 18]. Causes of altered neurovascular coupling phenomenon are subject to further investigations.

These studies have provided valuable insights into the physiological and neuronal bases of fMRI decreases. However, more systematic studies on the relationships between neuronal activity, hemodynamic responses (CBV, CBF, and $CMRO_2$), and fMRI BOLD decreases in subcortical structures during seizures and under normal conditions must be undertaken to better understand the relationship between neuroimaging signals and brain electrical activity.

Acknowledgements

This work was supported by NIH Grants R01 NS049307 (H.B.), R01 MH067528 (F.H.), P30 NS052519 (F.H.) and by the Betsy and Jonathan Blattmachr family.

References

1. Ogawa S et al (1990) Brain magnetic resonance imaging with contrast dependent on blood oxygenation. Proc Natl Acad Sci U S A 87(24):9868–9872
2. Huettel SA (2004) Non-linearities in the blood-oxygenation-level dependent (BOLD) response measured by functional magnetic resonance imaging (fMRI). Conf Proc IEEE Eng Med Biol Soc 6:4413–4416
3. Logothetis N et al (2001) Neurophysiological investigation of the basis of the fMRI signal. Nature 412(6843):150–157
4. Smith AJ et al (2002) Cerebral energetics and spiking frequency: the neurophysiological basis of fMRI. Proc Natl Acad Sci U S A 99(16): 10765–10770
5. Herman P et al (2009) Cerebral oxygen demand for short-lived and steady-state events. J Neurochem 109(suppl 1):73–79
6. Sanganahalli BG et al (2009) Oxidative neuroenergetics in event-related paradigms. J Neurosci 29(6):1707–1718
7. Motelow JE, Blumenfeld H (2009) Functional neuroimaging of spike-wave seizures. Methods Mol Biol 489:189–209
8. Shulman RG, Rothman DL, Hyder F (2007) A BOLD search for baseline. Neuroimage 36(2): 277–281
9. Shmuel A et al (2006) Negative functional MRI response correlates with decreases in neuronal activity in monkey visual area V1. Nat Neurosci 9(4):569–577
10. Barks SK, Parr LA, Rilling JK (2013) The default mode network in chimpanzees (Pan troglodytes) is similar to that of humans. Cereb Cortex
11. Yakushev I et al (2013) Metabolic and structural connectivity within the default mode network relates to working memory performance in young healthy adults. Neuroimage 79:184–190
12. Danielson NB, Guo JN, Blumenfeld H (2011) The default mode network and altered consciousness in epilepsy. Behav Neurol 24:55–65
13. Raichle ME et al (2001) A default mode of brain function. Proc Natl Acad Sci U S A 98(2):676–682
14. Bai X et al (2010) Dynamic time course of typical childhood absence seizures: EEG, behavior, and functional magnetic resonance imaging. J Neurosci 30(17):5884–5893
15. Berman R et al (2010) Simultaneous EEG, fMRI, and behavior in typical childhood absence seizures. Epilepsia 51(10):2011–2022
16. Englot DJ et al (2008) Remote effects of focal hippocampal seizures on the rat neocortex. J Neurosci 28(36):9066–9081
17. Mishra AM et al (2011) Where fMRI and electrophysiology agree to disagree: corticothalamic and striatal activity patterns in the WAG/Rij rat. J Neurosci 31(42):15053–15064

18. Schridde U et al (2008) Negative BOLD with large increases in neuronal activity. Cereb Cortex 18(8):1814–1827
19. Suh M et al (2006) Neurovascular coupling and oximetry during epileptic events. Mol Neurobiol 33(3):181–197
20. Bahar S et al (2006) Intrinsic optical signal imaging of neocortical seizures: the 'epileptic dip'. Neuroreport 17(5):499–503
21. ILAE (1981) Proposal for revised clinical and electroencephalographic classification of epileptic seizures. From the Commission on Classification and Terminology of the International League Against Epilepsy. Epilepsia 22(4):489–501.
22. Berg AT et al (2010) Revised terminology and concepts for organization of seizures and epilepsies: report of the ILAE Commission on Classification and Terminology, 2005-2009. Epilepsia 51(4):676–685
23. Blumenfeld H (2005) Cellular and network mechanisms of spike-wave seizures. Epilepsia 46(suppl 9):21–33
24. Blumenfeld H (2002) The thalamus and seizures. Arch Neurol 59(1):135–137
25. Blumenfeld H, McCormick DA (2000) Corticothalamic inputs control the pattern of activity generated in thalamocortical networks. J Neurosci 20(13):5153–5162
26. Kostopoulos GK (2001) Involvement of the thalamocortical system in epileptic loss of consciousness. Epilepsia 42(suppl 3):13–19
27. McCormick DA, Contreras D (2001) On the cellular and network bases of epileptic seizures. Annu Rev Physiol 63:815–846
28. Williams D (1953) A study of thalamic and cortical rhythms in petit mal. Brain 76(1):50–69
29. Buzsaki G, Kaila K, Raichle M (2007) Inhibition and brain work. Neuron 56(5):771–783
30. Hyder F et al (2006) Neuronal-glial glucose oxidation and glutamatergic-GABAergic function. J Cereb Blood Flow Metab 26(7):865–877
31. Riera JJ et al (2008) The micro-architecture of the cerebral cortex: functional neuroimaging models and metabolism. Neuroimage 40(4): 1436–1459
32. Shulman RG et al (2004) Energetic basis of brain activity: implications for neuroimaging. Trends Neurosci 27(8):489–495
33. Attwell D, Laughlin SB (2001) An energy budget for signaling in the grey matter of the brain. J Cereb Blood Flow Metab 21(10):1133–1145
34. Hyder F et al (2001) Quantitative functional imaging of the brain: towards mapping neuronal activity by BOLD fMRI. NMR Biomed 14(7–8):413–431
35. Ogawa S et al (1998) On the characteristics of functional magnetic resonance imaging of the brain. Annu Rev Biophys Biomol Struct 27: 447–474
36. Hyder F, Rothman DL, Shulman RG (2002) Total neuroenergetics support localized brain activity: implications for the interpretation of fMRI. Proc Natl Acad Sci U S A 99(16): 10771–10776
37. Hoge RD, Pike GB (2001) Oxidative metabolism and the detection of neuronal activation via imaging. J Chem Neuroanat 22(1–2): 43–52
38. Kennan RP, Zhong J, Gore JC (1994) Intravascular susceptibility contrast mechanisms in tissues. Magn Reson Med 31(1):9–21
39. Ogawa S et al (1993) Functional brain mapping by blood oxygenation level-dependent contrast magnetic resonance imaging. A comparison of signal characteristics with a biophysical model. Biophys J 64(3):803–812
40. Nersesyan H et al (2004) Relative changes in cerebral blood flow and neuronal activity in local microdomains during generalized seizures. J Cereb Blood Flow Metab 24(9):1057–1068
41. Martin C et al (2006) Haemodynamic and neural responses to hypercapnia in the awake rat. Eur J Neurosci 24(9):2601–2610
42. Logothetis NK (2002) The neural basis of the blood-oxygen-level-dependent functional magnetic resonance imaging signal. Philos Trans R Soc Lond B Biol Sci 357(1424):1003–1037
43. Shmuel A et al (2002) Sustained negative BOLD, blood flow and oxygen consumption response and its coupling to the positive response in the human brain. Neuron 36(6): 1195–1210
44. Shulman GL et al (1997) Common blood flow changes across visual tasks: II. Decreases in cerebral cortex. J Cogn Neurosci 9(5):648–663
45. Harel N et al (2002) Origin of negative blood oxygenation level-dependent fMRI signals. J Cereb Blood Flow Metab 22(8):908–917
46. Ingvar M, Siesjo BK (1983) Local blood flow and glucose consumption in the rat brain during sustained bicuculline-induced seizures. Acta Neurol Scand 68(3):129–144
47. Pereira de Vasconcelos A, Ferrandon A, Nehlig A (2002) Local cerebral blood flow during lithium-pilocarpine seizures in the developing and adult rat: role of coupling between blood flow and metabolism in the genesis of neuronal damage. J Cereb Blood Flow Metab 22(2):196–205
48. van Eijsden P et al (2004) In vivo 1H magnetic resonance spectroscopy, T2-weighted and diffusion-weighted MRI during lithium-pilocarpine-induced status epilepticus in the rat. Brain Res 1030(1):11–18
49. Schwartz TH (2005) The application of optical recording of intrinsic signals to simultaneously

acquire functional, pathological and localizing information and its potential role in neurosurgery. Stereotact Funct Neurosurg 83(1):36–44
50. Schwartz TH (2003) Optical imaging of epileptiform events in visual cortex in response to patterned photic stimulation. Cereb Cortex 13(12):1287–1298
51. Schwartz TH, Bonhoeffer T (2001) In vivo optical mapping of epileptic foci and surround inhibition in ferret cerebral cortex. Nat Med 7(9):1063–1067
52. Nersesyan H et al (2004) Dynamic fMRI and EEG recordings during spike-wave seizures and generalized tonic-clonic seizures in WAG/Rij rats. J Cereb Blood Flow Metab 24(6):589–599
53. Stefanovic B et al (2005) Hemodynamic and metabolic responses to activation, deactivation and epileptic discharges. Neuroimage 28(1):205–215
54. Slaght SJ et al (2004) On the activity of the corticostriatal networks during spike-and-wave discharges in a genetic model of absence epilepsy. J Neurosci 24(30):6816–6825
55. Hui KK et al (2000) Acupuncture modulates the limbic system and subcortical gray structures of the human brain: evidence from fMRI studies in normal subjects. Hum Brain Mapp 9(1):13–25
56. Moeller F et al (2008) Changes in activity of striato-thalamo-cortical network precede generalized spike wave discharges. Neuroimage 39(4):1839–1849
57. Moeller F et al (2008) Simultaneous EEG-fMRI in drug-naive children with newly diagnosed absence epilepsy. Epilepsia 49(9):1510–1519
58. Choi JK et al (2006) Brain hemodynamic changes mediated by dopamine receptors: role of the cerebral microvasculature in dopamine-mediated neurovascular coupling. Neuroimage 30(3):700–712
59. Shih YY et al (2009) A new scenario for negative functional magnetic resonance imaging signals: endogenous neurotransmission. J Neurosci 29(10):3036–3044
60. Shih YY et al (2011) Striatal and cortical BOLD, blood flow, blood volume, oxygen consumption, and glucose consumption changes in noxious forepaw electrical stimulation. J Cereb Blood Flow Metab 31(3):832–841
61. Haider B et al (2006) Neocortical network activity in vivo is generated through a dynamic balance of excitation and inhibition. J Neurosci 26(17):4535–4545
62. Steriade M, Nunez A, Amzica F (1993) A novel slow (<1 Hz) oscillation of neocortical neurons in vivo: depolarizing and hyperpolarizing components. J Neurosci 13(8):3252–3265
63. Englot DJ et al (2009) Cortical deactivation induced by subcortical network dysfunction in limbic seizures. J Neurosci 29(41):13006–13018

Chapter 12

Simultaneous Multi-Wavelength Optical Imaging of Neuronal and Hemodynamic Activity

Hongtao Ma, Mingrui Zhao, Samuel Harris, and Theodore H. Schwartz

Abstract

Functional brain imaging techniques, such as functional magnetic resonance imaging, infer changes in underlying neural activity from perfusion-related signals and can be sampled from large areas of cortex to examine hemodynamic network activity. However, measurement of the underlying neuronal activity in a correspondingly widespread network is not available for comparison. In order to accurately understand the origins of these hemodynamic signals and the mechanisms of neurovascular coupling, it is critical to be able to measure widespread neuronal activity simultaneously and homotopically with the vascular signal. For this reason, we designed a simultaneous multi-wavelength optical imaging system that can provide high-resolution spatiotemporal information of neuronal and hemodynamic activities, and their inter-relationship. We have demonstrated that this technique is an effective method with which to study neurovascular coupling during spontaneous brain activity, and pathological conditions such as epilepsy.

Key words Neurovascular coupling, Epilepsy, Intrinsic optical imaging, Voltage-sensitive dye imaging, Calcium imaging

1 Introduction

Neurovascular coupling refers to the relationship between local neural activity and subsequent changes in cerebral blood flow (CBF). The change in neuronal activity will adjust the local blood flow with optimal magnitude and spatial location through a complex sequence of coordinated events involving neurons, glia, and the regulation of vascular flow. Many perfusion-based functional brain imaging techniques, such as functional magnetic resonance imaging (fMRI), rely on this coupling to infer the changes in the underlying neural activity, which lead to the resulting hemodynamic response. The study of neurovascular coupling is essential to the correct interpretation of brain imaging data. Extensive research on the role of different cellular types (e.g., neuron, astrocytes, pericytes) and physiological state (e.g., temperature, pH, O_2, NO) on changes in CBF have elucidated the complexity of the

Mingrui Zhao et al. (eds.), *Neurovascular Coupling Methods*, Neuromethods, vol. 88,
DOI 10.1007/978-1-4939-0724-3_12, © Springer Science+Business Media New York 2014

neurovascular coupling process. However, the clinical application of hemodynamic-based functional imaging signals is primarily concerned with the temporal and spatial precision of the imaging results. To investigate this, a method that can record both neuronal dynamics and hemodynamic changes with equivalent spatiotemporal resolution as well as spatial coverage is needed. fMRI, as well as other hemodynamic techniques such as single-photon emission computed tomography (SPECT), are currently being used not only to investigate cognitive processes but also pathophysiology such as epilepsy and Alzheimer's disease for clinical use.

Currently, a large number of methods are available to record hemodynamic changes on a wide variety of scales, from individual blood cell movement to the whole brain, such as fMRI, positron emission tomography (PET), SPECT, near-infrared spectroscopy (NIRS), optical coherence tomography (OCT), intrinsic optical imaging, and two-photon optical imaging. There are also a number of methods available to record the voltage change or electric current on a wide variety of scales from single ion channel neuronal activity to whole brain, such as magnetoencephalography (MEG), electroencephalography (EEG), local field potential (LFP), voltage sensitive dye imaging (VSDI), calcium imaging, multi-unit activity, single unit activity, intracellular recording, and patch clamp techniques. Among all these methods, the only way to acquire simultaneous information about hemodynamics and electrophysiology with similar spatiotemporal resolution is to use epi-illumination and the resulting absorption and reflection of intrinsic light signals or chemical-mediated fluorescent light signals such as optical imaging of intrinsic signal (ORIS) and either voltage or calcium imaging. Other techniques can be used, which will be illustrated in this textbook, however, simultaneous acquisition of both electrical and hemodynamic signals with similar spatial sampling and scale is the ultimate goal addressed in this chapter.

Optical imaging of intrinsic signals: ORIS is a technique developed by Grinvald and co-workers [1–4]. ORIS is the visualization of changes of intrinsic optical properties of neural tissues, in particular, light reflection induced by neuronal activity. The surface of the brain is illuminated with light of a specific wavelength and subsequent reflection recorded with a camera. Depending on the recorded wavelength, the ORIS can reflect different physical processes. At 530, 546, and 570 nm, where the oxy-hemoglobin and deoxy-hemoglobin absorb light equally i.e., isosbestic wavelengths, ORIS reflects the changes in total hemoglobin (Hbt) concentration, or cerebral blood volume (CBV). At wavelengths >600 nm, in which light absorption by deoxy-hemoglobin is much greater than oxy-hemoglobin, the ORIS mainly reflects oximetry. At wavelengths >780 nm, the ORIS mainly reflects ion and water movement and the resulting light scatter in neurons and glial and swelling in the extra- and intracellular space [2, 5, 6]. ORIS can offer a spatial resolution of ~50 μm, a spatial coverage of millimeters to centimeters, with a fairly high-temporal resolution limited only by the data sampling rate, although the signal itself generally changes on the order of hundreds of milliseconds.

Voltage sensitive dye (VSD) imaging: VSDs are a kind of molecular probe that binds to the membrane and changes its absorption or fluorescence in response to membrane-associated voltage changes. Optical imaging with VSD permits visualization of neuronal activity with a submillisecond temporal resolution and a spatial resolution of 50–100 μm [7]. In single cell imaging, the VSD signal is analogous to an intracellular electrical recording [8–12]. When recording from cortical tissue, the optical signal represents the average membrane potential change, in both pre- and post-synaptic neuronal elements. The depolarization of neighboring glial cells may also contribute to the VSD signal. During in vivo preparations, fluorescence dye is usually selected to record cortical dynamics, with recent development of new blue dyes permitting real time monitoring of neuronal activity [13–15].

Calcium imaging: Calcium dyes are molecules that can respond to the binding of Ca^{2+} by changing their fluorescent properties. Calcium imaging can probe various aspects of cortical dynamics in vivo, including neuronal network activation [16], and glial and metabolic signaling [17]. Although the underpinning of calcium imaging is the Ca^{2+} change, the fast change of the calcium signal is similar to the VSD in single cell imaging [18]. Wide field and single photon calcium imaging has also been widely used to represent population neuronal activity [19, 20] with similar spatiotemporal resolution to VSD, in addition to similar spatial coverage. The difficulty with calcium imaging has been the inability to load large areas of adult cortex with calcium dyes. Although technique to overcome this have recently been developed involving multiple injections, limitations still exist. In this chapter we introduce another technique with powerful possibilities to bulk load the adult cortex for calcium imaging.

Furthermore, we will provide detailed descriptions of two methods for combining ORIS with either VSD or calcium imaging to obtain concurrent spatiotemporal measures of hemodynamic changes and neuronal activity.

2 System Design

The major technical challenge in simultaneous ORIS and neuronal activity imaging is the separation of the signals from both modalities. Figure 1a summarizes the most popular wavelengths employed in ORIS, calcium imaging, and VSD imaging. Note that the

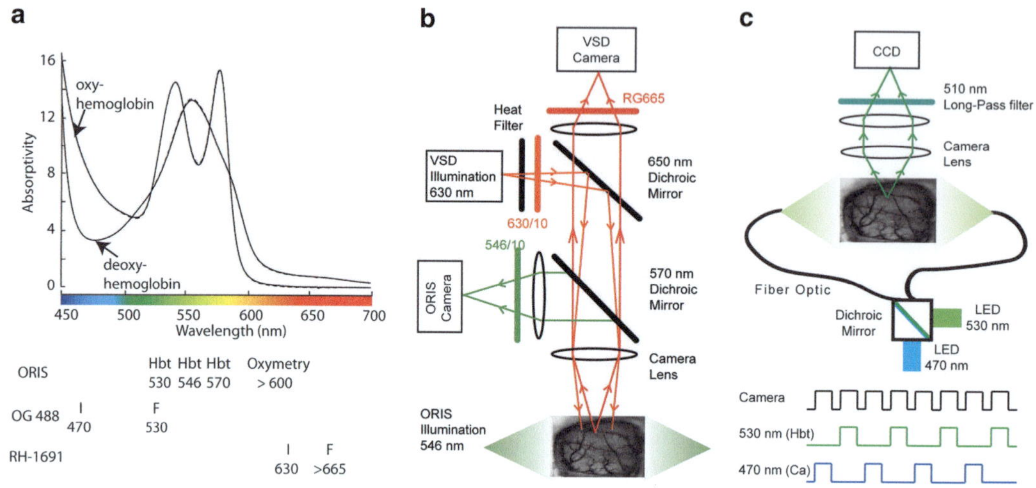

Fig. 1 Two designs of simultaneous multi-wavelength optical imaging system. (**a**) The absorption curve of oxy- and deoxy-hemoglobin. The commonly used wavelengths in ORIS, calcium dye (Oregon Green 488 BAPTA-1, AM), and voltage-sensitive dye (RH-1691) is listed. (**b**) Diagram of "spatial separation" setup for simultaneous VSD and ORIS imaging. Two illumination sources are used for VSD and IOS respectively. The IOS illumination is filtered using a 546 ± 10-nm filter and directed towards the cortex using two fiber optic cables. The IOS is reflected towards the camera via a 570-nm dichroic mirror. A 546 ± 10-nm filter is placed in front of the camera. The VSD illumination is filtered with a heat filter and a VSD filter (630 ± 10 nm) and reflected onto the cortex via a 650-nm dichroic mirror. The fluorescence is filtered with a long-pass filter, RG-665, before being focused onto the VSD camera. (**c**) Diagram of "temporal separation" setup for simultaneous calcium dye and ORIS imaging. A 470- and a 530-nm LED are coupled with a dichroic mirror to serve as the illumination source. Two LEDs are sequentially triggered by the camera frame indicator. A 510-nm long-pass filter is placed before the camera to filter out the calcium illumination (470 nm)

fluorescence for calcium dye Oregon Green 488 overlies the ORIS wavelengths that reflect Hbt changes (530 nm). In turn, the VSD illumination wavelength (new blue dye, RH-1691) overlies ORIS wavelengths that reflect oximetry changes (630 nm).

In traditional imaging techniques, halogen light is usually employed as the illumination source. Illumination at a specific wavelength is achieved using a filter placed after the illumination source or before the recording device. For ORIS, the illumination is targeted using fiber-optic guides to achieve a dark field illumination (Fig. 1b). For in vivo VSDI, the illumination is reflected using a dichroic mirror to form coaxial illumination (Fig. 1b), which is required due to the use of fiber-optic light guides that result in a significant loss of power. We designed a new system that combines these traditional imaging modalities to perform simultaneous VSD/ORIS imaging (Fig. 1b) [21]. Two halogen lights are filtered with 630 and 546 nm to serve as VSD and ORIS illumination sources, respectively. A low-speed video camera is employed to record ORIS (Imager 3001, Optical Imaging, Rehovot, Israel), with a temporal resolution of 33 Hz, a spatial resolution of

248×160 pixels with each pixel sampling an area of 40×40 μm. The VSD images were captured using a MiCAM ULTIMA CMOS camera (SciMedia, Costa Mesa, CA). The camera has a temporal resolution of 1,000 Hz and a spatial resolution of 100×100 pixels, with each pixel sampling an area of 50×50 μm. We performed ORIS imaging at 546 nm, an isosbestic wavelength of hemoglobin, to provide a measure of total hemoglobin (Hbt), which is equivalent to CBV if hematocrit remains constant and proportional to CBF [22, 23]. This specific isosbestic wavelength, of which there are several, was chosen to minimize overlap with the VSD absorption. ORIS imaging of oximetry could not be achieved with our setup because these wavelengths are mainly absorbed by the VSD dye rather than blood cells. A 570-nm dichroic mirror is the key component in this setup, as it separates the ORIS and VSD wavelengths into two different cameras (Fig. 1b). Since this technique uses filters and dichroic mirrors to separate wavelengths distributed at different spatial locations on the spectrum, we describe this as a "spatial separation" technique.

Correspondingly a "temporal separation" technique is also available. Through development of LED based illumination techniques, fast switching between multiple wavelengths can be achieved and overlapping wavelengths can be separated in the temporal domain. Bouchard and colleagues first applied this technique during simultaneous imaging of ORIS (530 nm) and calcium signals (Oregon Green 488 BAPTA-1 AM, illumination at 470 nm and signal at >510 nm) [20]. Figure 1c demonstrates the "temporal separation" setup for simultaneous ORIS and calcium imaging. Two LEDs, 470 and 530 nm (Thorlabs) were employed as the illumination source for calcium-sensitive dye and Hbt, respectively. The LEDs are coupled with a dichroic mirror and illumination directed using fiber-optic light guides. A 510-nm long-pass filter was placed before the camera (Dalsa camera in Imager 3001, Optical Imaging) to prevent the calcium illumination contamination, but permit both calcium dye fluorescence and Hbt signal. The ultra-fast multispectral switching between the two LEDs was time-locked to camera frames (Fig. 1c). A Master-9 was used to couple the frame indicator from the camera and the LEDs.

2.1 Animal Preparation

All experimental procedures were approved by the Weill Cornell Medical College Animal Care and Use Committee following NIH guidelines. Adult male Sprague–Dawley rats (250–380 g) were anesthetized with isoflurane at 70 % N_2:30 % O_2, 4 % induction and 1.5–2 % maintenance. Body temperature was maintained at 37 °C with a regulated heating blanket (Harvard Apparatus, Holliston, MA). The heart rate, pO_2, and $ETCO_2$ were carefully monitored with a small animal system Capnograph (Surgivet, Waukesha, WI) and were maintained stable throughout the experiment (heart rate: 250–300 pulse/min, $pO_2 > 90$ %, $ETCO_2 \sim 25–28$ mmHg). The head was fixed in a stereotaxic frame. A ~5 × 8-mm cranial window was opened over one or both hemispheres, between lambda and bregma, centered over the somatosensory cortex. For VSD imaging, the dura was carefully removed. A temporary well was built around the craniotomy window with Vaseline oil. The voltage-sensitive dye RH-1691 solution (1 mg/mL in 0.9 % NaCl solution), was applied to the exposed cortex for ~90 min. After staining, the cortex was washed with saline for ~10 min to remove unbound dye. For calcium imaging, convection-enhanced delivery was employed to stain a large area of cortex with calcium dye (Oregon Green 488 BAPTA-1 AM, 100 mg in 5-μL of 20 % Pluronic F-127 in DMSO (w/v) and then in 50-μL of artificial cerebrospinal fluid (ACSF)). The dye was injected via a glass electrode controlled by a micro pump (UMP-3). Eight microlitres of dye was continuously injected 1 mm below the surface at a speed of 100 μL/min to replace the cerebrospinal fluid cerebral with dye solution. The animal will rest 1 h for the cell to load calcium dye. The exposed window was covered with silicon oil (12,500 centistoke) to preserve the moisture of the cortex.

To record spontaneous activity, a glass microelectrode (2–4 MΩ) filled with saline was positioned ~500 μm deep from the surface. Seizure-like discharges were induced by injecting 4-AP (Sigma, 15 mM, 0.5 μL) 300–500 μm into the cortex, through a glass microelectrode using a Nanoject II injector (Drummond Scientific, Broomall, PA). A second glass microelectrode (2–4 MΩ) filled with saline was positioned <0.5 mm from the 4-AP electrode to record the LFP. The LFP was amplified and band-pass filtered (1–500 Hz) using a DAB-S system and digitized at 1,000 Hz by a CED Power 1401. The LFP data were recorded with a PC running Spike2 software (Cambridge Electronic Design, Cambridge, UK).

3 Results

3.1 Simultaneous Total Hemoglobin and VSD Imaging with "Spatial Separation"

We have applied the "spatial separation" techniques with simultaneous VSD/ORIS imaging during 4-AP induced ictal events. Figure 2a shows the images of the exposed cortex stained with RH-1691. The field of view of VSD is smaller than Hbt and the distribution of the light intensity between both signals is different. This arises due to the VSD signal being related to the staining of the dye and ORIS being associated with the reflected and absorbed intrinsic signal. The time course of the VSD recorded from a small region of interest (ROI) is similar to that of the LFP recorded in the same location. The slow ORIS signal cannot reflect the fast neuronal dynamics and shows a smooth curve (Fig. 2b). However, while the ORIS spreads smoothly and homogeneously, the VSD signal shows marked variability in the spatial spread of each wave (Fig. 2c). The two signals are mostly, though not precisely,

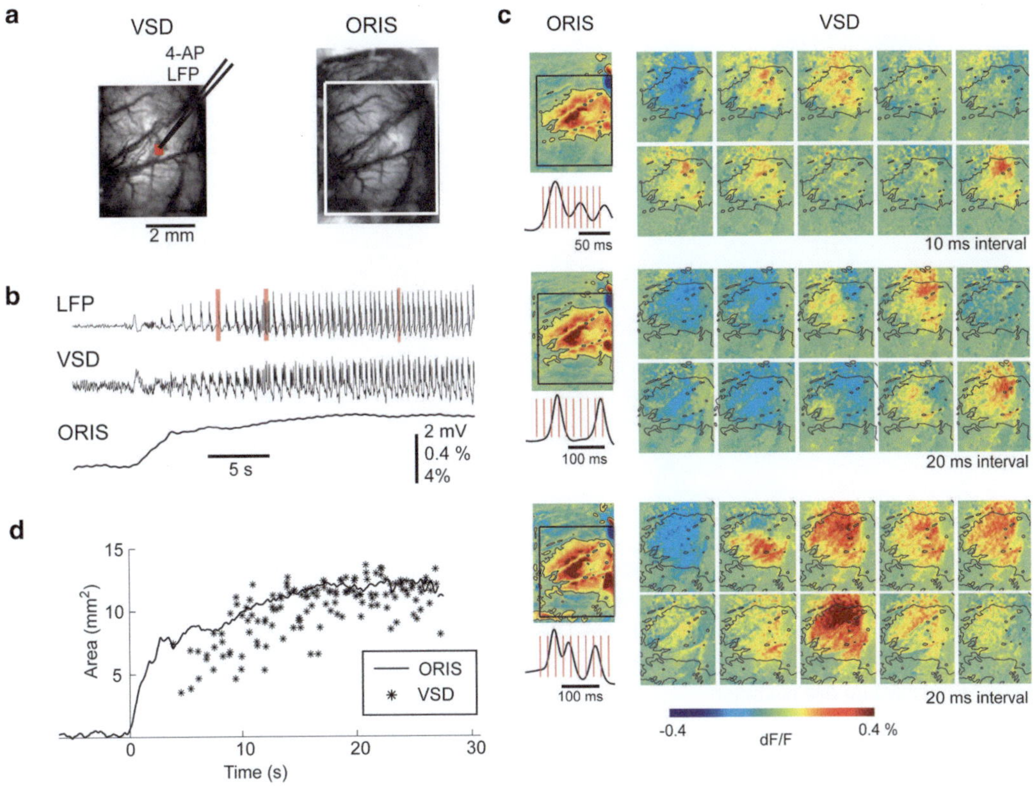

Fig. 2 Simultaneous VSD and ORIS of ictal event with "spatial separation" system. (**a**) The field of view imaged with VSD and ORIS. A *small red box* on the VSD image indicates an ROI from which the VSD and ORIS time-courses are derived. The *large red box* on the ORIS image indicates the area recorded with the VSD camera. (**b**) The simultaneously recorded LFP, VSD, and IOS traces from the ROI. Three *red shaded rectangles* indicate time periods of data shown in (**c**). (**c**) The spatial propagation of IOS and VSD signal at each of the three time periods shown in (**b**). *Vertical red lines* in magnified VSD trace (*inset*, corresponding to time-course from *red rectangles* in (**b**)) indicate the time points for the VSD images on the *right*. The IOS data is taken from the same time period. The spatial extent of the IOS signal is thresholded and overlaid on the VSD data as a *black outline*. Note, the IOS and VSD areas do not perfectly overlap. (**d**) Progressive change in the area of the IOS compared with the VSD data in a single seizure. The IOS data is represented as a smooth increase in area that subsequently plateaus. The maximal area of each VSD spike is indicated with an *asterisk*. Note that during the initial evolution of the seizure, the VSD area is smaller than the IOS area but that the correspondence improves as the seizure evolves. However, while the IOS area increases smoothly over time, the VSD areas are extremely variable. Although the IOS data blurs the spatial inhomogeneity of the VSD data, the IOS signal is a reasonable approximation of the maximal spread of membrane potential increases during later seizure evolution

overlapping at their lateral extent, and both signals spread to slightly different areas of cortex at their margins (Fig. 2c, d). These results indicate that the hemodynamic signal, although highly spatially correlated to the maximal area of membrane potential increase during the middle stages of seizure evolution, is a distinct and independent event with unique regulatory mechanisms.

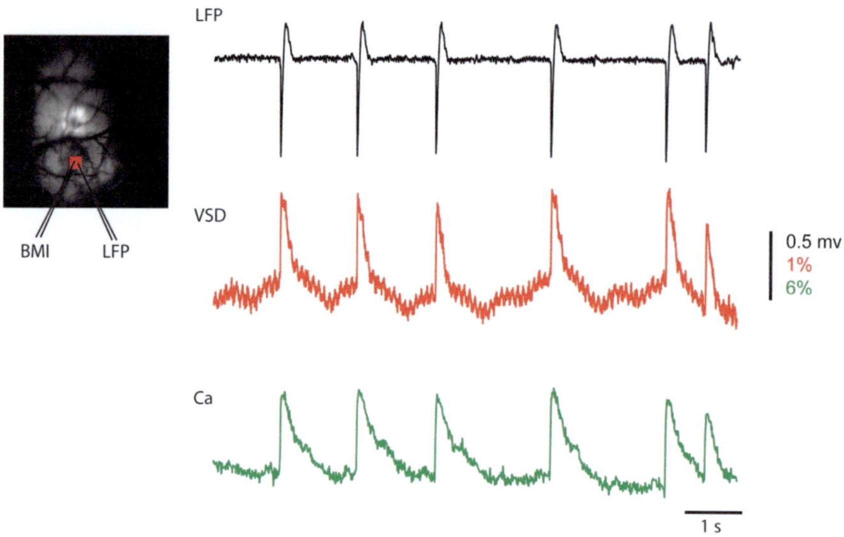

Fig. 3 Simultaneous VSD and calcium imaging during interictal events. The *left* shows the field of view. The bicuculline application electrode and LFP electrode are placed close to each other. A *small red box* over on the image indicates a ROI covering both electrodes. LFP, VSD, and calcium dye signals derived from the ROI are shown on the *right*. Note that the VSD and calcium waveforms are highly correlated with LFP although the decay of the calcium dye is longer

3.2 Simultaneous Total Hemoglobin and Calcium Imaging with "Temporal Separation"

The similarity between VSD and calcium signal:
Although calcium dyes have been widely used to study neuronal activity with single cell resolution, there is less data on wide field imaging of bulk loaded cortex where the calcium signal represents neuronal activity at a lower resolution of a few hundred microns and a few thousand neurons. We first compared the calcium signal with the VSD signal using simultaneous calcium and VSD imaging (Fig. 3). During bicucculine-induced interictal events, the VSD and calcium waveforms are highly correlated during individual spikes, in terms of both onset time and amplitude. Our data suggests that the calcium signal reflects neuronal electrical activity at a spatial resolution of the order of a hundred microns (Fig. 3).

The sensitivity of calcium dye and ORIS imaging:
Figure 4 demonstrates the application of "temporal separation" in simultaneous calcium and ORIS imaging during spontaneous cortical activity. Figure 4a shows images of the exposed cortex injected with fluorescent calcium dye (Oregon Green 488 BAPTA-1, AM). The ORIS reflectance (530 nm) can be recorded from both the cortex and the surrounding bone but the fluorescence from the calcium dye (510 nm) can only be recorded from the cortex where the dye has stained. The waveform of the calcium dye closely reflects the LFP recording, indicating that the calcium change is related to the electrical activity (Fig. 4b). However, the ORIS signal does not directly reflect the dynamics of the LFP, because the

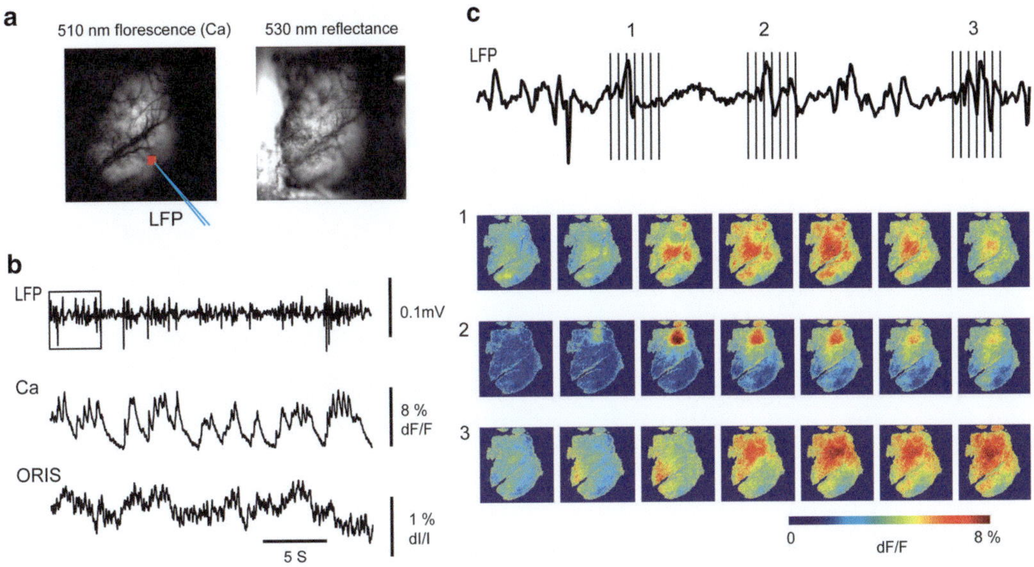

Fig. 4 Simultaneous calcium and ORIS of spontaneous activity with "temporal separation" system. (**a**) The field of view under calcium imaging and ORIS. A *small red box* on the calcium image indicates a ROI from which calcium and ORIS tracings are derived. Note that the bone is bright under ORIS but not under calcium imaging. (**b**) The simultaneously recorded LFP, calcium, and ORIS traces from ROI. A *black box* over the LFP indicates a section magnified in (**c**). Note that LFPs clearly share similar dynamics with the calcium waveform but not that of the ORIS waveform. (**c**) The spatial propagation of the calcium signal during three spontaneous events. *Vertical bars* over the LFP indicate the time points of calcium images shown below. Note that the initiation site and spatial propagation are different among each of the three events. The low signal-to-noise ratio of the calcium dye permits resolution of single events without imaging and provide the ability for creating an optical electrocorticography signal

amplitude of the spontaneous events is not adequately large to be recorded from the intrinsic signal, given the low signal-to noise-ratio, without averaging over multiple events. Since spontaneous brain activity is not repetitive but rather stochastic and variable, events cannot be averaged. Since the signal-to-noise ratio of the calcium dyes is much larger, individual spontaneous fluctuations in the LFP signal can be measured reliably with the calcium signal (Fig. 4c). These results suggest that the calcium dye imaging, with its higher signal-to-noise ratio is adequately sensitive to resolve spontaneous fluctuation in the surface electrocorticography (ECoG) and can be used as optical ECoG.

Application of simultaneous calcium dye and ORIS imaging on acute seizure:
Simultaneous calcium dye imaging and ORIS can also be applied to study neurovascular coupling dynamics during acute seizures. Unlike the VSD signal during a seizure (Fig. 2), which only has fast signal reflecting rapid fluctuations in membrane potential, the calcium signal clearly has two separate components, a slow signal and a fast signal. Using a simple threshold of 1 Hz, it is possible to

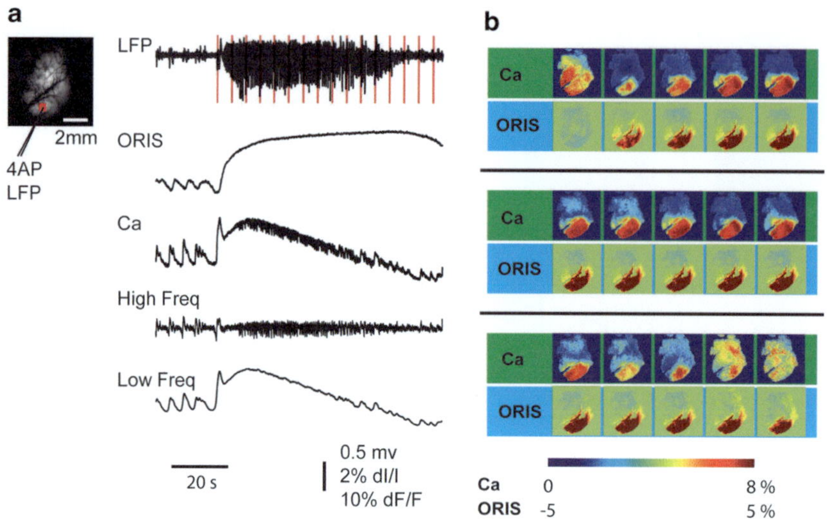

Fig. 5 Simultaneous calcium and ORIS of ictal activity with "temporal separation" system. (**a**) The waveform of an ictal event. A *small red box* indicates a ROI over the 4-AP/LFP electrode, from which the calcium and ORIS tracings are derived. The ORIS show a smooth waveform that does not reflect the dynamics of the LFP. The calcium waveform has two components: high-frequency activity riding on a low-frequency wave. The high-frequency component (>1 Hz), which shows a waveform similar to the LFP, may thus reflect neuronal activity, with the low-frequency component (<1 Hz) possibly associated with the glial activation. The *red bars* over the LFP trace indicate the time points for the images in (**b**). (**b**) The spatial propagation of calcium and ORIS signal. The first initiating spike propagates to a larger area than the subsequent spikes as the seizure propagates. This is seen in the calcium signal but not recorded in the hemodynamic ORIS signal. As the seizure evolves, the high-frequency calcium signal spread to a similar degree as the hemodynamic signal, however, the low-frequency calcium signal propagates further, thus demonstrating a reduced neurovascular drive

create separate maps of slow versus fast calcium transients (Fig. 5). Using simultaneous ORIS of CBV, in this example, it can be seen that the rapid transients have a higher spatial correlation with the CBV signal than the slower transients, which propagate to a larger area of cortex. Using this technique, it is possible to examine neurovascular coupling mechanisms for synaptic activity (fast activity) compared with glial waves and intracellular calcium accumulation (slower activity) simultaneously.

4 Discussion

We have characterized the "spatial separation" and "temporal separation" techniques for simultaneous multi-wavelength imaging of neuronal and hemodynamic activity during both spontaneous and pathological brain activity. These techniques can be allied to the study of neurovascular coupling mechanisms in a variety of brain states in both the temporal and the spatial domain.

4.1 Choice of Optimal Setup

An important question is how to choose the optimal set-up, which of course depends on the question being posed by the investigator. For the illumination device, we recommend the use of LEDs over incandescent (i.e., bulb based) illumination for a number of reasons. Foremost, incandescent light may generate significant heat during operation. Convection may occur in the air surrounding the light bulb and affect the stability of the illumination. Housing of the illumination source would therefore be required to prevent such convection. In contrast, LEDs operate at a much lower temperature and aluminum heat sinks are sufficient to keep these cool during operation. The size of a heat sink is much smaller than a light guard and allows more flexibility in set-up design. Moreover, LEDs have ultra-fast time responses to electrical current. LEDs offer strobe frequencies up to ~10 kHz with stable illumination power, which facilitates the simultaneous recording of multiple wavelengths with high-temporal resolution. The temporal resolution is then limited by the acquisition rate of the imaging devices. Wavelength switching with incandescent light can only be achieved using filter wheels, which may in turn generate mechanical-based artifacts during operation. Thus, LEDs are ideal illumination sources that can be applied to both spatial and temporal separation, with incandescent light being primarily suitable for the former. How does one decide between using spatial or temporal separation? We prefer the latter over the former as cost is reduced with fewer cameras required. Although we have only demonstrated "temporal separation" in simultaneous imaging of calcium and one ORIS wavelength, more wavelengths could be added to this design without difficulty (e.g., a second ORIS wavelength for oximetry at 610 nm). The spatial separation should be used when the temporal separation is not possible.

4.2 Difference Between Voltage and Calcium Dye

VSD signals represent the average membrane voltage changes in all stained cells, including neurons, glia, and blood vessels. However, the temporal response of glia and vessels is very slow compared to that of neurons. The VSD signal waveform is highly correlated, both temporally and morphologically, with LFPs and the temporal rapidity clearly indicates neuronal population activity. The magnitude of the VSD signal is strongly related to the cell membrane area, and hence the signal primarily reflects subthreshold synaptic activity in superficial cortical layers, and to less degree suprathreshold action potentials, but cannot differentiate between the two events [24]. Similarly, in vivo calcium dyes also reflect various aspects of cortical dynamics, including neuronal network activation [16] and glial and metabolic signaling [17]. When imaged at low-spatial resolution, the calcium signal mainly reflects strong synaptic activation in the superficial cortical layers, because the fluorescence collection attenuates markedly as a function of distance from the recording plane [19, 25]. Thus, VSD and Calcium reflect very similar aspects of cortical activity as evidenced by analogous waveforms when recorded simultaneously during low-frequency activity (Fig. 3). During long duration high-frequency activity, like ictal events, a profound slow component was observed with calcium dye but not VSD. The contribution of glial activation to VSD signals is thought to be small [16, 26] possibly due to glial responses being weak (depolarized only 1–7 mV in response to visual stimulation) [27]. However, the glial contribution to calcium signals is significant, as shown by Schulz and colleagues who demonstrated glial activation to produce prolonged low-frequency activity in calcium signals during electrical forepaw stimulation [19], similar to that observed here. Therefore, calcium-derived data is a mixture of high-frequency neuronal activity and low-frequency glial activity. A high-pass filter (1 Hz) is sufficient to remove the glial component (Fig. 5), thus calcium dyes potentially offer more data for investigating neurovascular coupling since the glial waves may be a major component of the vascular signal and thus to understand its etiology, one requires a mechanism for measuring neuronal as well as glial activity.

References

1. Bonhoeffer T, Grinvald A (1996) Optical imaging based on intrinsic signals. The methodology. In: Toga AW, Mazziota JC (eds) Brain mapping. The methods. Academic Press, San Diego, pp 55–99
2. Frostig RD, Lieke EE, Ts'o DY, Grinvald A (1990) Cortical functional architecture and local coupling between neuronal activity and the microcirculation revealed by in vivo high-resolution optical imaging of intrinsic signals. Proc Natl Acad Sci U S A 87(16):6082–6086
3. Grinvald A, Lieke E, Frostig RD, Gilbert CD, Wiesel TN (1986) Functional architecture of cortex revealed by optical imaging of intrinsic signals. Nature 324(6095):361–364
4. Grinvald A, Shoham D, Shmuel A, Glaser D, Vanzetta I, Shtoyerman E, Slovin H, Sterkin A, Wijnbergen C, Hildesheim R, Arieli A (1999) In-vivo optical imaging of cortical architecture and dynamics. In: Windhorst U, Johansson H (eds) Modern techniques in neuroscience research. Springer, New York
5. Narayan SM, Santori EM, Blood AJ, Burton JS, Toga AW (1994) Imaging optical reflectance in rodent barrel and forelimb sensory cortex. Neuroimage 1(3):181–190. doi:10.1006/nimg.1994.1003, pii: S1053-8119(84)71003-2
6. Narayan SM, Santori EM, Toga AW (1994) Mapping functional activity in rodent cortex using optical intrinsic signals. Cereb Cortex 4(2):195–204
7. Cohen LB, Salzberg BM, Grinvald A (1978) Optical methods for monitoring neuron activity. Annu Rev Neurosci 1:171–182
8. Ross WN, Salzberg BM, Cohen LB, Davila HV (1974) A large change in dye absorption during the action potential. Biophys J 14(12):983–986. doi:10.1016/S0006-3495(74)85963-1, pii: S0006-3495(74)85963-1
9. Salzberg BM, Davila HV, Cohen LB (1973) Optical recording of impulses in individual neurones of an invertebrate central nervous system. Nature 246(5434):508–509
10. Salzberg BM, Grinvald A, Cohen LB, Davila HV, Ross WN (1977) Optical recording of neuronal activity in an invertebrate central nervous system: simultaneous monitoring of several neurons. J Neurophysiol 40(6):1281–1291
11. Grinvald A, Ross WN, Farber I (1981) Simultaneous optical measurements of electrical activity from multiple sites on processes of cultured neurons. Proc Natl Acad Sci U S A 78(5):3245–3249
12. Grinvald A, Manker A, Segal M (1982) Visualization of the spread of electrical activity in rat hippocampal slices by voltage-sensitive optical probes. J Physiol 333:269–291
13. Shoham D, Glaser DE, Arieli A, Kenet T, Wijnbergen C, Toledo Y, Hildesheim R, Grinvald A (1999) Imaging cortical dynamics at high spatial and temporal resolution with

novel blue voltage-sensitive dyes. Neuron 24(4):791–802

14. Grinvald A, Hildesheim R (2004) VSDI: a new era in functional imaging of cortical dynamics. Nat Rev Neurosci 5(11):874–885
15. Kee MZ, Wuskell JP, Loew LM, Augustine GJ, Sekino Y (2008) Imaging activity of neuronal populations with new long-wavelength voltage-sensitive dyes. Brain Cell Biol 36(5–6):157–172. doi:10.1007/s11068-009-9039-x
16. Grewe BF, Helmchen F (2009) Optical probing of neuronal ensemble activity. Curr Opin Neurobiol 19(5):520–529. doi:10.1016/j.conb.2009.09.003, pii: S0959-4388(09)00125-1
17. Nedergaard M, Rodriguez JJ, Verkhratsky A (2010) Glial calcium and diseases of the nervous system. Cell Calcium 47(2):140–149. doi:10.1016/j.ceca.2009.11.010, pii: S0143-4160(09)00191-2
18. Michel K, Michaelis M, Mazzuoli G, Mueller K, Vanden Berghe P, Schemann M (2011) Fast calcium and voltage-sensitive dye imaging in enteric neurones reveal calcium peaks associated with single action potential discharge. J Physiol 589(pt 24):5941–5947. doi:10.1113/jphysiol.2011.219550, pii: jphysiol.2011.219550
19. Schulz K, Sydekum E, Krueppel R, Engelbrecht CJ, Schlegel F, Schroter A, Rudin M, Helmchen F (2012) Simultaneous BOLD fMRI and fiber-optic calcium recording in rat neocortex. Nat Methods 9(6):597–602. doi:10.1038/nmeth.2013, pii: nmeth.2013
20. Bouchard MB, Chen BR, Burgess SA, Hillman EM (2009) Ultra-fast multispectral optical imaging of cortical oxygenation, blood flow, and intracellular calcium dynamics. Opt Express 17(18):15670–15678. doi:10.1364/OE.17.015670, pii: 185149
21. Ma H, Zhao M, Schwartz TH (2012) Dynamic neurovascular coupling and uncoupling during ictal onset, propagation, and termination revealed by simultaneous in vivo optical imaging of neural activity and local blood volume. Cereb Cortex. doi:10.1093/cercor/bhs079, pii: bhs079
22. Kennerley AJ, Berwick J, Martindale J, Johnston D, Papadakis N, Mayhew JE (2005) Concurrent fMRI and optical measures for the investigation of the hemodynamic response function. Magn Reson Med 54(2):354–365
23. Sheth SA, Nemoto M, Guiou M, Walker M, Pouratian N, Toga AW (2004) Linear and nonlinear relationships between neuronal activity, oxygen metabolism, and hemodynamic responses. Neuron 42(2):347–355
24. Slovin H, Arieli A, Hildesheim R, Grinvald A (2002) Long-term voltage-sensitive dye imaging reveals cortical dynamics in behaving monkeys. J Neurophysiol 88(6):3421–3438
25. Engelbrecht CJ, Gobel W, Helmchen F (2009) Enhanced fluorescence signal in nonlinear microscopy through supplementary fiber-optic light collection. Opt Express 17(8):6421–6435, pii: 178935
26. Chemla S, Chavane F (2010) Voltage-sensitive dye imaging: technique review and models. J Physiol Paris 104(1–2):40–50. doi:10.1016/j.jphysparis.2009.11.009, pii: S0928-4257(09)00091-6
27. Kelly JP, Van Essen DC (1974) Cell structure and function in the visual cortex of the cat. J Physiol 238(3):515–547

Chapter 13

Noninvasive Laser-Induced Photoacoustic Tomography for Functional Imaging

Huabei Jiang

Abstract

While brain imaging and electrophysiology play a central role in neuroscience research and in the evaluation of neurological disorders, a single noninvasive modality that offers both high spatial and temporal resolution is currently not available. In this chapter, we describe an emerging optical imaging technique called photoacoustic tomography (PAT) which can noninvasively track brain hemodynamics with both high spatial and temporal resolution, and at a depth that is clinically relevant. We present in detail photoacoustic reconstruction methods and imaging systems. Using a rodent model of acute focal epilepsy, we demonstrate the noninvasive yet whole surface and depth capabilities of the PAT system that allow to actually see what is happening during ictogenesis in terms of seizure onset and spread. Both seizure onset and propagation are tomographically detected at a spatial resolution of 150 μm and a temporal resolution of 300 ms. Dynamical changes of vasculature during epileptiform events are also detected with high spatiotemporal resolution. Together, these findings suggest that PAT represents a powerful tool for noninvasively mapping seizure onset and propagation patterns, and the "functional" connectivity within epileptic brain networks.

Key words Functional imaging, Neurovascular coupling, Epilepsy, Photoacoustic tomography, Imaging methods, Reconstruction algorithms

1 Introduction

It has been known that optical contrast is highly sensitive to neuronal activity [1, 2]. The optical contrast is largely due to the changes in blood volume and blood oxygenation. When used intraoperatively, Haglund et al. have shown that optical imaging is able to provide dynamic (surface) maps of functional and epileptiform activity in rat, monkey, and human somatosensory cortices with high spatial and temporal resolution [3, 4]. Recently near-infrared spectroscopy/topography (NIRS) has been used to obtain cerebral blood volume and blood oxygenation in epilepsy patients [5–7]. These studies have suggested that NIRS can be used to localize epileptic foci and even distinguish between different seizure types. However, the applications of these surface imaging/topography

Mingrui Zhao et al. (eds.), *Neurovascular Coupling Methods*, Neuromethods, vol. 88,
DOI 10.1007/978-1-4939-0724-3_13, © Springer Science+Business Media New York 2014

methods are limited because no depth information is available. Such information is important to accurately localize the epileptic foci. Depth information can be obtained by using tomographic reconstruction methods such as diffuse optical tomography (DOT) [8–11], but DOT has relatively low spatial resolution (2–4 mm) due to strong tissue scattering. Laser-induced photoacoustic tomography (PAT) retains the desired high optical contrast while providing much better spatial resolution (10–150 μm, depth-dependent, and adjustable with ultrasound frequency) than pure optical tomographic methods by detecting much less-scattering ultrasonic waves. Basically, PAT combines the advantages of both optical and ultrasound imaging in a single modality, yet overcomes the limitations associated with optical and ultrasound.

PAT is based on the photoacoustic (PA) effect, which was first described in 1880 by Alexander Graham Bell. In PAT, a short-pulsed laser source is typically used to irradiate the tissue of interest. The absorption of the laser pulses gives rise to a rapid temperature rise and subsequent thermoelastic expansion of the irradiated tissue volume. The pressure distribution induced by tissue expansion prompts acoustic wave propagation towards tissue surfaces where they are detected by one or an array of ultrasound transducers. The acquired acoustic data by ultrasound transducers are then used to recover the PA images qualitatively or quantitatively. PAT can capture tissue mechanical properties such as ultrasound velocity, optical properties such as absorption and scattering coefficients, and physiological or functional parameters such as deoxyhemoglobin (HbR), oxyhemoglobin (HbO_2), and water (H_2O).

In this chapter, we focus our attention to the fundamentals of PAT and its application for functional imaging of epilepsy. Specifically, we describe two typical reconstruction methods used in PAT, followed by the description of PAT imaging instrumentation. Experiments using tissue-like phantoms are presented for validation of PAT imaging methods. In vivo applications for functional imaging of epileptic foci and network and neurovascular coupling are detailed, and conclusions are made in the end of the chapter.

2 Methods

2.1 Reconstruction Algorithm

2.1.1 Delay-and-Sum Beam Forming Algorithm

To form the PA image of an object, an image reconstruction algorithm is required. A simple, yet very effective method is the delay-and-sum beam forming algorithm that is commonly used in radar signal processing [12]. The image expression in the case of near-field for photoacoustic imaging can be stated as

$$S^f(t) = \frac{\sum_i w_i^f S_i\left(t + \delta_i^f\right)}{\sum_i w_i^f} \quad (1)$$

where $S^f(t)$ is the image output at a particular focus point f, $S_i(t)$ is the time signal from the ith receiver, δ_i^f is the delay applied to this signal, and w_i^f is an amplitude weighting factor, which is used to enhance the beam shape, to reduce sidelobe effects, or to minimize the noise level. The summed signal is typically normalized to make the output independent of the actual set of transducers. This delay-and-sum algorithm has been tested and evaluated using considerable phantom and in vivo experiments [12, 13].

2.1.2 Finite Element-Based Algorithm

While the delay-and-sum algorithm described above is simple to implement, it does not account for the diffraction effect that is essential to the PA waves. In particular, a major assumption has to be made in this and other linear algorithms seen in the literature (e.g., backprojection [14] or filtered backprojection [15]) that biological tissues are acoustically homogeneous which may not be true in reality. This assumption will certainly affect the accuracy in image reconstruction. In addition, acoustic properties cannot be reconstructed because of this assumption. Importantly, functional parameters such as HbR, HbO_2, and H_2O cannot be obtained with these linear methods.

To overcome the above-mentioned limitations, we have developed a nonlinear reconstruction algorithm, which is based on the finite element (FE) solution to the full PA wave equation without the homogeneous acoustic property assumption made thus far in the literature [16, 17]. Finite element method (FEM) has been a powerful numerical method for solving the Helmholtz-type equation, because of its computational efficiency and unrivaled ability to accommodate tissue heterogeneity and geometrical irregularity as well as allow complex boundary conditions and source representations. The FE-based reconstruction algorithm is implemented based on a dual meshing method, which allows us to use a large mesh for accurate forward solution of the PA wave equation subject to the well-known radiation or absorbing boundary conditions, while a much smaller mesh is used for the inverse solution. This reconstruction approach in PAT is an iterative Newton method with combined Marquardt and Tikhonov regularizations that can provide stable inverse solutions. The approach uses the hybrid regularizations-based Newton method to update an initial optical and acoustic property distribution iteratively in order to minimize an object function composed of a weighted sum of the squared difference between computed and measured data. In addition, the adjoint sensitivity method has also been incorporated in the algorithm, which is able to reduce the computational cost dramatically for calculating the Jacobian matrix involved in the nonlinear inverse procedures. Together with an iterative Newton method, this nonlinear algorithm is able to precisely solve the Helmholtz wave equation and fulfill reliable inverse computation for an arbitrary measurement configuration.

Here we detail the FE-based algorithm. The PA wave equation in frequency domain is written if a heterogeneous acoustic field is considered,

$$\nabla^2 P(r,\omega) + k_0^2 (1+O) P(r,\omega) = ik_0 v_0 \beta \Psi(r) / C_{\mathrm{p}} \tag{2}$$

where P is the pressure wave in frequency domain; $k_0 = \omega / v_0$ is the wave number described by the angular frequency, ω, and the speed of acoustic wave in a reference or coupling medium, v_0; $\Psi(r)$ is the absorbed optical energy density that is the product of optical absorption coefficient, μ_{a} and optical fluence or photon density; and O is a coefficient that depends on both acoustic speed and attenuation as follows:

$$O = v_0^2 / v^2 - 1 + iAv_0 / k_0 v^2 \tag{3}$$

where v is the speed of acoustic wave in the scattering medium/tissue and A is the acoustic attenuation coefficient. Expanding acoustic pressure, P, as the sum of coefficients multiplied by a set of basis functions, ψ_j: $P = \sum P_j \psi_j$, the finite element discretization of the Helmholtz wave Eq. 2 can be written as

$$\sum_{j=1}^{N} P_j \left[\nabla \psi_j \cdot \nabla \psi_i - k_0^2 (1+O) \psi_j \psi_i - \oint \left(\eta \psi_j + \gamma \frac{\partial^2 \psi_j}{\partial \varphi^2} \right) \psi_i ds \right] \tag{4}$$
$$= -ik_0 c_0 \beta \Psi / C_{\mathrm{P}} \psi_i$$

where the following second-order absorbing boundary conditions have been applied [18]:

$$\nabla P \cdot \hat{n} = \eta P + \gamma \frac{\partial^2 P}{\partial \varphi^2} \tag{5}$$

where $\eta = (-ik_0 - 3/2\rho + i3/8k_0\rho^2)/(1 - i/k_0\rho)$ and $\gamma = (-i/2k_0\rho^2)/(1 - i/k_0\rho)$; N is the total number of nodes of the finite element mesh, $\langle(\cdot)\rangle$ indicates the integration over the problem domain, and $\oint$ expresses the integration over the boundary. In both the forward and inverse calculations, the unknown coefficients, O and Ψ, need to be separated into real (O_{R} and Ψ_{R}) and imaginary (O_{I} and Ψ_{I}) parts, both of which are expanded in a similar fashion to P as a sum of unknown parameters multiplied by a known spatially varying basis function. The matrix form of Eq. 4 is expressed as follows:

$$[A]\{P\} = \{B\} \tag{6}$$

where

$$A_{ij} = \langle \nabla \psi_j \cdot \nabla \psi_i \rangle - k_0^2 \langle \psi_j \psi_i \rangle - k_0^2 \left\langle \sum_k O_{\mathrm{R},k} \psi_k \psi_j \psi_i \right\rangle$$

$$- ik_0^2 \left\langle \sum_l O_{\mathrm{I},l} \psi_l \psi_j \psi_i \right\rangle - \oint (\eta \psi_j + \gamma \frac{\partial^2 \psi_j}{\partial \varphi^2}) \psi_i ds$$

$$B_i = -Ik_0 c_0 \beta \left\langle \sum_k \Psi_{\mathrm{R},k} \psi_k \psi_i \right\rangle / C_\mathrm{p} + k_0 c_0 \beta \left\langle \sum_l \Psi_{\mathrm{I},l} \psi_l \psi_i \right\rangle / C_\mathrm{p}$$

$$\{P\} = \{P_1, P_2, \ldots, P_N\}^\mathrm{T}$$

To form an image from a presumably uniform initial guess of the optical and acoustic property distribution, we use iterative Newton's method to update O_R, O_I, Ψ_R, and Ψ_I from their starting values. In this method, we Taylor expand P about an assumed (O_R, O_I, Ψ_R, Ψ_I) distribution, which is a perturbation away from some other distribution, $\left(\tilde{O}_\mathrm{R}, \tilde{O}_\mathrm{I}, \tilde{\Psi}_\mathrm{R}, \tilde{\Psi}_\mathrm{I}\right)$, such that a discrete set of P values can be expressed as

$$P\left(\tilde{O}_\mathrm{R}, \tilde{O}_\mathrm{I}, \tilde{\Psi}_\mathrm{R}, \tilde{\Psi}_\mathrm{I}\right) = P\left(O_\mathrm{R}, O_\mathrm{I}, \Psi_\mathrm{R}, \Psi_\mathrm{I}\right) + \frac{\partial P}{\partial O_\mathrm{R}} \Delta O_\mathrm{R} + \frac{\partial P}{\partial O_\mathrm{I}} \Delta O_I + \frac{\partial P}{\partial \Psi_\mathrm{R}} \Delta \Psi_\mathrm{R} + \frac{\partial P}{\partial \Psi_\mathrm{I}} \Delta \Psi_\mathrm{I} + \cdots \tag{7}$$

where $\Delta O_\mathrm{R} = \tilde{O}_\mathrm{R} - O_\mathrm{R}$, $\Delta O_\mathrm{I} = \tilde{O}_\mathrm{I} - O_\mathrm{I}$, $\Delta \Psi_\mathrm{R} = \tilde{\Psi}_\mathrm{R} - \Psi_\mathrm{R}$, and $\Delta \Psi_\mathrm{I} = \tilde{\Psi}_\mathrm{I} - \Psi_\mathrm{I}$. If the assumed optical and acoustic property distribution is close to the true profile, the left-hand side of Eq. 7 can be considered as true data (observed or measured), and the relationship can be truncated to yield

$$J \Delta \chi = \boldsymbol{P}^\mathrm{o} - \boldsymbol{P}^\mathrm{c} \tag{8}$$

where

$$J = \begin{bmatrix} \partial P_1 / \partial O_{\mathrm{R},1} & \cdots & \partial P_1 / \partial O_{\mathrm{R},K} & \partial P_1 / \partial O_{\mathrm{I},1} & \cdots & \partial P_1 / \partial O_{\mathrm{I},L} & \partial P_1 / \partial \Psi_{\mathrm{R},1} & \cdots & \partial P_1 / \partial \Psi_{\mathrm{R},K} & \partial P_1 / \partial \Psi_{\mathrm{I},1} & \cdots & \partial P_1 / \partial \Psi_{\mathrm{I},L} \\ \partial P_2 / \partial O_{\mathrm{R},1} & \cdots & \partial P_2 / \partial O_{\mathrm{R},K} & \partial P_2 / \partial O_{\mathrm{I},1} & \cdots & \partial P_2 / \partial O_{\mathrm{I},L} & \partial P_2 / \partial \Psi_{\mathrm{R},1} & \cdots & \partial P_2 / \partial \Psi_{\mathrm{R},K} & \partial P_2 / \partial \Psi_{\mathrm{I},1} & \cdots & \partial P_2 / \partial \Psi_{\mathrm{I},L} \\ \vdots & \ddots & \vdots & \vdots & \ddots & \vdots & \vdots & \ddots & \vdots & \vdots & \ddots & \vdots \\ \partial P_M / \partial O_{\mathrm{R},1} & \cdots & \partial P_M / \partial O_{\mathrm{R},K} & \partial P_M / \partial O_{\mathrm{I},1} & \cdots & \partial P_M / \partial O_{\mathrm{I},L} & \partial P_M / \partial \Psi_{\mathrm{R},1} & \cdots & \partial P_M / \partial \Psi_{\mathrm{R},K} & \partial P_\mathrm{M} / \partial \Psi_{\mathrm{I},1} & \cdots & \partial P_M / \partial \Psi_{\mathrm{I},L} \end{bmatrix}$$

$$\Delta \chi = \left(\Delta O_{\mathrm{R},1}, \Delta O_{\mathrm{R},2}, \ldots, \Delta O_{\mathrm{R},K}, \Delta O_{\mathrm{I},1}, \Delta O_{\mathrm{I},2}, \ldots, \Delta O_{\mathrm{I},L}, \Delta \Psi_{\mathrm{R},1}, \Delta \Psi_{\mathrm{R},2}, \ldots, \Delta \Psi_{\mathrm{R},K}, \Delta \Psi_{\mathrm{I},1}, \Delta \Psi_{\mathrm{I},2}, \ldots, \Delta \Psi_{\mathrm{I},L} \right)^\mathrm{T}$$

$$\boldsymbol{P}^\mathrm{o} = \left(P_1^\mathrm{o}, P_2^\mathrm{o}, \ldots, P_M^\mathrm{o} \right)^\mathrm{T}$$

$$\boldsymbol{P}^\mathrm{c} = \left(P_1^\mathrm{c}, P_2^\mathrm{c}, \ldots, P_M^\mathrm{c} \right)^\mathrm{T}$$

and P_i^o and P_i^c are measured and calculated based on the assumed (O_R, O_I, Ψ_R, Ψ_I) distribution data for $i = 1, 2, \ldots, M$ locations. $O_{\mathrm{R},k}$ ($k = 1, 2, \ldots, K$) and $O_{i,l}$ ($l = 1, 2, \ldots, L$) are the reconstruction parameters for the acoustic property profile, while $\Psi_{\mathrm{R},k}$ ($k = 1, 2, \ldots, K$) and

$\Psi_{I,l}$ (l = 1, 2, …, L) are the reconstruction parameters for the optical property profile. In order to realize an invertible system of equations for $\Delta\chi$, Eq. 8 is left multiplied by the transpose of $\mathbf{J}$ to produce

$$\left(J^{T}J+\lambda I\right)\Delta\chi = J^{T}\left(P^{o}-P^{c}\right) \tag{9}$$

where regularization schemes are invoked in order to stabilize the decomposition of $J^{T}J$; J is the identity matrix and λ is the regularization parameter determined by combined Marquardt and Tikhonov regularization schemes. We have found that when $\lambda = (P^{o} - P^{c}) \times \text{trace}\ [J^{T}J]$, the reconstruction algorithm generates best results for PAT image reconstruction. The process now involves determining the calculated scattering field data and Jacobian matrix using dual meshing scheme coupled with adjoint sensitivity method. The reconstruction algorithm here uses the hybrid regularization-based Newton method to update an initial (guess) optical and acoustic property distribution iteratively via the solution of Eqs. 6 and 9 so that an object function composed of a weighted sum of the squared difference between computed and measured acoustic pressures for all acoustic frequencies and optical wavelengths can be minimized.

2.2 PAT Instrumentation

The PAT imaging setups can be classified as single transducer scanning system [15, 16], transducer array system [19–21], backward-mode planar scanning system [22], and hybrid or other forms of scanning systems [23, 24]. The core elements required for a PAT imaging system are a pulsed laser and ultrasound transducer(s). Q-switched Nd:YAG or Ti:Sapphire laser in the visible and near-infrared (NIR) wavelengths is the commonly used light source for generating acoustic waves. The laser pulses (several to 10 ns width) can be delivered via free space or an optical fiber to tissues. The choices of the wavelength and pulse energy depend on the absorption of the object being imaged. The acoustic transducer commonly used in PAT is the piezoelectric sensor, such as PVDF film. Recently transparent planar Fabry-Perot polymer film ultrasound sensor and hydrophone are also used to accurately record PA signals [22].

A typical single transducer scanning PAT system is shown in Fig. 1. In this system, pulsed light from an Nd:YAG, Ti:Sapphire, or diode laser is coupled into the phantom via an optical subsystem to generate acoustic pressure wave. A transducer (1 MHz or higher central frequency) is used to receive the acoustic signals. The transducer and the phantom/tissues are immersed in a water tank. A rotary stage rotates the receiver relative to the center of the tank. One set of data is taken at 120/360 positions when the receiver is scanned circularly over 360°. The complex wavefield signal is first amplified by a preamplifier and then amplified further by a pulser/receiver. A data acquisition board converts it into digital one which is fed to a computer.

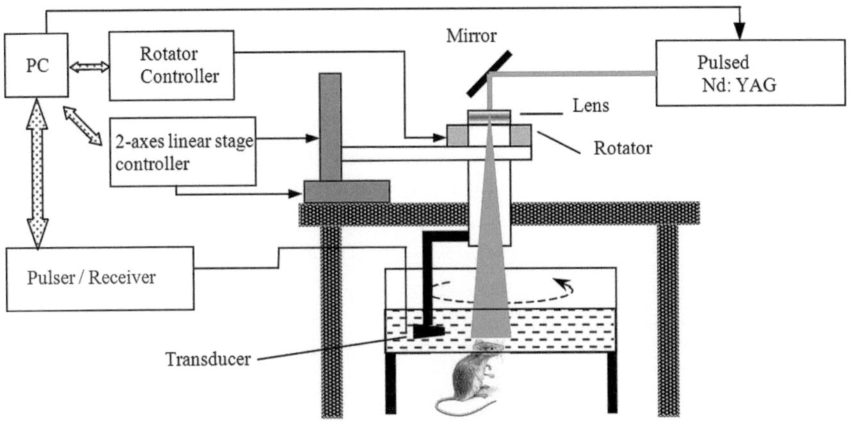

Fig. 1 Schematic of a typical PAT system. *BS* beam splitter, *PC* personal computer

The entire data acquisition is realized through C programming. In this system, data collection for a total of 120 measurements requires about 2 min.

In the typical PAT setup shown above, the mechanical scanning significantly reduces the speed of data acquisition. The use of a transducer array has clear acquisition time advantages over a mechanically scanned system, making real-time data acquisition possible. In addition, a transducer array-based system may eliminate the use of water tank and allow the direct contact of the array with the skin (of course ultrasound gel needs to be applied as coupling medium). The major drawback of an array-based system, however, is its significantly added cost. We will detail a transducer array-based PAT system in Sect. 3.2.

3 In Vivo Applications

3.1 Introduction

Epilepsy is a common, chronic, neurological disorder characterized by seizures. Three percent of people will be diagnosed with epilepsy at some time in their lives [25]. Indeed, approximately 50 million people worldwide have epilepsy, and 20–30 % of these patients are refractory to all forms of medical treatment [26]. Seizures are transient signs and symptoms of abnormal, excessive, or hypersynchronous neuronal activity in the brain [27]. In most cases, seizures are controlled, although not cured, with anticonvulsant medication. For those patients with medically intractable focal epilepsy, the best treatment option is resective brain surgery [28–30]. Although several factors can impact the success of epilepsy surgery, the primary reason of failure is the incomplete mapping of the local epilepsy network which results in incomplete resection of epileptogenic foci [31, 32]. Much of the attention on epilepsy surgery has been directed at identifying single neuronal populations.

This approach has, in many cases, led to failed surgical outcomes, because seizures typically involve groups of neurons interacting both locally and across several cortical and subcortical brain regions. A better understanding of the regional interactions occurring at the site of seizure onset and spread may provide important insights about the pathophysiology of seizures and aid with accurate brain mapping and resection of the epileptic focus.

Much of our understanding about focal seizure circuitry comes from electrophysiological recording methods. Although electrophysiology is currently the "gold standard" in mapping the epileptic focus, it is often inadequate to define the boundary of the epilepsy circuitry due to spatial sampling limitations and volume conduction. What we truly desire is a brain mapping modality, which could give a high-resolution real-time spatial and temporal "read out" of the dynamics of cortical processing and seizures. Previous attempts to track from moment to moment populations of neurons participating in an epileptiform event had not been possible with any technique or tool because of spatial and/or temporal sampling limitations. Here, PAT was employed to image seizures in an experimental acute bicuculline methiodide model of focal epilepsy. Bicuculline is a light-sensitive competitive antagonist of GABAA receptors that mimics focal epilepsy when applied to brain tissue. During focal application of bicuculline into the brain cortex, brains were imaged noninvasively with a transducer array-based PAT system that has three orders of magnitude higher temporal resolution and fourfold higher spatial resolution relative to a single transducer scanning PAT system. The high spatiotemporal properties of this PAT imaging tool allowed for identification of both bicuculline-elicited focal seizure onset and spread. Off-line, we employed measures of brain connectivity to further identify the functional anatomy implicated in focal cortical seizures. The high spatial and temporal sampling of the fast PAT system allowed for the complete functional mapping of an epileptiform event in vivo. It also allowed for mapping an epileptiform event at depths well below the cortex.

3.2 Results

3.2.1 PAT Imaging

Light from a Ti:Sapphire laser tunable (690–950 nm) was delivered through the skull to the brain through an optical fiber (Fig. 1a). The energy of each laser pulse was detected by a photodiode for calibration. A 192-element full-ring transducer array was used to capture the photoacoustic (PA) signals generated by the laser light. The 192-channel data acquisition system consisted of preamplifiers, secondary stage amplifiers (for optimizing the signal-to-noise ratio), and a 3:1 electronic multiplexer coupled with a 64-channel analog-to-digital converter. The estimated PAT data acquisition system speed was 0.33 s/frame. Data acquisition speed was limited in part by the 10 Hz laser repetition. The spatial resolution of the PAT system was inversely related to the bandwidth of the ultrasonic transducer. Each ultrasonic detector had a 5-MHz central frequency and a 70 % nominal bandwidth with a diameter of 6 mm (Blatek, Inc., PA, USA). This allowed for imaging objects of 50 μm in diameter with a spatial resolution of 150 μm.

Thirty minutes following the microelectrode placement, PAT imaging was performed to generate baseline tissue absorption maps, and visualize micro blood vessels and morphological cortical landmarks (Fig. 2b, c). Subsequently, rats ($n = 10$) received 10 μL of 1.9 mM bicuculline methiodide and 10 μL normal saline into the left [−1.0AP, 2.5ML, 2.4DV] and right [−1.0AP, 2.5ML, 0.4DV] parietal cortex, respectively, based on coordinates from a rat brain atlas [33]. Immediately following the focal infusions, PAT imaging was repeated to visualize changes in the injection site tissue absorption. An increase in tissue absorption was observed in the bicuculline methiodide cortical injection site and surrounding region (Fig. 3c, left), but not in the saline injection site or surrounding region (Fig. 3c, right). PAT scanning for subcortical changes was also performed. Slices 5, 7, and 9 (Fig. 3d) are the images obtained 3, 5, and 7 mm below the scalp, respectively. Figure 3e is the three-dimensional rendering of the epileptic foci obtained from different tomographic layers. Each image exhibited the variable patterns of seizure onset and propagation both at the cortical and subcortical regions. Seizures were confirmed by concomitant time-locked PAT/video-electroencephalography (Fig. 3a). Experiments were repeated for each animal at 2-h intervals.

3.2.2 Real-Time Monitoring of Epileptic Events

Epileptiform events were recorded with PAT from a focal region of interest (ROI) of ~2×3 mm (Fig. 4a). We observed a significant optical absorption change directly associated with deoxygenation at a wavelength of 755 nm (Fig. 4c). Furthermore, at the seizure onset at 1 min 6.267 s, the seizure focus measured 0.2053 mm^2 and increased to 0.5004 mm^2 as the electrographic seizure time series increased in frequency and amplitude (Fig. 4b). Corresponding rate changes (0.33 s) spike and wave discharges and PA images were observed in each experimental trial. The seizure onset dynamics captured photoacoustically are best appreciated in movies displaying how the seizure was generated and varied over time. PAT images suggest that in addition to ictal spread from the primary focus, homotopic foci are seen in the contralateral parietal cortex (Fig. $5a_{1-4}$). The PA signal in the contralateral foci was smaller in magnitude and delayed in time compared to the signal recorded from the primary seizure foci. We also observed a region of inverted optical signal immediately surrounding the seizure onset zone (Fig. 5b, left). We performed analysis of the optical signal at a distance of at least 2 mm away from the edge of the focus. Our results demonstrate that the inverse optical signal was inversely related to the optical signal recorded from the focus (Fig. 5b, right).

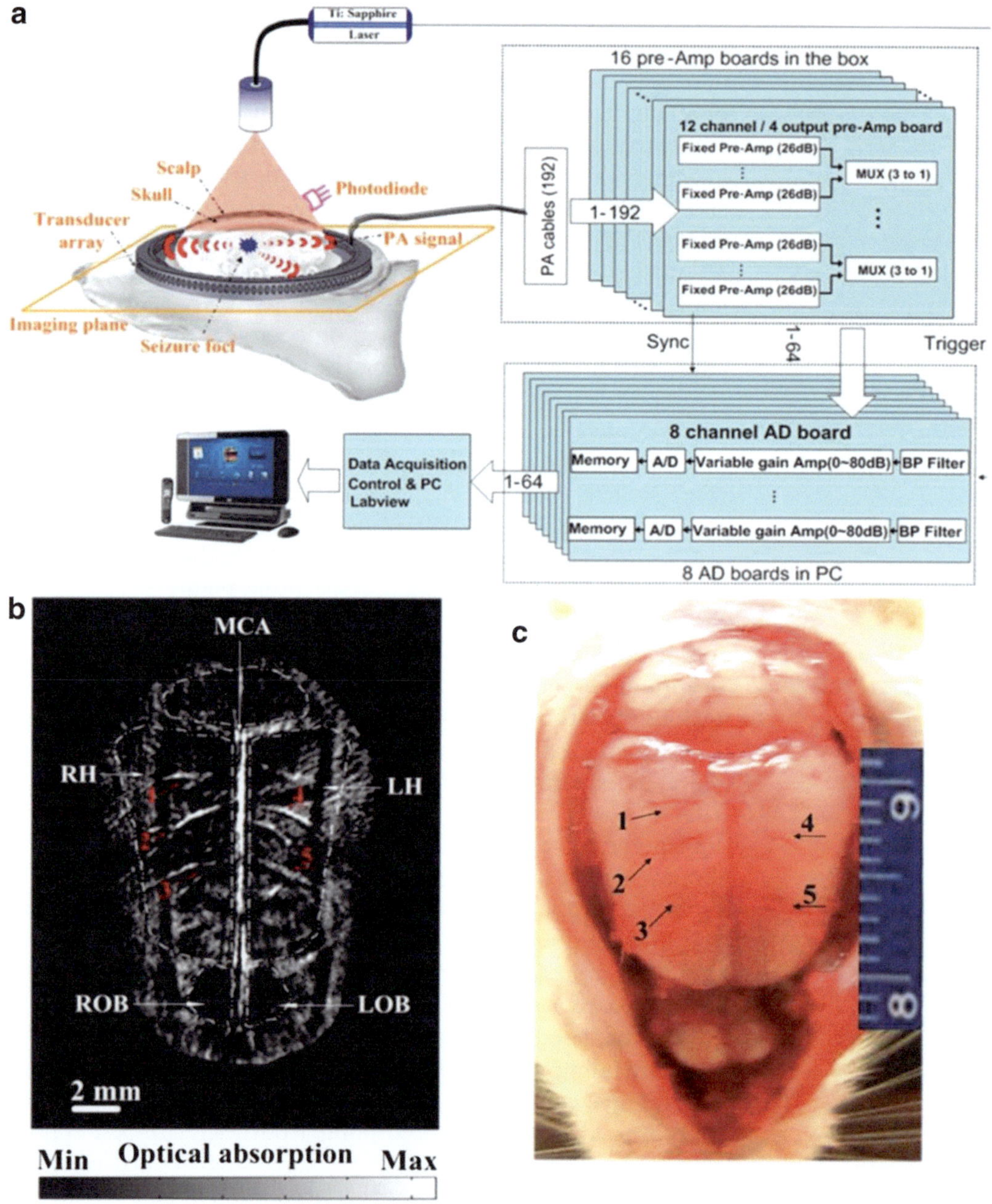

Fig. 2 Real-time PAT system for seizure dynamics. (**a**) Schematic of the real-time PAT system. A 192-element full-ring transducer array was used to capture the PA signal during seizure onset. Using 64-channel parallel data acquisition and 3:1 electronic multiplexing, imaging speed was up to 0.33 s/frame. The spatial resolution of this imaging system was ~150 μm. (**b**) Noninvasive PAT imaging of a rat brain in vivo with the skin and skull intact. *MCA* middle cerebral artery, *RH* right hemispheres, *LH* left hemispheres, *LOB* left olfactory bulbs, *ROB* right olfactory bulbs. (**c**) Open-skull photograph of the rat brain surface acquired after the PAT experiment. Numbers 1–5 indicate the corresponding blood vessels in the PAT image and rat brain photograph

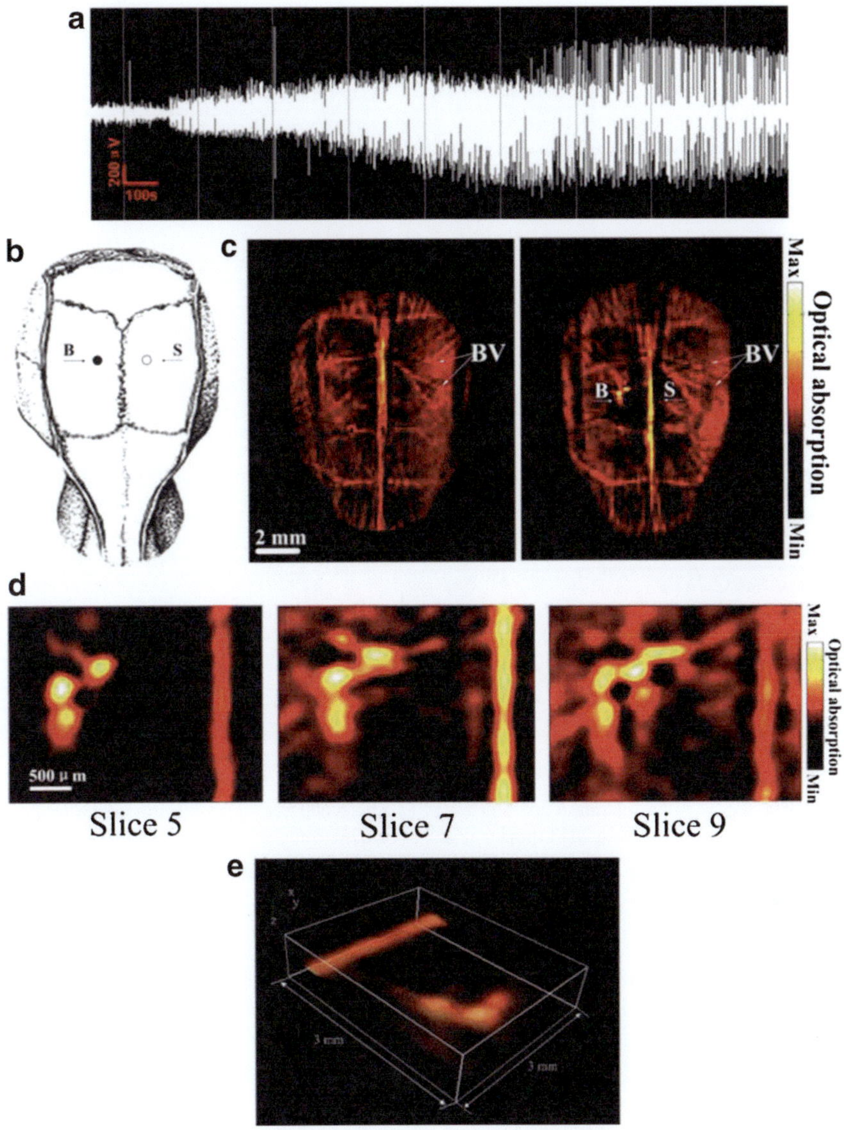

Fig. 3 Noninvasive epileptic foci localization. (**a**) EEG recordings of seizure onset after bicuculline methiodide injection. (**b**) Schematic showing the location of BMI and saline injection (control). *B* BMI injection, *S* saline injection. (**c**) Reconstructed PAT image before (*left*) and after (*right*) the bicuculline methiodide injection. Bicuculline methiodide and saline were injected into the left and right parietal neocortex, respectively. A significant increase of optical absorption is seen in the region of the bicuculline methiodide injection, while no absorption contrast is observed in the region of the saline injection relative to its surroundings. (**d**) Series of PAT images from three representative transverse planes parallel to the skin surface with 755 nm wavelength. *Slices 5, 7, and 9 are the images obtained 3, 5, and 7 mm under the skin, respectively*. The images show different spatial patterns at the seizure foci at different tomographic layers. (**e**) Three-dimensional rendering of the epileptic foci from different tomographic layers

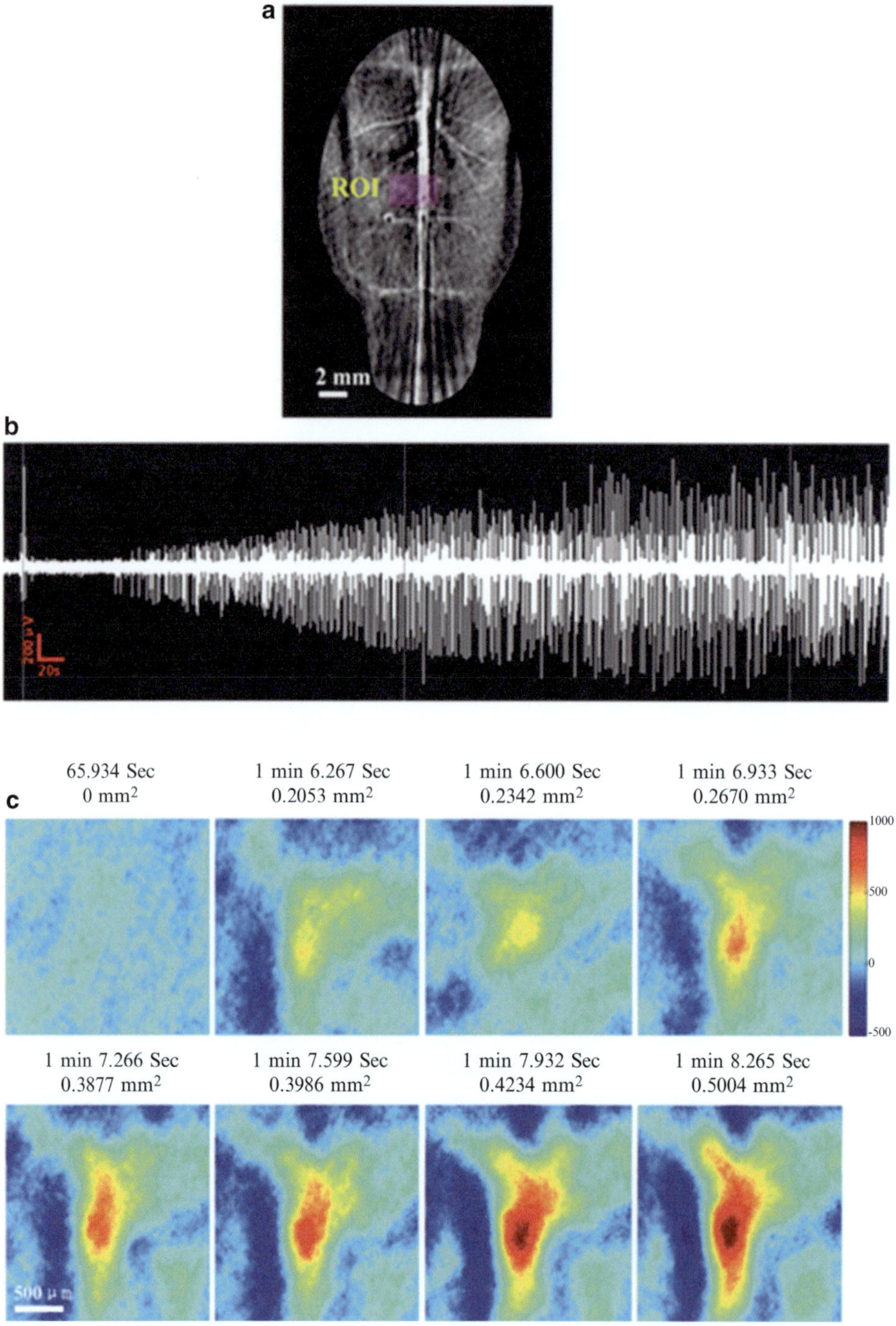

Fig. 4 Real-time monitoring of ictal onset. (**a**) A PAT image showing the cerebral vasculature and the position of BMI. *ROI* region of interest; scale bar: 2 mm. (**b**) EEG recordings showing the seizure onset about 1 min after BMI. (**c**) PAT images recording the ictal onset in real time. At 1 min 6.267 s the area of the focus was 0.2053 mm^2. The area of the focus then increased in size over the next 2 s to 0.5004 mm^2, corresponding to an increase in the amplitude of the EEG spikes. The area of the seizure focus was derived from the PAT images by thresholding to a pixel value one standard deviation above the pixel values from the area of the focus during the control conditions. Scale bar: 500 μm

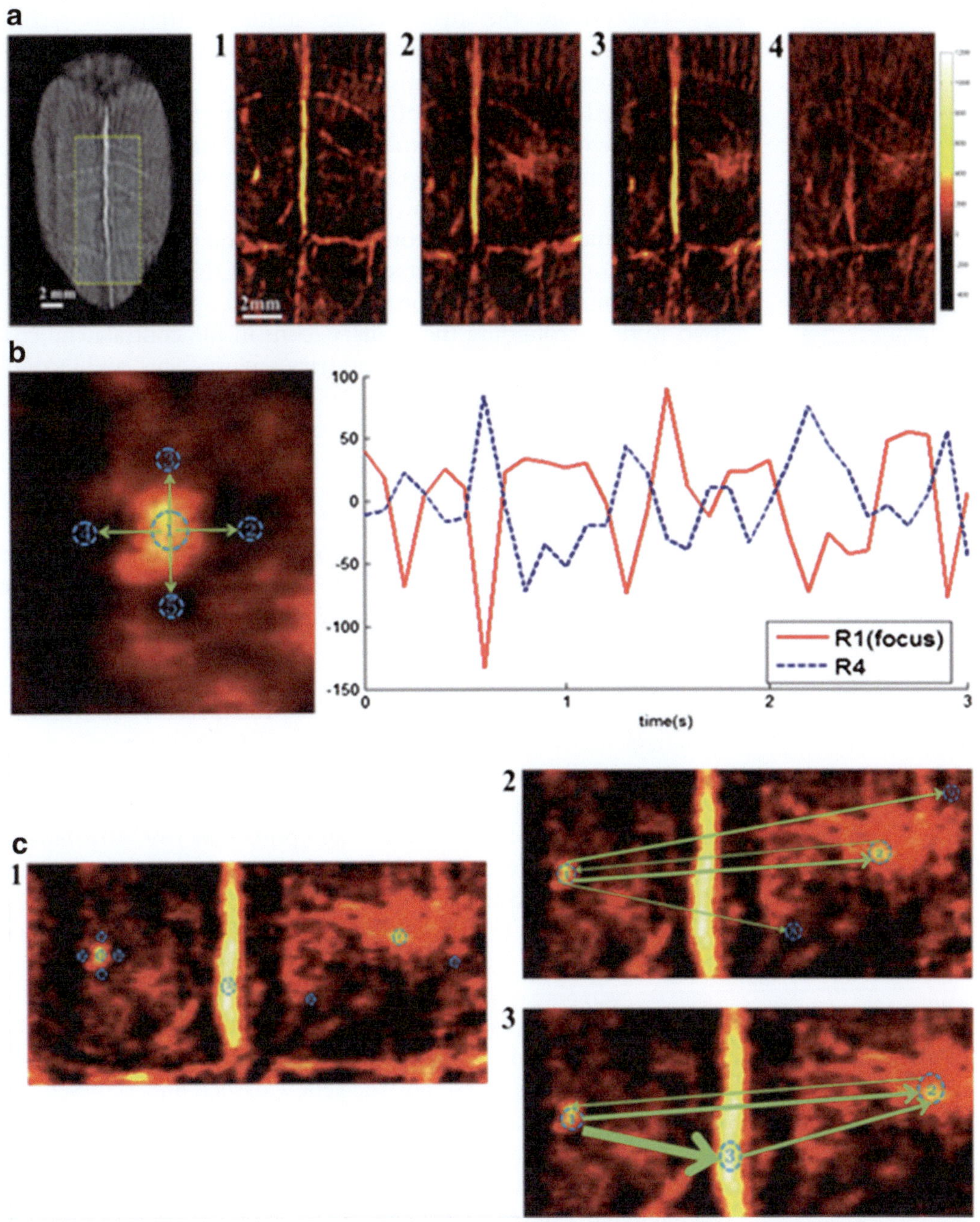

Fig. 5 In vivo maps of contralateral homotopic seizure foci. (**a**) *Left*, the PAT image of the rat brain; the *yellow dotted rectangle* shows the region of interest for analysis; *right*, noninvasive imaging of the primary and mirror foci in contralateral homotopic cortex. A1–4 shows how secondary epileptic foci were generated and diminished. The PA signal in the secondary foci was smaller in magnitude and delayed in time compared to the signal recorded from the primary foci. (**b**) *Left*, Granger analysis of the image around the foci and their immediate secondary foci; *right*, PA signal from the foci (*red*) and secondary area (*blue*) during an ictal event. (**c**) Pairwise Granger analysis of the seizure foci, contralateral homotopic foci, and middle cerebral artery. (**c1**) PAT image showing the nine selected regions of interest; (**c2**) *arrows* indicate Granger connectivity maps between the primary and secondary foci; (**c3**) connectivity maps between the primary and secondary foci, and the middle cerebral artery

To evaluate the dynamic interactions within the seizure circuitry and capture the associated networks we used the frequency domain Granger causality [34–36]. PAT time series data was selected from 200 sets of PAT images sampled at 3 Hz within the seizure onset period. At each time point we chose nine ROIs (Fig. $5c_1$) and used the averaged optical absorption within each region for the network analysis. Nine sets of time series analysis were classified into three groups including the primary ictal onset area and corresponding four surrounding ROIs including the primary or initiating seizure focus, contralateral homotopic foci, and two ROIs surrounding the primary focus in the contralateral cortex. The middle cerebral artery, primary seizure focus, and secondary homotopic seizure focus were also analyzed. Results from group 1 analysis demonstrate the causal influence from the primary focus to the surrounding ROIs (Fig. 5b, left). The negative zero-lag correlation (−0.68) between both the primary and secondary foci showed that neural activities within these two regions changed in opposite direction over time. We postulate that the inverted change in optical signal was likely due to the opposite oxygen delivery or blood flow [37]. Figure $5c_2$ shows the Granger causality analysis for group 2 data where we see that the primary and secondary contralateral brain foci influenced each other mutually with a stronger influence of the primary focus relative to the contralateral focus. Moreover, we suggest that the primary seizure focus may influence the contralateral foci as previously suggested using optical intrinsic imaging in epilepsy [38] and microelectrode recordings of hippocampus local field potentials [39, 40]. Finally, from group 3 analyses (Fig. $5c_3$) we found that the primary focus strongly influenced the hemodynamic change in the middle cerebral artery, which in turn showed influence on the contralateral homotopic foci.

3.2.3 Dynamical Changes of Vasculature During Interictal Discharges

Figure 6a shows image of the rat cortex vasculature through the intact scalp and skull. The red arrow indicates the microvascularity along the direction marked by the yellow dashed line, representing a cross-sectional scanning using a 50 MHz ultrasound transducer. Positive and negative acoustic peaks induced by a 25 μm blood vessel were sorted out as target signals to track the change of the vessel diameter. Typical PA signals from the targeted vessel at different times show apparent vessel vasomotion by ~40 % during the interictal discharges (Fig. 6b). Local field potential recordings (Fig. 6c) showed that interictal spikes had a strong correlation with the discrete change in vessel size observed photoacoustically (Fig. 6d; error bars ± s.d. were calculated from 20 consecutive spikes).

Finally, we conducted multispectral ultrahigh-resolution PAT of a blood vessel using the 50 MHz transducer (the blood vessel is located at the cortex about 1.5 mm below the skin as indicated by yellow arrow in Fig. 6a). In this case, 100 mg/kg pentylenetetrazol

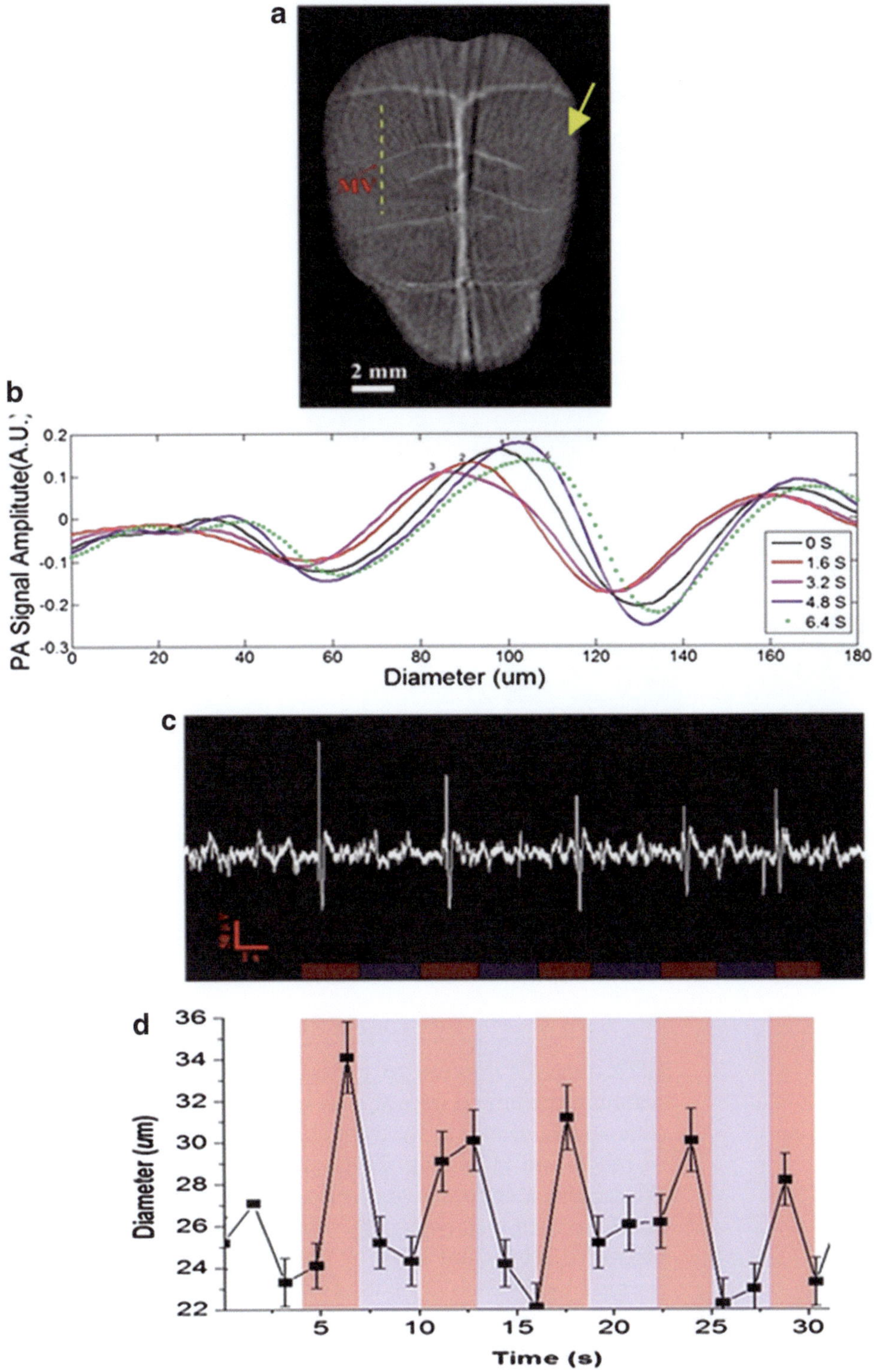

Fig. 6 Changes of blood vessel diameter during interictal onset. (**a**) PA image of the rat cortex vasculature with the intact scalp and skull. The *red arrow* indicates the microvascularity (MV) along the *yellow dashed line* direction for a cross-sectional scanning using a 50 MHz ultrasound transducer. (**b**) Typical photoacoustic signals from the targeted vessel at different times. (**c**) EEG recordings show the interictal spikes. (**d**) Change of the vessel size captured by PAT. There was a clear correlation between the interictal spikes and the changes of the vessel size. Error bars (±s.d.) were calculated from 20 consecutive spikes

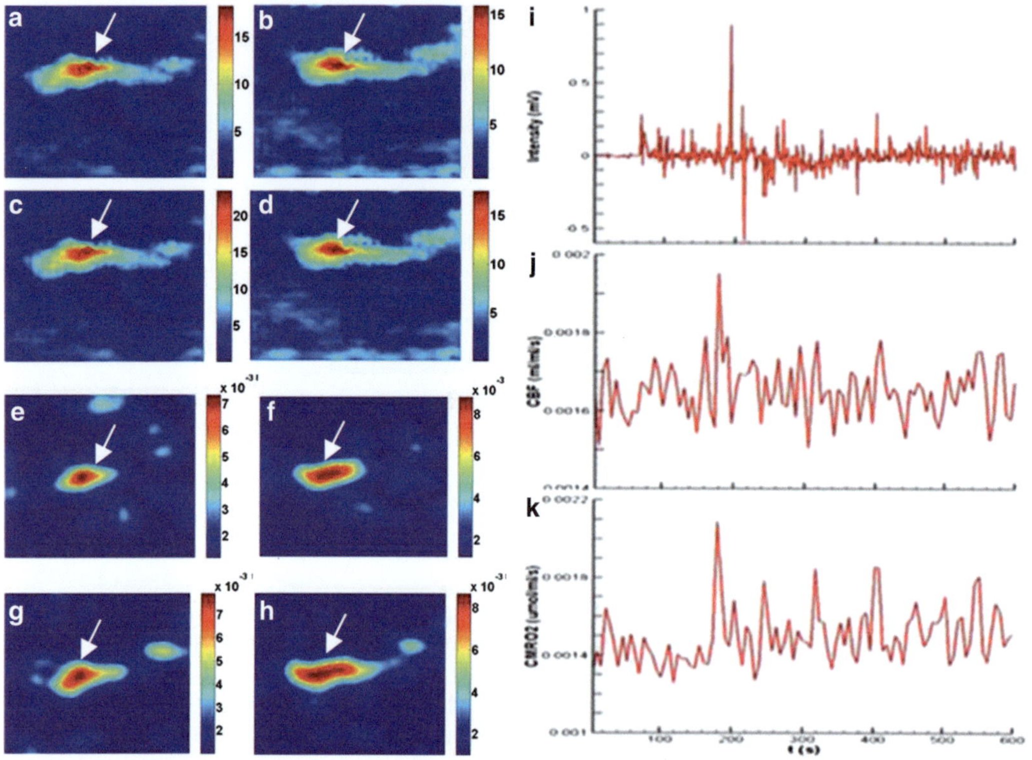

Fig. 7 Dynamic sagittal images of HbO_2, HbR, CBF, and $CMRO_2$ during ictal onset. (**a**) HbO_2 image at 48 s. (**b**) HbO_2 image at 330 s. (**c**) HbR image at 48 s. (**d**) HbR image at 330 s. (**e**) CBF image at 48 s. (**f**) CBF image at 48 s. (**g**) $CMRO_2$ image at 48 s. (**h**) $CMRO_2$ image at 330 s. (**i**) EEG recoding in 600 s. (**j**) Average CBF *within the blood vessel* in 600 s. (**k**) Average $CMRO_2$ *within the blood vessel* in 600 s. The units for HbO_2, HbR, CBF, $CMRO_2$, and EEG are, respectively, μM, μM, mL/mL/s, μM/mL/s, and mV

(PTZ) was IP injected to generate generalized seizure [41]. By scanning the pulsed laser at 730 and 1,064 nm alternatively over a time course of 600 s, we were able to obtain the dynamic images of HbO_2 and HbR (Fig. 7a–d; arrow indicates the blood vessel). We are also able to derive cerebral blood flow (CBF) and cerebral rate of oxygen metabolism ($CMRO_2$) images (Fig. 7e–h) using the recovered HbO_2 and HbR by adapting a model of oxygen transport in blood vessels (*see* Jiang [42], pp. 298–302, for description). We can observe significant changes in blood vessel size from the CBF and $CMRO_2$ images at the two representative instants (Fig. 7e–h). Most interestingly, hemodynamic "spikes" (Fig. 7j, k) similar to EEG spikes (Fig. 7i) are clearly notable in plots of average CBF and $CMRO_2$ within the blood vessel. While the hemodynamic spiking has lower frequency than the EEG spiking, it is also noted that the hemodynamic spiking occurred about 20 s earlier than the EEG spiking.

3.3 Discussion and Conclusions

The main finding is that PAT imaging is a powerful tool for noninvasively mapping seizure dynamics with both high spatial and temporal resolution. This study investigated the hemodynamic changes during focal seizure onset and propagation in an acute model of focal epilepsy. The experimental results suggest that the increase in local and surrounding brain tissue absorption was due to the focal bicuculline methiodide-induced seizure rather than other factors such as stab cortex tissue wound injection. Epileptiform events, including interictal spikes, and ictal onset and spread were identified in near real time. This study demonstrates solid experimental evidence for millisecond temporal resolution, micron scale imaging, and centimeters depth, of focal seizure onset and spread by a noninvasive technique.

A key advantage of the noninvasive PAT system for epileptiform event monitoring is the real-time imaging ability. Seizures are generated by complex interactions of a large group of neurons in which the population of neurons involved varies largely from moment to moment [37]. By achieving 1,000 times faster data acquisition of the transducer array-based PAT imaging system relative to a single transducer scanning PAT system [43], we were able to observe that the spatial aspects of the seizure focus varied over time as suggested by the changes in hemodynamics (Fig. 4c). At present, the imaging speed was limited only by the 10 Hz laser pulse-repetition-rate. Since considerably faster lasers of ~500 Hz pulse-repetition-rates are now commercially available, the temporal resolution of PAT will soon reach a level of a few milliseconds. This, coupled with the compactness of an optical system, will make it possible to bring a PAT system to the bedside for real-time and chronic monitoring of seizures in infants or during extra-operative epilepsy brain mapping, for example.

One hypothesis is that long-standing epilepsy may lead to the development of secondary epileptogenic regions located at a site distant from the original focus, a factor that may reduce the likelihood of successful epilepsy presurgical planning [44]. The ability to identify the epileptic focus and associated network may lead to better epilepsy surgery outcomes for many individuals. Various methods such as cross correlation and Granger causality have been explored for assessing the dynamic directional relationships among brain regions using time series data [45, 46]. We used Granger causality as a way to quantify the dynamic interactions among the primary seizure focus, secondary seizure foci, and middle cerebral artery based on the time series photoacoustic data. Specifically, Granger causality allowed for assessment of the magnitude and direction of temporal relationships during overlapping PAT time series windows. Results further exemplified the temporal aspects of seizure onset, seizure secondary spread, and seizure termination. Collectively, PAT image analysis and tools for effective connectivity may result in a better understanding of epileptic networks at high spatiotemporal resolution.

Little data are available regarding a link between vascular changes and seizures [47]. We found that the vasomotor phenomena of micro blood vessels correlate with interictal spike and wave discharges (Fig. 5d). Interictal spikes generated a strong cerebral metabolic change that induced cerebral vessel dilation and a focal increase in CBF provided oxygenated hemoglobin to hypermetabolic neurons, with a corresponding increase in total hemoglobin [48]. Vasomotion during seizure onset and spread was clearly caused by the oscillation in vessel diameter and the expansion of the vessel cross section. We reasoned that the dilation resulted in an active cerebral microvasculature response to increased metabolic activity accompanying the seizure [49]. The combination of multi-scale imaging ability and endogenous hemoglobin contrast makes PAT a promising tool for imaging microvasculature during seizure onset. Future quantitative PAT studies will allow us to resolve many of the hemodynamic components, including changes in level of hemoglobin oxygenation (i.e., deoxy- and oxyhemoglobin), CBF, and the rate of cerebral oxygen metabolism, for a detailed understanding of the neurovascular coupling phenomenon, both during and in between seizures [50, 51].

The noninvasive yet whole surface and depth capabilities of the PAT system allowed to actually see what is happening during ictogenesis in terms of seizure onset and spread. The challenge of mapping focal epilepsy stems from the observation that the pathology associated with focal epilepsy is often distributed across a number of brain sites. Seizures in animal models and in people often have a multifocal or broadly synchronized onset. The implication of these observations is that each of the sites could act independently to initiate a seizure or, potentially, to drive another site into a seizure. This study lends support to the theory that seizure onset and spread involve a rich interplay between multiple cortical and subcortical foci during the onset and spread of focal epilepsy. The findings are timely in the sense that the neuroscience community is questioning the long-held dogma that seizure onset involves some sort of single focus or epicenter in favor of the emerging thought that seizures instead involve multiple cortical and subcortical foci.

References

1. Grinvald A, Frostig R, Lieke E, Hildesheim R (1988) Optical imaging of neuronal activity. Physiol Rev 68:1285–1366
2. Hill D, Keynes R (1949) Opacity changes in stimulated nerve. J Physiol 108:278–281
3. Haglund MM, Ojemann GA, Hochman DW (1992) Optical imaging of epileptiform and functional activity in human cerebral cortex. Nature 358:668–671
4. Haglund MM, Hochman DW (2004) Optical imaging of epileptiform activity in human neocortex. Epilepsia 45:43–47
5. Steinhoff BJ, Herrendorf G, Kurth C (1996) Ictal near infrared spectroscopy in temporal lobe epilepsy: a pilot study. Seizure 5:97–101
6. Sokol D, Markand O, Daly E, Luerssen T, Malkoff M (2000) Near infrared spectroscopy (NIRS) distinguishes seizure types. Seizure 9: 323–327
7. Watanabe E, Maki A, Kawaguchi F, Yamashita Y, Koizumi H, Mayanagi Y (2000) Noninvasive cerebral blood volume measurement during seizures using multichannel near infrared spectroscopic topography. J Biomed Opt 5:287–290

8. Jiang H, Paulsen KD, Osterberg UL, Pogue BW, Patterson MS (1996) Optical image reconstruction using frequency-domain data: simulations and experiments. J Opt Soc Am A 13:253–266
9. O'Leary M, Boas D, Li X, Chance B, Yodh A (1996) Fluorescence lifetime imaging in turbid media. Opt Lett 21:158–160
10. Bluestone A, Abdoulaev G, Schmitz C, Barbour R, Hielscher A (2001) Three-dimensional optical tomography of hemodynamics in the human head. Opt Express 9: 272–286
11. Boas DA, Brooks DH, Miller EL, DiMarzio CA, Kilmer M, Gaudette RJ, Zhang Q (2001) Imaging the body with diffuse optical tomography. IEEE Signal Process Mag 18:57–75
12. Hoelen CGA, de Mul FFM (2000) Image reconstruction for photoacoustic scanning of tissue structures. Appl Optics 39:5872–5883
13. Manohar S, Vaartjes SE, Hespen JCG, Klaase JM, Engh FM, Steenbergen W, Leeuwen TG (2007) Initial results of in vivo non-invasive cancer imaging in the human breast using near-infrared photoacoustics. Opt Express 15: 12277–12285
14. Kruger RA, Liu P, Appledorn CR (1995) Photoacoustic ultrasound (PAUS)—reconstruction tomography. Med Phys 22:1605
15. Wang X, Xu Y, Xu M, Yokoo S, Fry ES, Wang LV (2002) Photoacoustic tomography of biological tissues with high cross-section resolution: reconstruction and experiment. Med Phys 29:2799
16. Yuan Z, Wu C, Zhao H, Jiang H (2005) Imaging of small nanoparticle-containing objects by finite-element-based photoacoustic tomography. Opt Lett 30:3054–3056
17. Jiang H, Yuan Z, Gu X (2006) Spatially varying optical and acoustic property reconstruction using finite-element-based photoacoustic tomography. J Opt Soc Am A 23:878–888
18. Jin J (2002) The finite element method in electromagnetics. Wiley, New York
19. Siphanto R, Thumma K, Kolkman R, Van Leeuwen T, De Mul F, Van Neck J, van Adrichem L, Steenbergen W (2005) Serial noninvasive photoacoustic imaging of neovascularization in tumor angiogenesis. Opt Express 13:89–95
20. Yin B, Xing D, Wang Y, Zeng Y, Tan Y, Chen Q (2004) Fast photoacoustic imaging system based on 320-element linear transducer array. Phys Med Biol 49:1339
21. Oraevsky AA, Andreev VA, Karabutov AA, Fleming RD, Gatalica Z, Singh H, Esenaliev RO (1999) Laser optoacoustic imaging of the breast: detection of cancer angiogenesis. In: BiOS'99 international biomedical optics symposium, International Society for Optics and Photonics, pp 352–363
22. Zhang E, Laufer J, Beard P (2008) Backward-mode multiwavelength photoacoustic scanner using a planar Fabry-Perot polymer film ultrasound sensor for high-resolution three-dimensional imaging of biological tissues. Appl Optics 47:561–577
23. Chen Z, Tang Z, Wan W (2007) Photoacoustic tomography imaging based on a 4f acoustic lens imaging system. Opt Express 15:4966–4976
24. Niederhauser JJ, Jaeger M, Lemor R, Weber P, Frenz M (2005) Combined ultrasound and optoacoustic system for real-time high-contrast vascular imaging in vivo. IEEE Trans Med Imaging 24:436–440
25. Hauser WA, Kurland LT (1975) The epidemiology of epilepsy in Rochester, Minnesota, 1935 through 1967. Epilepsia 16:1–66
26. Hauser WA (2008) Seizure disorders: the changes with age. Epilepsia 33:6–14
27. Fisher RS, Boas WE, Blume W, Elger C, Genton P, Lee P, Engel J Jr (2005) Epileptic seizures and epilepsy: definitions proposed by the International League Against Epilepsy (ILAE) and the International Bureau for Epilepsy (IBE). Epilepsia 46:470–472
28. Birbeck GL, Hays RD, Cui X, Vickrey BG (2002) Seizure reduction and quality of life improvements in people with epilepsy. Epilepsia 43:535–538
29. Duncan JS, Sander JW, Sisodiya SM, Walker MC (2006) Adult epilepsy. Lancet 367:1087–1100
30. Berg AT, Langfitt JT, Spencer SS, Vickrey BG (2007) Stopping antiepileptic drugs after epilepsy surgery: a survey of US epilepsy center neurologists. Epilepsy Behav 10:219–222
31. Engel J Jr (2004) The goal of epilepsy therapy: no seizures, no side effects, as soon as possible. CNS Spectr 9:95
32. Jeha LE, Najm I, Bingaman W, Dinner D, Widdess-Walsh P, Lüders H (2007) Surgical outcome and prognostic factors of frontal lobe epilepsy surgery. Brain 130:574–584
33. Paxinos G, Watson C (1998) A stereotaxic atlas of the rat brain. Academic, New York
34. Brovelli A, Ding M, Ledberg A, Chen Y, Nakamura R, Bressler SL (2004) Beta oscillations in a large-scale sensorimotor cortical network: directional influences revealed by Granger causality. Proc Natl Acad Sci U S A 101:9849–9854
35. Granger CWJ (1969) Investigating causal relations by econometric models and cross-spectral methods. Econometrica 37:424–438
36. Wang X, Chen Y, STEVEN LB, Ding M (2007) Granger causality between multiple interdependent neurobiological time series: blockwise

versus pairwise methods. Int J Neural Syst 17: 71–78

37. Schwartz TH, Bonhoeffer T (2001) In vivo optical mapping of epileptic foci and surround inhibition in ferret cerebral cortex. Nat Med 7:1063–1067
38. Khalilov I, Holmes GL, Ben-Ari Y (2003) In vitro formation of a secondary epileptogenic mirror focus by interhippocampal propagation of seizures. Nat Neurosci 6:1079–1085
39. Cadotte AJ, DeMarse TB, Mareci TH, Parekh MB, Talathi SS, Hwang DU, Ditto WL, Ding M, Carney PR (2010) Granger causality relationships between local field potentials in an animal model of temporal lobe epilepsy. J Neurosci Methods 189:121–129
40. Zhou J, Talathi SS, Cadotte A, Liu Z, Holmes GL, Carney PR (2009) The effects of hippocampus Ca1 single neuron firing properties on interictal spike patterns during seizure onset in a rat model of temporal lobe epilepsy. Epilepsia 50:468–469
41. André V, Pineau N, Motte JE, Marescaux C, Nehlig A (2001) Mapping of neuronal networks underlying generalized seizures induced by increasing doses of pentylenetetrazol in the immature and adult rat: ac-Fos immunohistochemical study. Eur J Neurosci 10:2094–2106
42. Jiang H (2010) Diffuse optical tomography: principles and applications. CRC Press, Boca Raton, FL
43. Zhang Q, Liu Z, Carney PR, Yuan Z, Chen H, Roper SN, Jiang H (2008) Non-invasive imaging of epileptic seizures in vivo using photoacoustic tomography. Phys Med Biol 53:1921
44. Cendes F, Cook M, Watson C, Andermann F, Fish D, Shorvon S, Bergin P, Free S, Dubeau F, Arnold D (1995) Frequency and characteristics of dual pathology in patients with lesional epilepsy. Neurology 45:2058–2064
45. Kamiński M, Ding M, Truccolo WA, Bressler SL (2001) Evaluating causal relations in neural systems: Granger causality, directed transfer function and statistical assessment of significance. Biol Cybern 85:145–157
46. Bressler SL, Tang W, Sylvester CM, Shulman GL, Corbetta M (2008) Top-down control of human visual cortex by frontal and parietal cortex in anticipatory visual spatial attention. J Neurosci 28:10056–10061
47. Ndode-Ekane X, Hayward N, Gröhn O, Pitkänen A (2010) Vascular changes in epilepsy: functional consequences and association with network plasticity in pilocarpine-induced experimental epilepsy. Neuroscience 166: 312–332
48. Ma H, Zhao M, Suh M, Schwartz TH (2009) Hemodynamic surrogates for excitatory membrane potential change during interictal epileptiform events in rat neocortex. J Neurophysiol 101:2550–2562
49. Myers RR, Intaglietta M (1976) Brain microvascular hemodynamic responses to induced seizures. Stroke 7:83–88
50. Yao L, Sun Y, Jiang H (2009) Quantitative photoacoustic tomography based on the radiative transfer equation. Opt Lett 34: 1765–1767
51. Yuan Z, Wang Q, Jiang H (2007) Reconstruction of optical absorption coefficient maps of heterogeneous media by photoacoustic tomography coupled with diffusion equation based regularized Newton method. Opt Express 15:18076–18081

Part IV

Neurovascular Coupling Techniques in Cerebrovascular Disease: Ischemia, Hemorrhage, and Spreading Depression

Chapter 14

Neurovascular Coupling in Relation to Cortical Spreading Depression

Henning Piilgaard and Martin Lauritzen

Abstract

Cortical spreading depression (SD) has been known for more than 70 years. It has been a unique method investigating basic neurophysiology. A growing body of evidence indicates SD is involved in clinical conditions such as migraine with aura, subarachnoid hemorrhage, and traumatic brain injury. In this chapter we describe the application of SD in studying neurovascular and neurometabolic coupling in rat cerebral cortex. Description of the different techniques (electrophysiology, laser-Doppler flowmetry, and oxygen-sensitive electrodes) is followed by in depth step-by-step instructions on how to use them. This chapter will also describe the vascular consequences of SD and how to elicit SD experimentally.

Key words Spreading depression, Neurovascular coupling, Neurometabolic coupling, Local field potential, Laser-Doppler flowmetry, Oxygen-sensitive electrodes, In vivo

1 Introduction (Cortical Spreading Depression: The Tsunami of the Brain)

Cortical spreading depression (SD) was first described by Leão and published in the Journal of Neurophysiology in 1944 [1]. During SD, neurons and glia depolarize, often associated with an initial transient of burst activity lasting a few seconds. This is followed by a complete loss of neuronal activity, which can last minutes before complete recovery occurs over several minutes [2]. SD is associated with a dramatic failure of brain ion homeostasis (interstitial K^+ rising from 3 to 60 mM; Na^+ decreasing from 150 to 60 mM and Ca^{2+} decreasing from 1.5 to 0.08 mM) [3], efflux of excitatory amino acids from nerve cells, including glutamate and aspartate [4, 5] and enhanced energy metabolism. SD has been induced in most gray-matter regions studied so far, including neocortex, hippocampus, and cerebellum of a variety of species [2]. Furthermore, human neocortical [6] and hippocampal [7] tissue in vivo and their counterparts studied in vitro [8, 9] support SD. In healthy cortex the electrophysiological footprint of SD is the loss of spontaneous electrical activity in the electroencephalogram (EEG) or

Mingrui Zhao et al. (eds.), *Neurovascular Coupling Methods*, Neuromethods, vol. 88,
DOI 10.1007/978-1-4939-0724-3_14, © Springer Science+Business Media New York 2014

electrocorticogram (ECoG) attended by a DC potential shift. This pattern moves over the cortex at a speed of 3 mm/min [10]. Repeated episodes of SD have been shown in slice preparations to decrease the paired pulse inhibition without altering the excitatory potential suggesting selectively the suppression of GABAergic function in the wake of SD [11]. Selective reduction of the inhibitory postsynaptic potential has been demonstrated in vivo in rats in the wake of SD [12]. Associated to the electrophysiological events are changes in CBF. In rats an initial vasoconstriction at onset of the depolarization can be seen but is usually short of duration [13]. During return of ionic gradients to normal values a hyperemia phase (100–200 % increase) is evident lasting for 1–2 min in anesthetized animals [12] followed by a prolonged oligemia phase with CBF reduction of 20–30 % lasting more than 1 h [14]. The oligemia phase can be ameliorated by cyclosporine A [15]. Cyclosporine A affects both the mitochondrial permeability transition pore and calcineurin pathways. Recent data additionally indicate that the oligemia is mediated by 20-HETE since HET0016 (*N*-hydroxy-*N'*-(4-*n*-butyl-2-methylphenyl)formandine) blocks the 20-HETE production due to SD and ameliorate the oligemia phase after SD [16]. Species differences have been reported for the SD-associated CBF response indicating that mouse cortex has a more pronounced initial vasoconstriction and a diminished if any hyperemia phases [17].

Following several episodes of SD, protein synthesis is inhibited [18] but repeated episodes of SD increase the immunohistochemical staining of glial fibrillary acidic protein in the rat cortex, lasting for up to 4 days [19]. The *c-fos* proto-oncogene and mRNA for nerve growth factor increases for up to 24 h after repeated SD episodes [20], as do several other markers for immediate early genes, even after single SD episodes [21]. Neuronal NO synthase (nNOS) activity in glial cells has been reported to increase after SD [22]. Additionally, studies have shown increased expression of heat- shock proteins, cyclooxygenase 2, and other growth factors/cytokines besides those already mentioned [23–26].

The burden of evidence is growing for SD taking place in a variety of clinical disorders. The first of them recognized being migraine with aura, where SD is likely to be the mechanism responsible for the aura phase [27]. The idea of SD as the underlying mechanism of migraine has been supported by neuroimaging studies recording changes in cerebral blood flow or the BOLD signal in migraine patients [27, 28].

Recent and ongoing studies including the invasive clinical studies by the COSBID group (Co-Operative Study on Brain Injury Depolarizations) have demonstrated that SD-like episodes are taking center stage in the progression of acute brain disorders like subarachnoid hemorrhage, malignant hemispheric stroke, intracerebral hemorrhage, and traumatic brain injury [6, 29–31].

Diaschisis is decreased metabolism and CBF in anatomically intact brain regions remote from a cerebral lesion or in other words a functional ablation. Diaschisis is often seen in acute brain disorders like stroke and there is evidence that SD induces this condition [32].

2 Materials

The materials and methods outlined in the following, will address the materials, equipment, and procedures for inducing single episodes of SD as well as performing experiments which allows for calculations of a neurovascular and a neurometabolic coupling.

2.1 Anesthesia

Male Wistar rats are anesthetized with Isoflurane (Abbott; induction: 5 %, surgery: 2 %) while ventilated with O_2 and air. This regimen is held throughout the initial procedures as described in Sect. 2.2. After surgery the anesthesia is changed to α-chloralose (1,2–0-[2,2,2-trichloro-ethylidene]-[alpha]-D-gluco-furanose) HBC complex (Sigma), dissolved in saline (0.5 g/mL), (bolus: 1.6 mL/kg i.v., continuous infusion: 1.1 mL/kg/h i.v.). This anesthetic regime is considered ideal for studies of neurovascular and neurometabolic coupling [33, 34]. The level of anesthesia is checked by observing arterial blood pressure during stimulation and by tail pinch. Experiments are performed after a post-operative recovery period of at least half an hour to obtain a stable level of anesthesia, a stable CBF baseline, and an arterial blood pressure of 90–120 mmHg, not varying more than 10 % in the same animal.

2.2 General Procedures

The head of the rat is placed in a stereotactic head holder. Lidocaine (5 mg/mL) is used at the contact spots for the ear pins. A femoral artery catheter is inserted for measuring arterial blood pressure and taking blood–gas samples. A femoral vein catheter is inserted for saline infusion.

The trachea is cannulated and rats are ventilated throughout experiments. Blood gas analysis is performed hourly to assure physiological conditions (ABL 715, Radiometer). Body temperature is maintained at 36–37 °C using a custom-made heating pad. Three craniotomies are performed with a dental drill; two over homologous regions of the somatosensory cortex, and one over the frontal cortex on the registration side for elicitation of CSD (Fig. 1). Wells are made around each craniotomy using 5 % agar in Ringer solution. The dura is carefully removed under a microscope. Any visible sign of brain damage or bleeding excludes the rat from the protocols. During and after removal of the dura, the craniotomy sites are continuously superfused with artificial cerebrospinal fluid (aCSF, composition in mmol/L: 2.0 glucose, 126.0 NaCl, 2.8 KCl, 22.0 $NaHCO_3$, 1.45 $CaCl_2$, 1.0 Na_2HPO_4, and 0.876 $MgCl_2$).

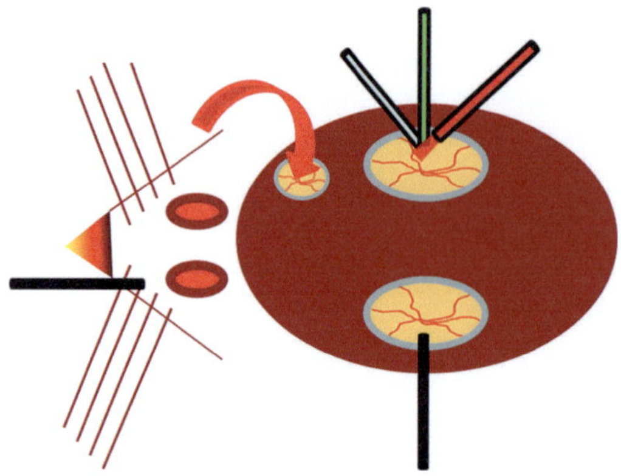

Fig. 1 Experimental design. The figure shows a rat brain from above. A stimulation electrode is inserted into the hiatus infraorbitalis on the *left side*. Measurements of oxygen tension in the tissue (*blue bar*), CBF (*red bar*), and ECoG/LFP (*green bar*) are performed over the primary somatosensory cortex on the right hemisphere (registration side). After 40-min incubation with the NMDA-receptor antagonist MK801 a stimulation electrode is placed in the primary somatosensory cortex of the stimulation side (*left side*) used for transcallosal stimulations. Single episodes of SD are elicited with a needle stab in the frontal craniotomy on the *right side*

2.3 Laser-Doppler Flowmetry

Laser-Doppler flowmetry (LDF) is a convenient technique for measuring changes in cerebral blood flow. One of the strongest advantages is the robustness of the technique. LDF was first used in 1972 as the technique for measurements of blood flow in the retina of the rabbit [35]. Monochromatic light is transmitted through a fiber and hits both stationary (i.e., tissue) and moving elements, mainly red blood cells (RBC). The interaction with the moving RBC causes a frequency shift (Doppler shift) that can be used for estimating the blood flow [36]. Via a receiving fiber both light with the original frequency and the Doppler shifted light enters the photodetector generating photocurrents. The depth of measurements can be modified both by the wavelength of the transmitted light and by the fiber separation between transmitting and receiving fiber. We use a 780-nm wavelength (red laser light) and a fiber separation of 250-μm (Periflux 4001 Master, Perimed, Jarfälla, Sweden), which corresponds to a recording depth of up to 1 mm into the cortex [37]. The signal is A/D converted and recorded using the CED 401-plus interface and CED Spike 2.6 software (10 Hz digital sampling rate). The laser-Doppler flowmetry monitor displays blood flow readings in arbitrary units that do not allow for measurement of CBF in terms of absolute values, but the method is valid in determining relative changes of CBF during moderate flow increases. LDF has been validated against other techniques for example [^{14}C]-Iodoantipyrine where it was found to have a tendency of overestimating flow at a very high flow [38].

2.4 Oxygen Electrode

A modified Clark-type polarographic oxygen microelectrode (OX-10, Unisense A/S, Aarhus, Denmark) with a guard cathode is used for measurements of tissue partial pressure of oxygen (tpO_2) [39]. The advantage of this electrode type is its small tip size (10 μm) and its built-in guard cathode, which removes all oxygen from the electrolyte reservoir. This enables us to measure tpO_2 over time (and among different treatment conditions) with excellent long-term stability (signal drift, 0–0.5 % h^{-1}). The field of sensitivity is a sphere of two times the tip diameter. Oxygen microelectrodes respond linearly to changes in oxygen concentration. The oxygen electrode is connected to a high-impedance Picoamperometer (PA 2000, Unisense A/S) sensing the currents of the oxygen electrode. Signals are A/D converted and recorded at a sampling rate of 10 Hz (Power 1401 and Spike 2.6; Cambridge Electronic Design).

2.4.1 Calibration of Oxygen Electrode

Since oxygen electrodes respond linearly to changes in oxygen concentration, a two-point calibration is sufficient to establish the range for a given electrode (measured in picoAmpere). Calibration of the oxygen electrode is performed both before and after each experiment to check that the electrode is responsive and not broken.

Values obtained from both the calibration procedure and measurements performed are expressed in picoAmpere (pA). A zero-oxygen calibration (pA_{zero}) is performed in an oxygen-free solution consisting of 0.1-M (+)-Sodium l-ascorbate and 0.1 M NaOH (solution can be made by applying 2 g of (+)-Sodium l-ascorbate to 100 mL of 0.1 mM NaOH).

In air-saturated saline a maximum value of oxygen tension (pA_{max}) at a pressure of 1 atm is obtained. To convert from pico Ampere to partial pressure of oxygen expressed in mmHg, put in the calibration values and data points into Eq. 1:

$$tpO_{2,data}(\text{mmHg}) = \frac{(pA_{data} - pA_{zero})}{(pA_{max} - pA_{zero})} \times 760\,\text{mmHg} \times 0.21 \qquad (1)$$

The pressure in the environment is set to 760 mmHg. Of this pressure 21 % is due to oxygen. $tpO_{2,data}$ corresponds to any given value obtained during experiment.

2.5 Electrocorticogram/Field Potential Recordings

Electrophysiological recordings are performed using glass microelectrodes filled with 2-M saline. Electrodes are custom made and have a tip diameter of less than 2 μm and an impedance of 2–3 MΩ. An Ag/AgCl ground electrode is placed in the neck muscle. The preamplified (×10) signal is A/D-converted, amplified, band pass filtered at 1–1,000 Hz, and digitized using the 1401-plus hardware (Cambridge Electronic Design, Cambridge, UK) connected to a PC running the Spike 2.6 software. Digital sampling rates for local field potentials (LFPs) and electrocorticogram (ECoG) are

5 kHz. DC potentials are recorded from the same glass electrode using a 0.1-Hz low-pass filter.

2.6 Avoiding Unwanted Episodes of SD

Both insertion of the stimulation electrode and direct cortical stimulation could potentially elicit episodes of SD on the stimulated side. To circumvent this problem, we apply the NMDA-receptor (ionotropic glutamate receptor) antagonist MK801 (Sigma–Aldrich) to the stimulated cortex in a concentration of 1 mM 40 min prior to placing the stimulation electrode. Stock solution is made in distilled water. For topical application stock solution is further diluted in aCSF to a final concentration of 1 mM. This procedure prevents the occurrence of unwanted SD [40, 41].

2.7 How to Elicit SD Experimentally

SD can be experimentally elicited in different ways, the most common methods being topical application (or microinjection) of KCl or a brief needle stab to the exposed cortex. One SD may develop differently, depending on the procedures used to elicit it. In the rat cerebral cortex, the profiles of DC potential and potassium shifts change as a function of depth and the mode of elicitation (needle stab versus KCl application). Recurrent SDs propagating from the site of KCl application have longer duration and lower frequencies in superficial (dendritic) layers than in deeper (cell-body) layers.

2.8 Stimulation Procedures

2.8.1 Infraorbital Nerve Stimulation

The infraorbital nerve is stimulated by placing custom-made electrodes as follows: the cathode in the hiatus infraorbitalis and the anode in the masticatory muscles. Stimulations are given as square wave pulses (0.1 ms, 1.5 mA).

2.8.2 Transcallosal Fiber Stimulation

A coated bipolar stainless steel electrode (SNEX 200, contact separation: 0.25 mm; RMI, Woodland Hills, CA, USA) is used for direct current (DC) stimulations (ISO-flex; A.M.P.I., Israel) given as square wave pulses (0.1 ms, 1.5 mA). We use stimulations delivered in trains lasting 1 s with frequencies ranging from ½ Hz up to 40 Hz, with inter-train intervals varying between 30 and 90 s depending on the stimulation frequency (be sure of return of baseline before stimulating again).

3 Procedures

3.1 Experiments Step by Step

1. Male Wistar rats are anesthetized with Isoflurane (Abbott; induction: 5 %, surgery 2 %) while ventilated with O_2 and air.
2. A femoral artery catheter is inserted for continuous measuring of arterial blood pressure and taking blood–gas samples.
3. A femoral vein catheter is inserted for saline infusion (and for later i.v. anesthesia).

4. The trachea is cannulated and the rat is ventilated throughout the experiment.
5. The head is placed in a stereotactic head holder. Lidocaine (5 mg/mL) is used at the contact spots for the ear pins.
6. Three craniotomies are performed with a dental drill: two over homologous regions of the somatosensory cortex, and one over the frontal cortex on the registration side are used for elicitation of SD (*see* Fig. 1).
7. Wells are made around each craniotomy using 5 % agar in Ringer solution.
8. In all three craniotomies the dura is now carefully removed and the exposed brain is superfused with a CSF.
9. After surgery the anesthesia is changed to α-chloralose (1,2–0-[2,2,2-trichloro-ethylidene]-[alpha]-D-gluco-furanose) HBC complex (Sigma), dissolved in saline (0.5 g/mL), (bolus: 1.6 mL/kg i.v., continuous infusion: 1.1 mL/kg/h i.v.).
10. The exposed brain on the stimulation side (left side) is incubated with 1 mM MK801 (NMDA receptor antagonist) for at least 40 min.
11. Infraorbital nerve stimulator is placed: the cathode in the hiatus infraorbitalis and the anode in the masticatory muscles (in our experiments on the left side).
12. Transcallosal fiber stimulator is placed 0.5-mm into the whisker-Barrel cortex of the primary somatosensory cortex at coordinates 3 mm posterior to the Bregma and 6 mm lateral to the midline on the stimulation side.
13. Oxygen electrode is placed on the registration side in a depth of 300–600 μm at coordinates of 3 mm posterior to the Bregma and 6 mm lateral to the midline.
14. Glass electrode is placed as close as possible to the oxygen electrode on the registration side.
15. Laser-Doppler probe is placed 0.3 mm above the pial surface juxtaposed to the oxygen electrode and the glass electrode on the registration side.
16. The presence of LFPs when stimulating infraorbital nerve (2 Hz) confirms the location of the registration electrode being in the barrel cortex.
17. Stimulations using the transcallosal fiber pathway can now be performed.
18. Single episodes of SD can now be elicited with a brief needle stab in the frontal craniotomy. Since the spread of SD over the cortex is approximately 3 mm/min within a few minutes the

described electrophysiological changes and associated events will arrive at the recording side.

19. If needed, additional transcallosal stimulations can be performed or additional SDs can be elicited.
20. Animal is euthanized.
21. Oxygen electrode is withdrawn and calibrated to assure reliable oxygen measurements and that the electrode is not broken.
22. Analysis can begin

3.2 Calculating Cerebral Metabolic Rate of Oxygen from tpO_2 and CBF Data

During experiments simultaneous recordings of tpO_2 and CBF are obtained. From these we calculate the cerebral metabolic rate of oxygen ($CMRO_2$) offline. In order to do so we used a steady-state model as described by Gjedde [42] (*see* Eq. 2), where P_{50} is the half-saturation tension of the oxygen–hemoglobin dissociation curve, h, the Hill coefficient of the same dissociation curve, C_a, the arterial oxygen concentration, and L the effective diffusion coefficient of oxygen in the brain tissue. For details about the theory behind the algorithm *see* [41]. The practical use of this algorithm will be described in the following.

$$\mathrm{tpO_2} = P_{50}\sqrt[h]{\frac{2C_a\mathrm{CBF}}{\mathrm{CMRO_2}} - 1} - \frac{2\,\mathrm{CMRO_2}}{3L} \quad (2)$$

1. From every animal entering the study an initial value of simultaneously obtained "baseline" tpO_2 and CBF are defined.
2. The constant L, the effective diffusion coefficient of oxygen in the brain tissue, has to be calculated. For this purpose values from the literature from animals being under α-chloralose anesthesia are entered into Eq. 2: CBF (53 mL/100 g per min), $CMRO_2$ (219 μmol/100 g per min) (both values reported by Zhu et al. [43]), P_{50} (36 mmHg), h (2.7), and C_a (8 μmol/mL).
3. P_{50} (36 mmHg), h (2.7), and C_a (8 μmol/mL) and the calculated L (μmol/100 g) are put into the algorithm.
4. Simultaneously obtained recordings of tpO_2 (mmHg) and CBF (relative to the baseline value) from the individual experiments are run through the algorithm.

3.3 Local Field Potentials

LFP represents, as the name indicates, the mean field potential generated by the slow components of synaptic and neural events in the vicinity of the recording electrode [44]. The recorded LFP is composed of two peaks: an initial negative peak (reflecting the excitatory postsynaptic potential, EPSP) followed by a positive peak (reflecting the inhibitory postsynaptic potential, IPSP). LFP is used as an indicator for synaptic activity. Evoked CBF increases correlates to the sum of neural activities (ΣEPSP, i.e., the negativity of LFPs multiplied by the stimulus frequency) suggesting that extracellular ion fluxes contribute to the coupling of brain activity to blood flow [45].

3.4 Data Analysis

Values are expressed as mean ± SEM. Differences between two groups are tested using student's *t*-test. We consider differences significant at $P < 0.05$. Linear regression with random effects (in SAS the procedure is named "proc mixed") can be used for characterization of the influence of cortical spreading depression on both evoked responses as function of stimulation frequency and on the neurometabolic and neurovascular coupling. All statistical tests are performed using SAS version 9.1.3 (SAS Institute Inc., Cary, NC, USA).

4 Experimental Variables

During experiments recordings are made of the ongoing CBF and tpO_2 (allowing also calculation of ongoing $CMRO_2$). Measurements of evoked responses are performed making it possible to establish a neurovascular coupling and a neurometabolic coupling. As measure of evoked CBF response (ΔCBF) either the area under the curve (AUC) or the peak amplitude can be used. Likewise for the evoked $CMRO_2$ response ($\Delta CMRO_2$) either the AUC or the peak amplitude can be used. As indicator of synaptic activity (i.e., neuronal activity) we use, as described earlier, the sum of the negative part of the LFP in a given stimulation train (ΣEPSP). This means that a neurovascular coupling can be established as ΣEPSP versus ΔCBF and a neurometabolic coupling can be established as ΣEPSP versus $\Delta CMRO_2$.

5 Typical Results

At the onset of CSD where the flattening of EEG/ECoG and the DC-shift is occurring, dramatically increases in CBF of up to 200–300 % and increases in $CMRO_2$ of up to 71 % appears lasting for a few minutes (Fig. 2). Following these early events an oligemia phase with 20–30 % reduction in CBF and a persistent increase in $CMRO_2$ of up to 8 % appears lasting for up to 2 h before returning to the original level (Fig. 3). Evoked responses are affected in the wake of SD. The EPSP of LFP returns to normal values within minutes. The IPSP is reduced for up to 2 h. Evoked CBF responses are reduced but evoked $CMRO_2$ are unaffected by SD. This translates into an impaired neurovascular coupling in the wake of SD but preserved neurometabolic coupling (Fig. 4) [12].

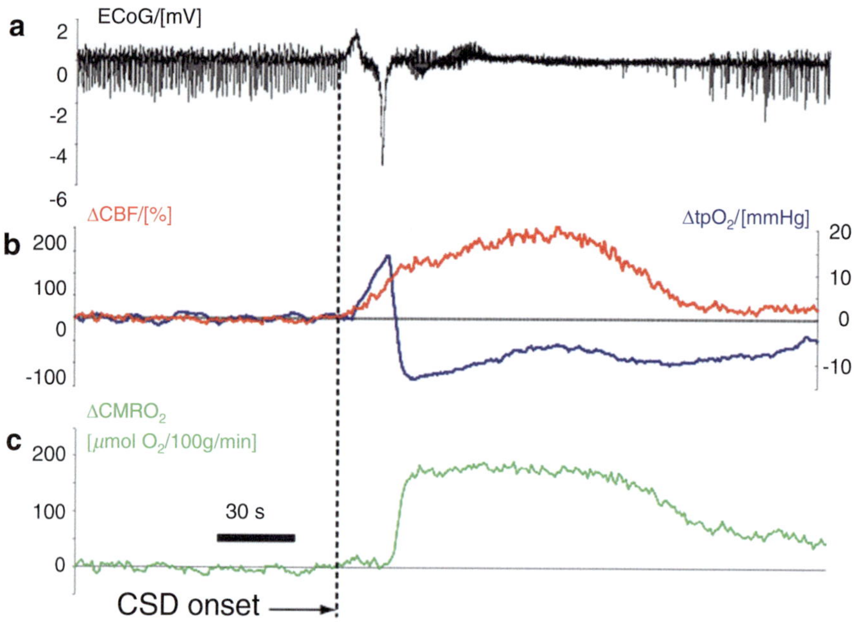

Fig. 2 Single episode of SD—the acute phase. The electrocorticogram shows disappearance of spontaneous activity and the coinciding DC-shift (**a**). During these events, CBF increases up to 200–300 % for minutes. The initial increase in tpO_2 mediated by the increase in CBF is not sufficient to maintain tpO_2 at steady-state level and a decrease is evident (**b**). Calculations of $CMRO_2$ reveal an increase in oxygen consumption of up to 71 % during SD (**c**). Figure adapted with permission from ref. [12]

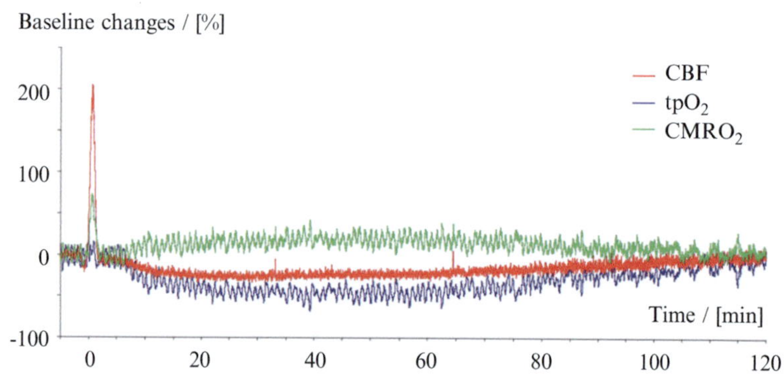

Fig. 3 In the wake of single episodes of SD. The initial 200 % increase in CBF and 71 % increase in $CMRO_2$ is followed by a prolonged oligemia phase. Also the tpO_2 is decreased in this time period. Calculations of $CMRO_2$ reveal persistent increase in oxygen consumption lasting for up to 2 h. Figure adapted with permission from ref. [12]

6 Troubleshooting

6.1 Oxygen Electrode

If during experiments measurements are getting unstable, the oxygen probe has to be checked if it has clotted at the tip or if it has a broken tip. The latter will also at the end of experiment be

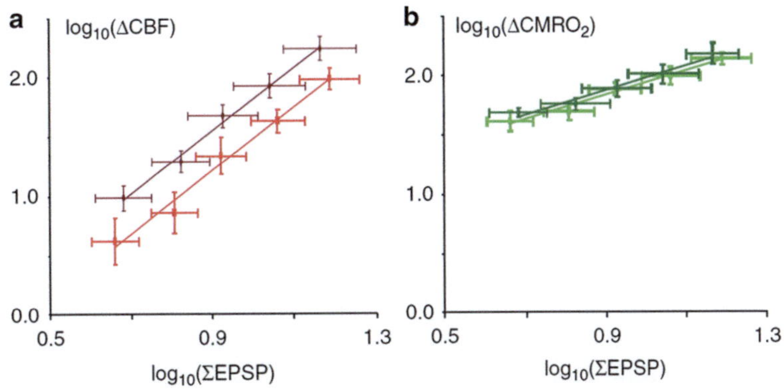

Fig. 4 Neurovascular and neurometabolic coupling: Single episodes of SD did not affect the slope of the neurovascular coupling but leads to a significant decrease in response level (**a**). The neurometabolic coupling was unaffected in the same period (**b**). Figure adapted with permission from ref. [12]

revealed because a broken electrode will have very much different calibration values.

6.2 Laser-Doppler Flowmetry

Pitfalls for LDF include movement artifacts. The LDF method is sensitive for vibrations [46] which we avoided by placing the rat in a stereotactic frame and by using a fixed holding arm for the LDF-probe securing the placing throughout the experiments. Also one has to be careful selecting a ROI (region of interest) for placing the probe that does not contain major vessels which would otherwise not reflect the microvasculature which we are interested in and the homogeneity of the sampling volume would not be equal throughout which is one of the basic assumptions on which LDF measurements rely.

6.3 Spreading Depressions

Potassium-induced SDs sometimes propagate independently in the superficial and deep layers [47]. In contrast, a SD elicited by needle stab usually invades the entire cortex with DC potentials that differ in shape from recurrent KCl-elicited SDs [47]. Furthermore, there appears to be a barrier for the vertical spread of SD between a depth of 800 and 1,200 μm in neocortex. Thus, the electrophysiological correlates of SD differ depending on the site of recording and the mode of elicitation in susceptible brain regions. Consequently, caution should be exercised and what is meant by "SD" always should be clearly defined when comparing results between groups.

7 Conclusions

Spreading depression is a unique method for challenging the neurovascular coupling. SD can be elicited in a variety of brain regions and in different species ranging from mice to man thereby being an excellent tool for translational studies into basic neurophysiology. The ever growing evidence of the presence and impact of SD on various clinical situations puts another great dimension to the use of this method. As described in this chapter, a huge number of factors (CBF, LFP, protein synthesis, and many more) have already been investigated associated to SD. The future will add even more pieces into the great puzzle of SD that has by now already been evolving for almost 70 years since its first discovery.

Acknowledgements

This study was supported by the NORDEA Foundation/Center for Healthy Aging, the Lundbeck Foundation via the Lundbeck Foundation Center for Neurovascular Signaling (LUCENS), the NOVO-Nordisk Foundation, the Danish Medical Research Council, and Foundation Leducq.

References

1. Leao AA (1944) Spreading depression of activity in the cerebral cortex. J Neurophysiol 7:359–390
2. Bures J, Buresova O, Krivanek J (1974) The mechanism and applications of Leão's spreading depression of electroencephalographic activity. Academia, Prague
3. Hansen AJ, Zeuthen T (1981) Extracellular ion concentrations during spreading depression and ischemia in the rat brain cortex. Acta Physiol Scand 113(4):437–445
4. Van Harreveld A, Fifkova E (1970) Glutamate release from the retina during spreading depression. J Neurobiol 2:13–29
5. Fabricius M, Jensen LH, Lauritzen M (1993) Microdialysis of interstitial amino acids during spreading depression and anoxic depolarization in rat neocortex. Brain Res 612:61–69
6. Mayevsky A, Doron A, Manor T, Meilin S, Zarchin N, Ouaknine GE (1996) Cortical spreading depression recorded from the human brain using a multiparametric monitoring system. Brain Res 740(1–2):268–274
7. Sramka M, Brozek G, Bures J, Nadvornik P (1977) Functional ablation by spreading depression: possible use in human stereotactic neurosurgery. Appl Neurophysiol 40:48–61
8. Avoli M, Drapeau C, Louvel J, Pumain R, Olivier A, Villemure JG (1991) Epileptiform activity induced by low extracellular magnesium in the human cortex maintained in vitro. Ann Neurol 30(4):589–596
9. Avoli M, Louvel J, Drapeau C, Pumain R, Kurcewicz I (1995) GABAA-mediated inhibition and in vitro epileptogenesis in the human neocortex. J Neurophysiol 73(2):468–484
10. Leao AAP, Morison RS (1945) Propagation of spreading cortical depression. J Neurophysiol 8:33–45
11. Kruger H, Luhmann HJ, Heinemann U (1996) Repetitive spreading depression causes selective suppression of GABAergic function. Neuroreport 7(15–17):2733–2736
12. Piilgaard H, Lauritzen M (2009) Persistent increase in oxygen consumption and impaired neurovascular coupling after spreading depression in rat neocortex. J Cereb Blood Flow Metab 29(9):1517–1527
13. Fabricius M, Akgoren N, Lauritzen M (1995) Arginine-nitric oxide pathway and cerebrovascular regulation in cortical spreading depression. Am J Physiol 269(1 pt 2):H23–H29
14. Lauritzen M, Jorgensen MB, Diemer NH, Gjedde A, Hansen AJ (1982) Persistent oligemia of rat cerebral cortex in the wake of spreading depression. Ann Neurol 12:469–474
15. Piilgaard H, Witgen BM, Rasmussen P, Lauritzen M (2011) Cyclosporine A, FK506, and NIM811 ameliorate prolonged CBF reduction and impaired neurovascular coupling after cortical spreading depression. J Cereb Blood Flow Metab 31(7):1588–1598
16. Fordsmann JC, Ko R, Choi HB, Thomsen KJ, Witgen BM, Mathiesen C et al (2013)

Increased 20-HETE synthesis explains reduced cerebral blood flow but not impaired neurovascular coupling after cortical spreading depression in rat cerebral cortex. J Neurosci 33(6):2562–2570

17. Ayata C, Shin HK, Salomone S, Ozdemir-Gursoy Y, Boas DA, Dunn AK et al (2004) Pronounced hypoperfusion during spreading depression in mouse cortex. J Cereb Blood Flow Metab 24(10):1172–1182
18. Mies G (1993) Inhibition of protein synthesis during repetitive cortical spreading depression. J Neurochem 60:360–363
19. Kraig RP, Dong LM, Thisted R, Jaeger CB (1991) Spreading depression increases immunohistochemical staining of glial fibrillary acidic protein. J Neurosci 11:2187–2198
20. Herrera DG, Maysinger D, Gadient R, Boeckh C, Otten U, Cuello AC (1993) Spreading depression induces c-fos-like immunoreactivity and NGF mRNA in the rat cerebral cortex. Brain Res 602:99–103
21. Kokaia Z, Gido G, Ringstedt T, Bengzon J, Kokaia M, Siesjo BK et al (1993) Rapid increase of BDNF mRNA levels in cortical neurons following spreading depression: regulation by glutamatergic mechanisms independent of seizure activity. Brain Res Mol Brain Res 19(4):277–286
22. Caggiano AO, Kraig RP (1998) Neuronal nitric oxide synthase expression is induced in neocortical astrocytes after spreading depression. J Cereb Blood Flow Metab 18(1): 75–87
23. Caggiano AO, Breder CD, Kraig RP (1996) Long-term elevation of cyclooxygenase-2, but not lipoxygenase, in regions synaptically distant from spreading depression. J Comp Neurol 376(3):447–462
24. Kunkler PE, Hulse RE, Kraig RP (2004) Multiplexed cytokine protein expression profiles from spreading depression in hippocampal organotypic cultures. J Cereb Blood Flow Metab 24(8):829–839
25. Nimura T, Weinstein PR, Massa SM, Panter S, Sharp FR (1996) Heme oxygenase-1 (HO-1) protein induction in rat brain following focal ischemia. Brain Res Mol Brain Res 37(1–2): 201–208
26. Plumier JC, David JC, Robertson HA, Currie RW (1997) Cortical application of potassium chloride induces the low-molecular weight heat shock protein (Hsp27) in astrocytes. J Cereb Blood Flow Metab 17(7):781–790
27. Lauritzen M (1994) Pathophysiology of the migraine aura. The spreading depression theory. Brain 117(pt 1):199–210
28. Hadjikhani N, Sanchez DR, Wu O, Schwartz D, Bakker D, Fischl B et al (2001) Mechanisms of migraine aura revealed by functional MRI in human visual cortex. Proc Natl Acad Sci U S A 98(8):4687–4692
29. Dreier JP, Woitzik J, Fabricius M, Bhatia R, Major S, Drenckhahn C et al (2006) Delayed ischaemic neurological deficits after subarachnoid haemorrhage are associated with clusters of spreading depolarizations. Brain 129: 3224–3237
30. Dohmen C, Sakowitz OW, Fabricius M, Bosche B, Reithmeier T, Ernestus RI et al (2008) Spreading depolarizations occur in human ischemic stroke with high incidence. Ann Neurol 63(6):720–728
31. Strong AJ, Fabricius M, Boutelle MG, Hibbins SJ, Hopwood SE, Jones R et al (2002) Spreading and synchronous depressions of cortical activity in acutely injured human brain. Stroke 33(12):2738–2743
32. Gold L, Lauritzen M (2002) Neuronal deactivation explains decreased cerebellar blood flow in response to focal cerebral ischemia or suppressed neocortical function. Proc Natl Acad Sci U S A 99(11):7699–7704
33. Bonvento G, Charbonne R, Correze JL, Borredon J, Seylaz J, Lacombe P (1994) Is alpha-chloralose plus halothane induction a suitable anesthetic regimen for cerebrovascular research? Brain Res 665(2):213–221
34. Cholet N, Seylaz J, Lacombe P, Bonvento G (1997) Local uncoupling of the cerebrovascular and metabolic responses to somatosensory stimulation after neuronal nitric oxide synthase inhibition. J Cereb Blood Flow Metab 17(11):1191–1201
35. Riva C, Ross B, Benedek GB (1972) Laser Doppler measurements of blood flow in capillary tubes and retinal arteries. Invest Ophthalmol 11:936–944
36. Frederiksson I, Fors C, Johansson J (2007) Laser Doppler flowmetry—a theoretical framework. Department of Biomedical Engineering, Linköping University. www.imt.se/bit/ldf/ldfmain.html
37. Fabricius M, Akgoren N, Dirnagl U, Lauritzen M (1997) Laminar analysis of cerebral blood flow in cortex of rats by laser-Doppler flowmetry: a pilot study. J Cereb Blood Flow Metab 17(12):1326–1336
38. Fabricius M, Lauritzen M (1996) Laser-Doppler evaluation of rat brain microcirculation: comparison with the [14C]-iodoantipyrine method suggests discordance during cerebral blood flow increases. J Cereb Blood Flow Metab 16(1):156–161

39. Revsbech NP (1989) An oxygen microsensor with a guard cathode. Limnol Oceanogr 34: 474–478
40. Hoffmeyer HW, Enager P, Thomsen KJ, Lauritzen MJ (2007) Nonlinear neurovascular coupling in rat sensory cortex by activation of transcallosal fibers. J Cereb Blood Flow Metab 27(3):575–587
41. Lauritzen M, Hansen AJ (1992) The effect of glutamate receptor blockade on anoxic depolarization and cortical spreading depression. J Cereb Blood Flow Metab 12(2): 223–229
42. Gjedde A (2006) Blood-brain transfer and metabolism of oxygen. In: Dermietzel R, Spray DC, Nedergaard M (eds) Blood-brain interfaces: from ontogony to artificial barriers. Wiley-VCH Verlag GmbH & Co. KGaA, Weinheim, pp 523–549
43. Zhu XH et al (2002) Development of O-17 NMR approach for fast imaging of cerebral metabolic rate of oxygen in rat brain at high field. Proc Natl Acad Sci USA 99: 13194–13199
44. Quian QR, Panzeri S (2009) Extracting information from neuronal populations: information theory and decoding approaches. Nat Rev Neurosci 10(3):173–185
45. Mathiesen C, Caesar K, Akgoren N, Lauritzen M (1998) Modification of activity-dependent increases of cerebral blood flow by excitatory synaptic activity and spikes in rat cerebellar cortex. J Physiol 512(pt 2):555–566
46. Oberg PA (1999) Tissue motion—a disturbance in the laser-Doppler blood flow signal? Technol Health Care 7(2–3):185–192
47. Richter F, Lehmenkuhler A (1993) Spreading depression can be restricted to distinct depths of the rat cerebral cortex. Neurosci Lett 152: 65–68

Chapter 15

Cerebral Blood Flow Imaging with Laser Speckle Contrast Imaging

S.M. Shams Kazmi, Lisa M. Richards, and Andrew K. Dunn

Abstract

Laser speckle contrast imaging (LSCI) is a technique for dynamic visualization of blood flow on tissue surfaces. LSCI produces images of blood flow in real-time using very simple instrumentation without the need for exogenous contrast agents, and has been widely used both in pre-clinical studies of neurological disease as well as clinical applications for monitoring of perfusion. One of the limitations of LSCI is the challenge of obtaining quantitative blood flow information. An extension to LSCI called multi-exposure speckle imaging (MESI) overcomes some of these limitations and enables repeated blood flow measurements to be performed reliably. This chapter describes MESI imaging and the use of LSCI in humans during neurosurgery.

Key words Blood flow, Laser speckle, Optical imaging

1 Introduction

Imaging methods for in vivo monitoring of blood flow dynamics are of great interest for a myriad of physiological and disease applications in both clinical and research settings. While MRI, CT, and ultrasound are commonplace in clinics for both hemodynamic and angiographic information, optical techniques based on dynamic light scattering (DLS) are often utilized for blood flow monitoring in pre-clinical research. DLS techniques include laser Doppler, speckle contrast imaging, and photon correlation spectroscopy [1, 2]. These essentially Doppler-based techniques are increasingly translating to the clinic as well. Although each of these techniques differs in their measurement geometry and analysis, each method is reliant on label-free intrinsic signals based on laser–tissue interactions. Among these techniques, laser speckle contrast imaging (LSCI) has emerged over the past decade as a powerful, yet relatively simple, method for imaging blood flow dynamics in real time. Full field imaging of blood flow based on laser speckle was first demonstrated by Fercher and Briers in 1981 [3], and this

Mingrui Zhao et al. (eds.), *Neurovascular Coupling Methods*, Neuromethods, vol. 88,
DOI 10.1007/978-1-4939-0724-3_15,

method was given the name LASCA (laser speckle contrast analysis). A significant limitation of early LASCA instruments however, was the need to record images on film and then process the images later, which resulted in low temporal resolution and large uncertainties in the speckle contrast values. With the advent of CCD cameras and modern computers in the 1990s, acquisition and processing of speckle images improved drastically and imaging of blood flow in tissues such as the skin and retina became possible. The rapid adoption of LSCI for physiological studies more recently is due to the ease and low cost of building an instrument as well as the ability to reliably quantify blood flow changes with excellent spatial and temporal resolution. Although measurements are depth integrated and weighted heavily towards superficial tissues, LSCI has been instrumental in pre-clinical studies of neurological disorders as well as clinical applications including dermatological, neurosurgical, and endoscopic studies.

2 Laser Speckle Contrast Imaging Basics

When an object is illuminated with coherent laser light, a speckle pattern forms at the camera due to the fact that the laser light reaching each pixel has traveled slightly different pathlengths, which then combine to form a random interference pattern. Temporal changes in the speckle pattern encode information about the motion of the scattering particles encountered in the sample. Particularly, when the scattering particles are in motion (i. e. blood cells) the speckle pattern fluctuates in time. Provided that the exposure time of the camera is longer than the time scale of the speckle intensity fluctuations, the camera integrates these variations resulting in the appearance of a blurring of the speckle pattern.

The degree of spatial blurring can be quantified by calculating the speckle contrast, K, defined as the ratio of the standard deviation, σ_s, to the mean intensity of pixel values, $<I>$, in a small region of the image [4],

$$K(T) = \frac{\sigma_s(T)}{\langle I \rangle} \tag{1}$$

where T is the exposure time of the camera. In areas of increased motion there is more blurring of the speckles resulting in a lower spatial contrast value. A spatially resolved map of local speckle contrast can be calculated from a raw speckle image by computing this ratio at each point in the image from the pixels in a surrounding NxN region, (typically $N=7$). Theoretically the speckle contrast values span the range between 0 and 1 when the speckle pattern is sampled properly [5, 6]. A speckle contrast of 1 indicates that there is no blurring of the speckle pattern and therefore, no motion, while a speckle contrast of 0 means that the scattering particles are moving fast enough to average out all of the speckles. Quantitative measurement of blood flow can be challenging due to the complex physics that relate the measured speckle contrast to the underlying blood flow. However, a number of technical advances have demonstrated significant improvements to the accuracy of speckle imaging.

The basic LSCI configuration consists of a laser whose beam is expanded and adjusted to illuminate the area of interest, which can vary from a few millimeters to several centimeters. The angle of the incident laser light ranges from near normal to oblique, and the wavelength is generally in the red to near infrared where tissue scattering dominates absorption. Diode lasers are widely available in this range (600–800 nm) and are the prevalent illumination sources for LSCI. While different wavelengths will lead to slightly different sampling depths, these differences are not large. Since LSCI is based purely on scattering, the wavelength is significantly less important than imaging methods reliant on absorption or fluorescence contrast. More importantly, the coherence length of the laser will govern the largest pathlength in the tissue that can provide quantitative flow information at a camera pixel. This can be optimized by selecting lasers with narrow spectral bandwidths relative to their central wavelength, denoted generally as single longitudinal mode or narrow linewidth lasers. The second main component is the camera on which light scattered from the sample is imaged to enable recording of the speckle pattern. The specifications of the cameras used for LSCI vary widely, but inexpensive cameras have been demonstrated to provide excellent blood flow images [7] and enable detailed physiological studies to be performed [6, 8, 9].

A significant limitation to LSCI is the shallow penetration depth of the images. Because of the illumination and detection geometry, detected photons sample only the superficial few hundred microns of tissue. Although skin and retinal blood flow monitoring does not require any surgical access [10, 11], LSCI is more commonly used in applications that require removal or thinning of overlying tissues such as the skull in brain [6] to obtain optical access to the tissue of interest.

A typical example of a raw speckle image of the rat cortex taken through a thinned skull, and the computed speckle contrast are shown in Fig. 1. The raw speckle image illustrates the grainy appearance of the speckle pattern, with some regions appearing more blurred than others. The speckle contrast image is computed from raw speckle image using Eq. 1 to generate a two-dimensional map of blood flow in the tissue. Areas of higher baseline flow, such as large vessels, have lower speckle contrast values and appear darker in the speckle contrast images. LSCI is sensitive to blood flow in visible surface vasculature as well as areas of tissue that do not contain visible vasculature (i.e., parenchyma). Speckle contrast values in areas containing no obvious surface vasculature contain information

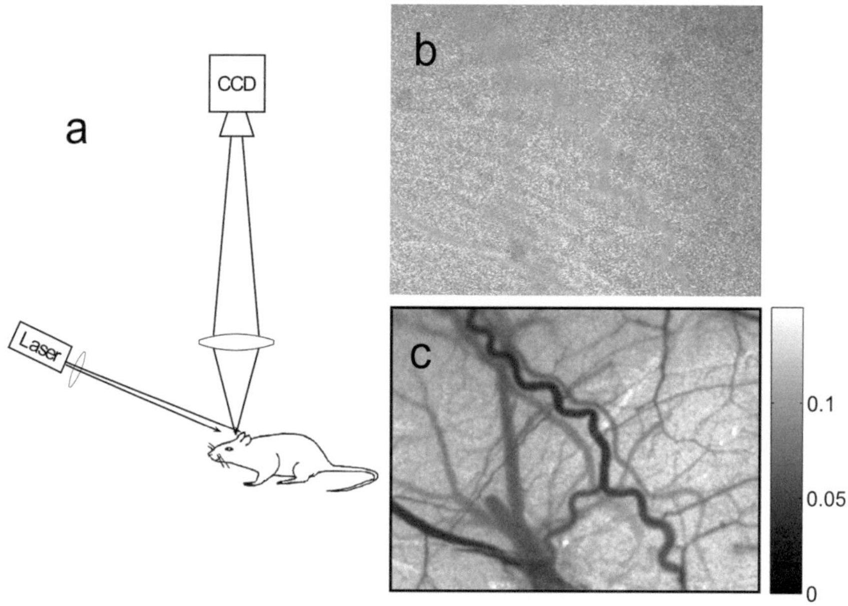

Fig. 1 (**a**) Basic laser speckle contrast imaging set up consisting of a laser diode and camera. A raw speckle image of the rat cortex, taken through a thinned skull, contains seemingly little information (**b**). However, when the speckle contrast is calculated using a sliding window (**c**), a tremendous amount of information is revealed about the motion of the scattering particles in the sample

about blood flow in the underlying microvasculature, even though the microvasculature is not spatially resolved. Parenchymal speckle contrast values are higher than areas containing large supply and draining vessels since overall blood flow is lower. However, changes in blood flow can still be detected [9, 12] and the speckle contrast increases drastically when flow is interrupted [13, 14].

Although as introduced, speckle contrast values, $K(T)$, are indicative of the level of motion in the sample, the flow estimates are limited to qualitative examinations. The relationship with the underlying blood flow is nonlinear and is currently being examined in the field of DLS. Obtaining quantitative flow measurements from speckle contrast images first involves accurately relating the speckle contrast values to the characteristic autocorrelation decay time of the speckles, τ_c. This is a unique process of working with time-integrated speckle images. The next step is to relate τ_c to the rate of DLS events, which quantifies the underlying flow or speed; a common process with all DLS techniques, including diffused correlation spectroscopy, laser Doppler flowmetry, and speckle techniques. In the seminal LSCI studies, a simplified model was proposed that related the speckle contrast values to the speckle correlation time, τ_c and speed of the scattering particles[3]. More recently a corrected version of this model was proposed by Bandyopadhyay et al.

$$K(T,\tau_c)=\left(\beta\frac{e^{-2x}-1+2x}{2x^2}\right)^{1/2} \tag{2}$$

where $x = T/\tau_c$. This model [15] and its earlier variants [3, 16, 17], each dependent on the derivational assumptions, have been demonstrated to be fairly accurate in its ability to predict relative blood flow changes and have been adopted for a wide range of blood flow imaging applications.

3 Multi-Exposure Laser Speckle Contrast Imaging

The flow estimates from traditional single exposure LSCI are relative and sensitive to a number of non-flow related parameters which affect the ability to establish a quantitative baseline [13, 18]. Recently, an extension to LSCI called Multi Exposure Speckle Imaging (MESI) was introduced that improves the quantitative accuracy of blood flow changes by enabling separation of non-flow related contributions from the measured speckle contrast [19]. Particularly, one of the limiting factors on the accuracy of blood flow changes measured with LSCI is the presence of light scattered from static (i.e., non-moving) tissue elements. Typically, mathematical models that relate speckle contrast to τ_c do not take into account the presence of a static scattering from non-moving tissues including thinned or intact skull. The MESI technique uses an improved model and instrumentation to more precisely extract the flow-related contribution to the observed speckle contrast and thus provides an improved assessment of flow [14].

The instrumentation for MESI image acquisition, although similar to traditional LSCI, requires control over both the camera exposure duration and the laser intensity. Typically, a laser diode illuminates the sample while the computer triggers 15 camera images spanning four decades of exposure while simultaneously adjusting the amplitude of the laser light in each exposure through an acousto-optic modulator (AOM) (Fig. 2a). By holding the total amount of laser light in each image constant, illumination variations do not impact the flow analysis over the multiple exposures.

Over very long exposure times, the speckle intensity fluctuations have sufficient time to blur completely, lowering the measured speckle contrast values. Conversely over very short exposures, the speckles will essentially be frozen and contrast values will be maximized (in theory close to 1). Thus, speckle contrast values are a strong function of the camera exposure duration, T, as illustrated in Fig. 2b, c.

Alternatively stated, lower flow regions have speckle fluctuations that are slow compared to the camera exposure and therefore result in no appreciable blurring. As the exposure duration is increased, areas with lower flow, such as small vessels, begin to manifest in the speckle contrast image (Fig. 2b). Therefore, a close relationship exists between the camera exposure and the underlying perfusion. The MESI model maps the dependence of the speckle contrast, K, on the exposure duration of the camera, T,

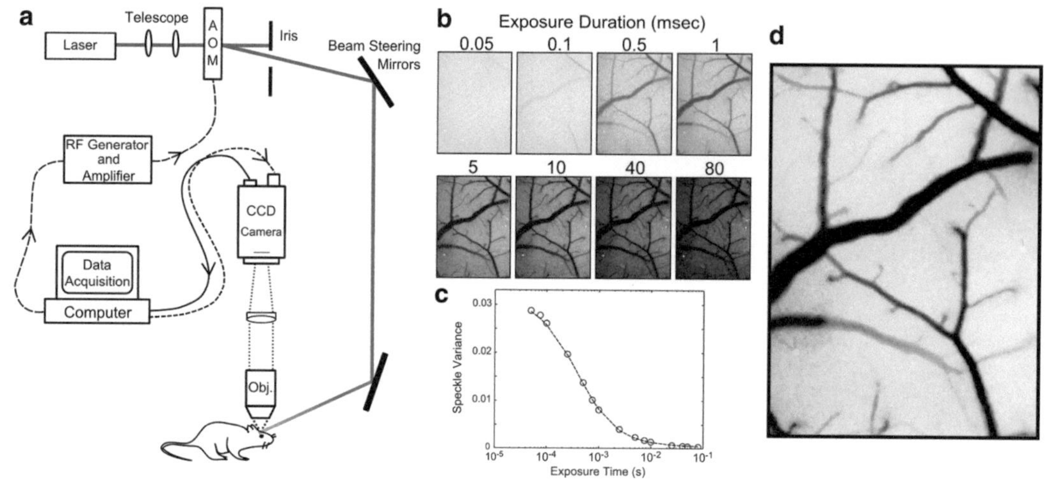

Fig. 2 (**a**) Schematic of combined multi-exposure speckle & RBC reflectance imaging. (**b**) Representative single exposure speckle contrast images (15 exposures total, 8 shown). (**c**) Speckle contrast (or variance) dependence on camera exposure duration. (**d**) MESI inverse correlation time image of flow computed from 15 exposures. Darker pixels are linearly representative of increasing flow

(Fig. 2c) to obtain an estimate of the correlation time of the speckles, τ_c. By doing so, the MESI technique more holistically samples the range of flow distributions in the tissue while more accurately extracting the flow-related contribution from the imaged speckles. Thus, a more reliable inverse correlation time image, derived from the analysis over the multiple exposures, can be generated (Fig. 2d).

Particularly, speckle contrast measurement at multiple exposure durations has been shown to improve the quantitative accuracy of flow changes (versus single exposure) in calibrated microfluidic phantoms [19] as well as in vivo [14]. Measurement of acute flow changes during middle cerebral artery occlusion demonstrate that single exposure LSCI measurements often underestimate large flow changes, while MESI accurately determined the flow deficits. The chronic accuracy and reproducibility of MESI flow measures have also been examined against high-frame rate red blood cell photography (e.g., RBC tracking) as an absolute flow calibration in mice over several days [20].

The MESI approach is an extension to the traditional theory relating τ_c to $K(T)$ that accounts for the presence of static scattering. The field and intensity autocorrelation functions of the speckle pattern, g_1 and g_2, respectively, are related through the Siegert relation. However, a modification [21, 22] must be employed in the presence of significant non-moving tissue components resulting in a relationship of the form

$$g_2(\tau) = 1 + A\beta\,|g_1(\tau)|^2 + B\beta\,|g_1(\tau)|, \tag{3}$$

where $A = I_f^{\,2}/(I_f + I_s)^2$ and $B = 2I_fI_s/(I_f + I_s)^2$. I_f and I_s are the intensities of dynamically and statically scattered light, respectively.

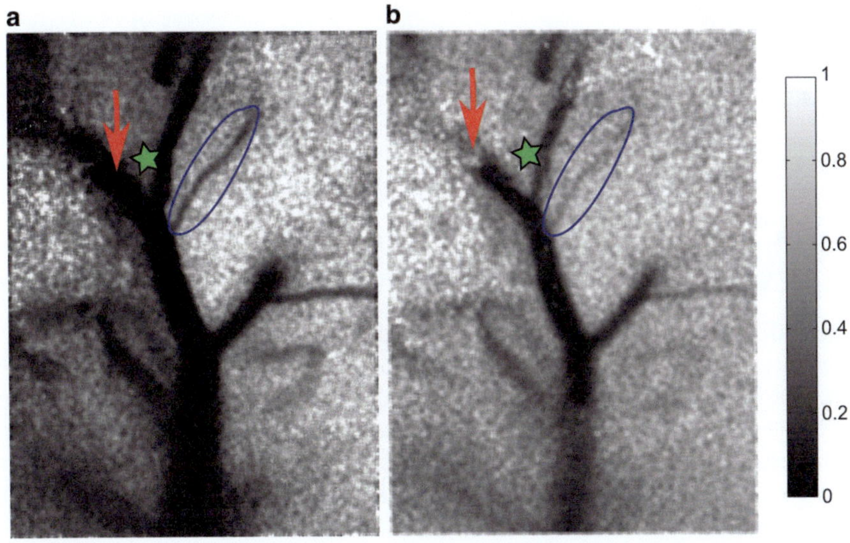

Fig. 3 Full field relative correlation time maps obtained using the (**a**) MESI technique (**b**) LSCI technique (5 ms exposure). Three corresponding regions marked in the figures illustrate the superior performance of the MESI technique. The boundary between the thin skull and the craniotomy indicated by the *red arrow* is clearly visible in (**b**), but not in (**a**). There is a clear change gradient in the region indicated by the *star* in (**b**), but this gradient is invisible in (**a**). The vessel circled is more visible in (**a**) versus (**b**). Relative correlation time estimates obtained using the MESI technique are not affected by the presence of thinned skull. Hence, similar estimates of blood flow changes are obtained across the boundary between the thin skull and craniotomy regions

By relating the normalized field autocorrelation function (g_1) to the normalized intensity autocorrelation function (g_2) and accounting for the mixing of static and dynamically scattered light a more robust expression of speckle variance is obtained,

$$K(T,\tau_c)=\left\{\beta\rho^2\frac{e^{-2x}-1+2x}{2x^2}+4\beta\rho(1-\rho)\frac{e^{-x}-1+x}{x^2}+v_{ne}+v_{noise}\right\}^{1/2}, \quad (4)$$

where $x=\frac{DT}{\tau_c}$, ρ is the fraction of light dynamically scattered, and β is again an instrument-dependent normalization factor. Residual variance that arises from scattered light in the absence of flow, v_{ne} (non-ergodic variance), and from noise sources, v_{noise}, is also modeled. The reliability of perfusion estimates from vessels traversing exposed tissue to thinned skull is depicted in Fig. 3 in terms of inverse correlation times (ICT) maps. The flow visibility is greatly improved with the MESI technique, highlighting a better decoupling of the motion from the speckle contrast analysis. These factors translated into better fidelity of MESI predicted flow dynamics observed in acute imaging settings [14].

This MESI model also includes accounting of the effects of multiple scattering inherent in its derivational assumptions, particularly relevant for interpreting perfusion information from tissues containing more than just an isolated capillary flow. As the

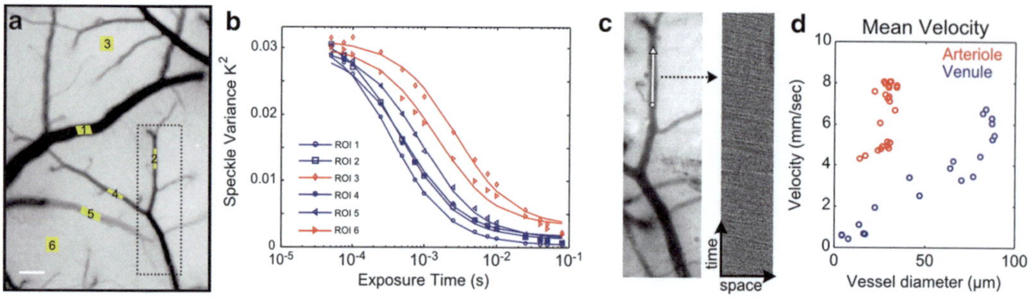

Fig. 4 (**a**) MESI ICT image of flow computed from 15 exposures. Scale bar = 150 μm. (**b**) MESI computed speckle visibility curves for selected ROIs in (**a**). Six Regions of Interest (ROIs) selected for correlation time computation. Four ROIs are selected from areas with large resolvable flow projections from vessels (*blue curves*) and two in areas without resolvable projections (*red curves*). (**c**) Green light illuminated reflectance image (*left*) corresponding to selected region in (**a**). Centerline (white, 100 μm) represents the region over which RBCs are tracked. RBC time-course (*right*) corresponding to selected vessel's centerline compiled from repeated photographs. (**d**) Extracted mean RBC velocity versus vessel diameter from RBC tracking

number of scattering events increases, the speckles observed at the camera are further decorrelated [21, 23, 24]. A unitless weighting term, *D*, proportional to the vessel caliber is used to inversely scale the correlation times to account for the disparity in the number of moving particles encountered between vessels of differing sizes. Contributions to speckles from parenchymal regions, however, cannot be decoupled or attributed to single vessels, and ICT can rather be interpreted as a regional perfusion index [9, 12, 25] integrating a host of unresolvable microvasculature Fig. 4a, b.

A common criticism of the LSCI technique has been the inability to establish a quantitative baseline [13, 18] due to the limited flow sensitivity with a single camera exposure along with the static and noise factors that further confound the motion contribution to the imaged speckles. This has not hindered some studies in showing CBF dynamics from collateral flow after occlusion [26] in a short-term chronic setting, but has limited the flow observations to largely qualitative dynamics. In dealing with these limitations to traditional LSCI, the advancement of using multiple exposure durations to obtain speckle contrast measurements with higher accuracy has been validated initially in microfluidic flow phantoms [19] and later examined in the presence of a thinned skull [14] in vivo. Noting the inherent improvement in flow sensitivity, the multi-exposure aspect is increasingly being adapted by several groups [27–29] for imaging vascular flows in varying applications.

For examining the long-term accuracy over multiple imaging sessions, dynamic high frame rate RBC photography [30, 31] has been used as a calibration procedure for MESI computed ICT in a chronic setting (Fig. 4c, d). Specifically, comparisons between changes in inverse correlation time ($1/\tau_c$) dynamics have been tested

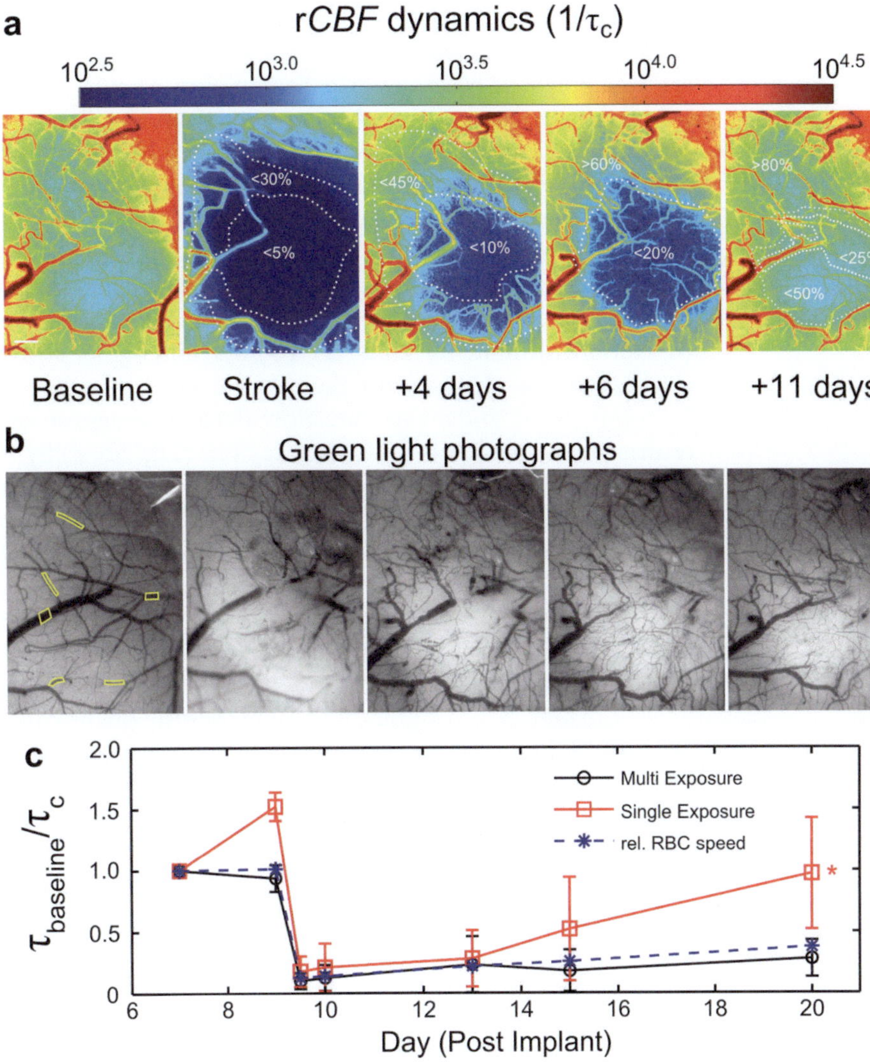

Fig. 5 (**a**) MESI inverse correlation time (ICT) images of perfusion before and after photothromboses. Logarithmic scale is utilized to capture occlusion as well as residual flow, redistribution, and remodeling. Higher ICT values signify higher flow. Regional outlines group flow dynamics with labels of percentage of baseline. Scale bar = 300 μm. (**b**) Corresponding green reflectance photography of the cortical surface. Labeled ROIs correspond to both single-exposure LSCI and MESI computed flow indices averaged in (**c**). (**c**) Average flow dynamics from MESI and LSCI versus RBC speed changes in six vessels normalized to the first imaging session. Asterisk denotes significant ($p < 0.05$, repeated measures ANOVA) difference between LSCI (5 ms) and RBC tracking flow predictions over the chronic period

against centerline RBC speed dynamics of surface microvessels in the rodent brain over multiple days.

Chronic imaging of cerebral blood flow is an important tool for investigating vascular remodeling following injury such as stroke. In Fig. 5c RBC tracking was conducted in selected vessels over five imaging sessions spanning nearly 2 weeks in a cranial window implanted mouse. A flow deficit by means of photothrombotic occlusion was also induced in a branch of the middle cerebral artery supplying a portion of the field of view. The density of redistributed flows was monitored over several days (Fig. 5a, b) with the regional perfusion levels also demarcated as a percentage of baseline. The MESI technique was useful in highlighting the changing vascular remodeling phases over the chronic period by following the influx and dispersion of new and returning flows. MESI flow estimates retained consistency with RBC tracking across the entire chronic period and registered a nearly threefold lower percent deviation with RBC tracking ($\Delta = 10 \pm 5\%$) and higher correlations ($r = 0.94 \pm 0.09$) than any of the single-exposures independently. In comparison, single-exposure flow dynamics suggest an errant return to baseline levels in selected regions (Fig. 5b, c) by the end of the chronic imaging period, along with early baseline inaccuracies. Over multiple animals [20], improved accuracy in longitudinal imaging is demonstrated between flow measurements derived from MESI and RBC tracking (~10 % deviation) versus single-exposure LSCI and RBC tracking (~24 % deviation) flow measures. Correlations of flow dynamics with RBC tracking were closer as well with MESI ($r = 0.88$) than LSCI ($r = 0.65$).

Ultimately, MESI significantly improves the quantitative accuracy of LSCI blood flow measurements in both acute and chronic imaging settings. The reliability and quantitative accuracy of between-subject comparisons remains to be confirmed and may be susceptible to morphological variations (i.e., vascular architecture variations) due to the spatially integrated imaging paradigm. Nevertheless, the improved quantitative accuracy extends LSCI beyond qualitative descriptions of changes in flow, necessary in physiological, preclinical, and clinical studies of hemodynamics.

4 Intraoperative Laser Speckle Contrast Imaging

In addition to its use in animal studies, the utility of LSCI for use during neurosurgery has been demonstrated in several studies. To be most useful to surgeons, LSCI hardware can be integrated into an existing neurosurgical microscope. By integrating the instrumentation into the current surgical workflow, additional setup time can be minimized and interruption to the surgical procedure during imaging is eliminated. This section describes one implementation of LSCI intraoperatively and highlights the instrumentation and data analysis for imaging of blood flow in humans during neurosurgical procedures.

4.1 Clinical Instrumentation

For use during neurosurgery, we integrated the LSCI hardware into a Zeiss OPMI Pentero neurosurgical microscope, which was convenient and easy to use for the surgeons [32–34]. A schematic

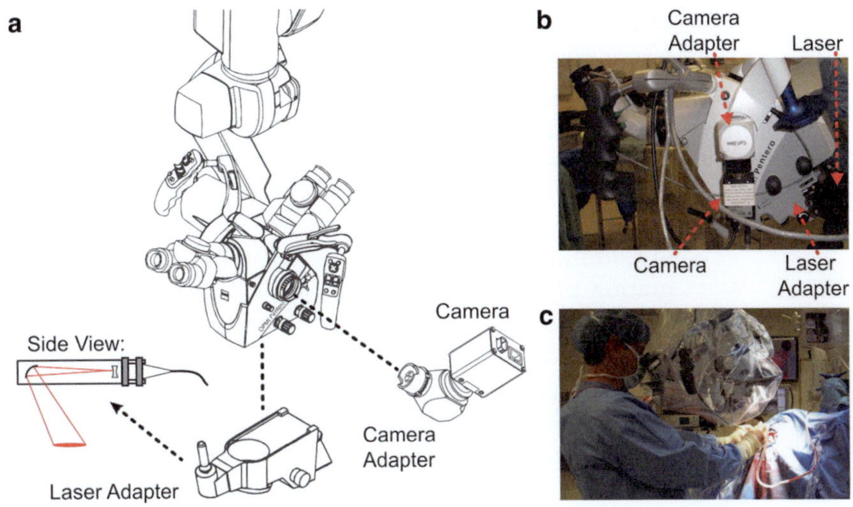

Fig. 6 (**a**) A schematic of the intraoperative instrumentation shows how the Zeiss OPMI Pentero Microscope is adapted with add-on attachments to measure cerebral blood flow intraoperatively using LSCI. Drawings are adapted from the *Zeiss OPMI Pentero Manual Issue 9.3*. (**b**) Photograph of the intraoperative instrumentation with all add-on components labeled. (**c**) The LSCI adapted microscope is covered in a sterile drape and used to assist with the tumor resection procedure without any interference from the additional hardware

of the adapted microscope is shown in Fig. 6a, with a corresponding photograph in Fig. 6b. For illumination of the cortical surface, a laser diode ($\lambda = 660$ nm, $P = 120$ mW) was incorporated into an add-on laser adapter, which was attached to the bottom of the microscope head. The laser beam travels linearly through the adapter until it encounters a curved mirror, which directs the beam towards the cortex as shown in the side view of Fig. 6a. The steering toggle on the laser adapter controls the angle of the mirror and consequently the position of the beam on the tissue surface. The laser power was measured to be 28 mW/cm^2 for the set up, which is far below the ANSI standard of 200 mW/cm^2 for maximum permissible exposure to a visible laser beam [35]. A camera (Basler A602f) was connected to the side viewing port on the microscope head using a c-mount camera adapter, which allowed use of the microscope's built-in imaging optics for focus and zoom controls. For this study, the microscope was set to maximum zoom and the camera field of view was ~2×1.5 cm. The addition of the laser and camera attachments did not interfere with sterile draping or with normal use of the microscope, as shown in Fig. 6c.

4.2 Clinical Procedure

During the pilot clinical study, ten patients were imaged with the adapted microscope during brain tumor resection surgeries performed at the NeuroTexas Institute in St. David's Hospital, located in Austin, TX. The clinical study was approved by the Institutional Review Boards of the University of Texas at Austin and St. David's

Hospital. Before each procedure, the laser and camera attachments were added to the microscope and sterile draping was performed. The timing of the LSCI imaging procedure was at the discretion of the surgeon, and was performed either before or after the tumor resection portion of the procedure. If the microscope was not already in use during the procedure, the surgeon positioned the microscope over the cortical area immediately prior to LSCI imaging, using the built-in xenon lamp to guide placement. The integration of the hardware into the surgical microscope allowed the surgeon to precisely control the positioning of the microscope head, which provided flexibility to accommodate for variability in the craniotomy location of each patient. Setup time was kept to less than 5 min, since the surgeons were familiar with the microscope.

After positioning the microscope, the built-in xenon lamp illumination was turned off and the laser diode was turned on for LSCI. The surgeon adjusted the focus and zoom controls as needed and flushed the cortical tissue with sterile saline to reduce specular reflections in the camera field of view. LSCI images were recorded for ~10–15 min using a camera exposure time of 5.0 ms. During image acquisition, the patient's electrocardiogram (ECG) signal was recorded using an existing anesthesia monitoring system in the operating room. The camera exposure signal was also recorded to allow location of each image in time relative to the cardiac cycle for post-process filtering. Images of baseline blood flow conditions were recorded for all patients, and in three cases, images were acquired before and after cortical stimulation with either bipolar cautery or the Ojemann Cortical Stimulator. After the surgical procedure, the LSCI hardware attachments were removed from the microscope, leaving it exactly as it was before the surgery. The add-on LSCI hardware can be easily portable to different operating rooms or different hospitals, as long as the surgical microscope is available.

4.3 Clinical Study Overview

The results from the pilot clinical study are promising and demonstrate the instrument's ability to visualize blood flow. For a subset of patients, images were acquired with the Zeiss built-in color camera under xenon lamp illumination for comparison with the speckle contrast images. Figure 7 shows the camera field of view recorded by the color camera along with corresponding representative speckle contrast images recorded from the same areas for three patients. The color photographs have been registered to match the orientation of the speckle contrast images, which shows excellent alignment of the anatomical vasculature in the color images and the blood flow maps in the LSCI images. In the LSCI images, the blood vessels appear dark, which indicates low contrast and high flow in those regions. This indicates that the vasculature in the field of view is unobstructed. The exception is the large vein in the center of Fig. 7f, which does not appear obstructed in the

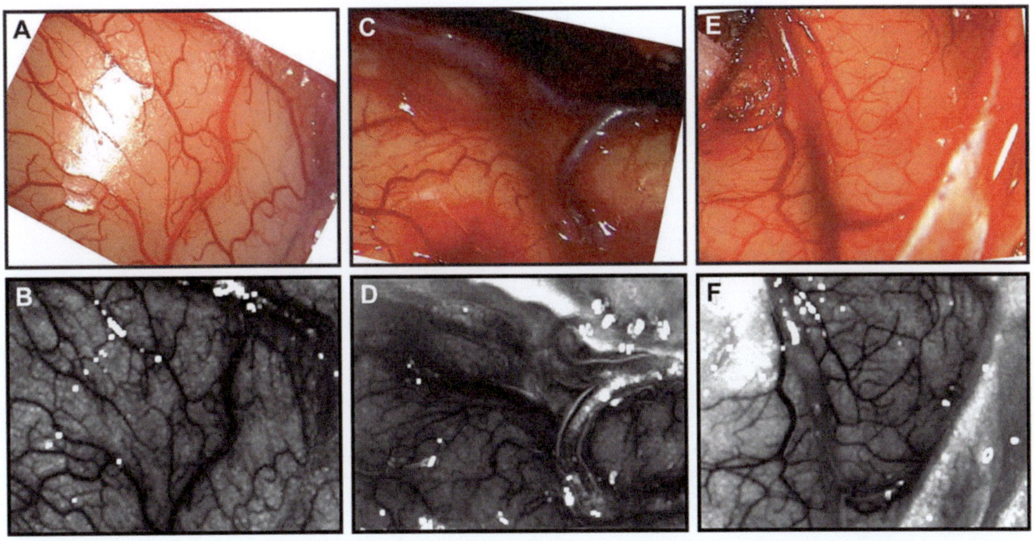

Fig. 7 (**a**, **c**, **e**) Color digital photographs taken with the built-in Zeiss color camera under the microscope's xenon lamp illumination from patients 7, 8, and 9. The photographs have been registered to match the orientation of the corresponding speckle contrast images (**b**, **d**, **f**), which are averaged over 10 frames with a field of view is ~2 × 1.5 cm. The *bright white* areas in both the color images and the speckle contrast images are regions of specular reflection

corresponding color image but has higher contrast and thus slower flow than surrounding vessels. The color images in Fig. 7c, e have blood pools present in the surgical field, which obstructs the view of the local vasculature in those regions. The corresponding LSCI images in Fig. 7d, f are not impacted by the presence of the blood pools, as flow is still observed in those regions. This ability to "see through" blood pools in the surgical field is an example of the improved visualization possible with LSCI and could be useful for identifying obstructed vessels in those regions.

4.4 Compensating for Physiological Motion Artifacts

One of the challenges involved with any in vivo imaging technique is compensating for physiological motion. Because LSCI is highly sensitive to motion, it is important to separate and remove any contributions to the signal that do not originate from blood flow alone. The cardiac cycle leads to unavoidable pulsatile variation in blood flow that is visible in most in vivo applications. Using the recorded ECG signal for each patient, we can account for fluctuations in blood flow that occur during the cardiac cycle using an ad hoc filter to improve visualization of functional changes in blood flow. There are also motion artifacts from tissue deformation during pulsation and respiration after the craniotomy procedure, which results in a constantly changing camera field of view. Image registration can be used to account for this physical motion of the brain and allows tracking of a specific tissue region over time.

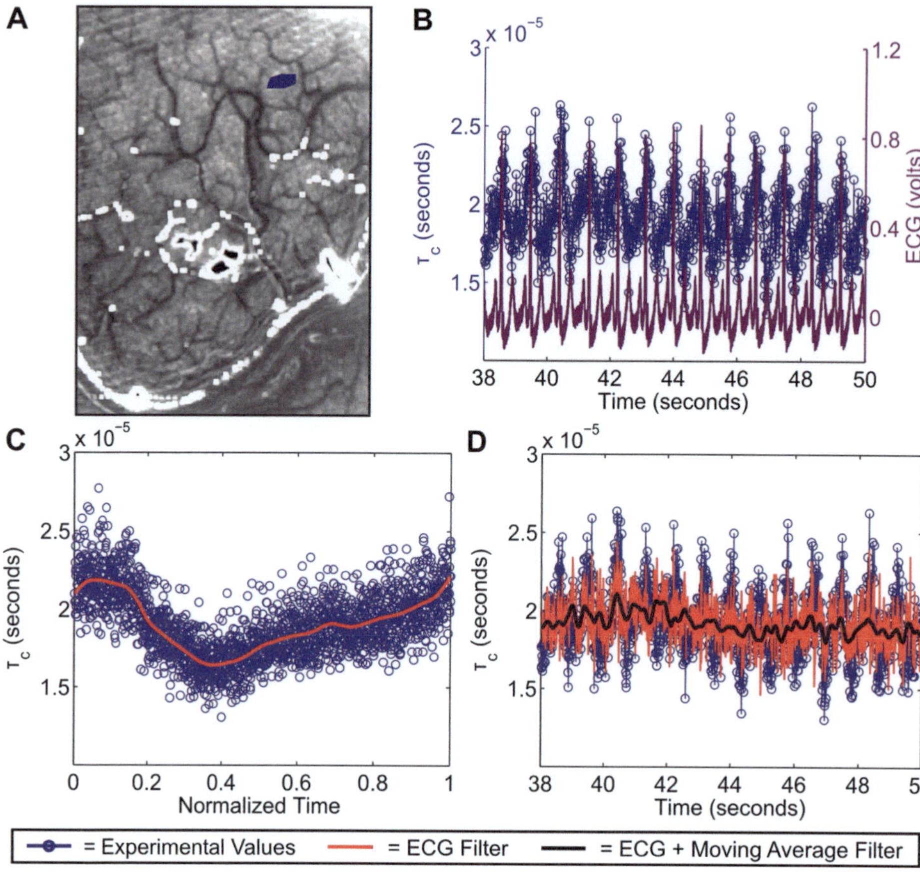

Fig. 8 An illustration of the ECG filtering process used to remove pulsatile motion artifacts. (**a**) The speckle contrast image from patient 4 is averaged over 10 frames with the region of interest (ROI) shown in blue and a field of view of ~2×1.5 cm. (**b**) The average correlation times τ_c computed from the ROI are co-localized in time with the ECG waveform recorded for the patient. (**c**) The ad hoc ECG filter shape is generated using the ROI average correlation times from 25 heartbeats and the normalized time of each acquisition. (**d**) The ECG filtered output is shown in *red* and the result after applying a small window moving average filter to the filtered output is shown in *black*. The correlation time data shown in (**b**), (**c**), and (**d**) is taken from the translation-registered image set

By matching the spatial location of images taken at different time points using image registration, blood flow can be monitored more accurately over time for distinct tissue regions.

4.5 Cardiac Filtering

To illustrate the motivation for cardiac filtering more clearly, representative results are shown in Fig. 8 from patient 4. Figure 8a shows the speckle contrast image with the region of interest (ROI) used for subsequent analysis in blue. The measured correlation time for the ROI is plotted on the same time-scale as the patient's ECG waveform in Fig. 8b, which illustrates the magnitude of the fluctuations observed within each heartbeat. These pulsatile changes in blood flow would be acceptable during procedures when large changes in blood flow are expected, such as aneurysm clipping or vessel bypass, as they would not hinder visualization of significant changes in flow. However, these fluctuations would mask small changes in blood flow, which would be an issue during functional mapping procedures where blood flow changes may be as small as 10 % from baseline values [36]. Thus, it is important to remove this artificial rise and fall of blood flow within each heartbeat to allow more accurate monitoring during any surgical procedure.

Because the fluctuations are synchronized in time with the cardiac cycle, we can use an ad hoc ECG filter based on the shape of the signal within each heartbeat to reduce the beat-to-beat variability. The filter design is similar to an adaptive filter technique for MRI developed by Deckers et al. [37], and has been previously described by our group [32]. The filter is designed using the images from 25 cardiac cycles and their corresponding acquisition time relative to the heartbeat. The correlation time measured from each image is assigned a "normalized time," which is the time of image acquisition relative to the nearest previous R peak in the ECG waveform. The correlation time data from the ROI is plotted against its normalized time and a large window moving average filter is used to generate the ad hoc ECG filter function, as shown in Fig. 8c. Thus, the filter function represents the average shape of the correlation time signal during the ECG cycle and will be used to suppress this expected shape in the rest of the image set. To apply this filter, the filter value for the corresponding normalized time is subtracted and the median of the ECG filter is added back for every measured correlation time. To further reduce noise and improve visualization, a small window moving average filter is applied to the ECG filtered result. The result after ECG filtering and after applying the small window moving average filter to the ECG filtered result is shown in Fig. 8d, which illustrates how ECG filtering reduces the variability within the cardiac cycle while retaining inherent physiological changes in flow. One note regarding Fig. 8 is that the correlation time data displayed is from the registered image set, and the procedure used for image registration will be described next.

4.6 Image Registration

Image registration was performed to account for large-scale physical motion of the tissue and to align the camera field of view for each image set. An open source software package for medical image registration called *Elastix* was used to register the image sets in post-processing [38]. This program is based on the Insight Segmentation and Registration Toolkit (ITK) and performs intensity-based image registration to match the position of a moving image to a fixed image using a given transformation type, similarity measure, and optimization procedure specified by the user. The type of transform used during registration is one of the most important parameters, as it determines what types of deformations are allowed in the spatial mapping between the moving and fixed images. Because the brain tissue deforms during physiological motion, a nonrigid transform would be required for an exact mapping of the images to account for tissue distortion during pulsation and respiration. However, LSCI records a 2D view of the 3D deformation, which limits the majority of tissue motion in the recorded image sets to the *x* and *y* directions. Thus, the translation transform was used to preserve the image integrity and decrease the complexity of the image registration procedure. To validate this choice, the B-spline nonrigid transform was used for a subset of patients and the results were directly compared to the translation transform results from the same patients. The mean percent difference between the translation and B-spline registered results across 16 ROIs from four patients was 2 %, which confirms that the translation transform is appropriate for these image sets. The similarity measure and the optimization procedure were set to mutual information and adaptive stochastic gradient descent, respectively, as good performance was observed with these settings. Registration was performed on the speckle contrast images and then the ECG filtering procedure was performed on the registered image set.

A side-by-side comparison of the original and registered image sets is shown for one baseline case and one cortical stimulation case in Fig. 9. The left column shows the LSCI image with the locations of four or five color-coded ROIs, the middle column shows the original time courses of each of the ROIs, and the right column shows the same time course after image registration. Figure 9a–c is the baseline case from patient 1 and Fig. 9d–f is the cortical stimulation case from patient 6. The two yellow ovals in Fig. 9d depict the regions where the Ojemann cortical stimulator contacted the cortex. The time courses in these plots have been ECG filtered to remove the cardiac motion artifact and are equivalent to the black curve from Fig. 8d. Looking first at the baseline case, the original time course (Fig. 9b) shows the cyan ROI in the vessel has both spike-like and broad increases in τ_c. Because the correlation time is inversely related to flow, these increases in τ_c correspond to a decrease in flow. The time course is significantly altered after registration (Fig. 9c) and appears much more stable over time compared to the original image set. Thus, these increases in τ_c were solely due to the ROI moving outside of the vessel and into the surrounding parenchyma as a result of tissue motion. For the parenchyma ROIs, the time courses appear very similar between the original and registered image sets, especially for the red and blue ROIs that are located far from surrounding vessels. For the green parenchyma ROI that is closer to nearby vessels, there is some observable change after registration, mostly increased stability and higher τ_c values that are less influenced by surrounding vessels. For the cortical stimulation case, it is clear that the registration makes a significant impact (Fig. 9f), with a reduction of spikes and noise in the registered results. Figure 9f shows a brief increase in flow in all regions followed by a return to baseline, with the cyan region farther from the stimulation location peaking later in time

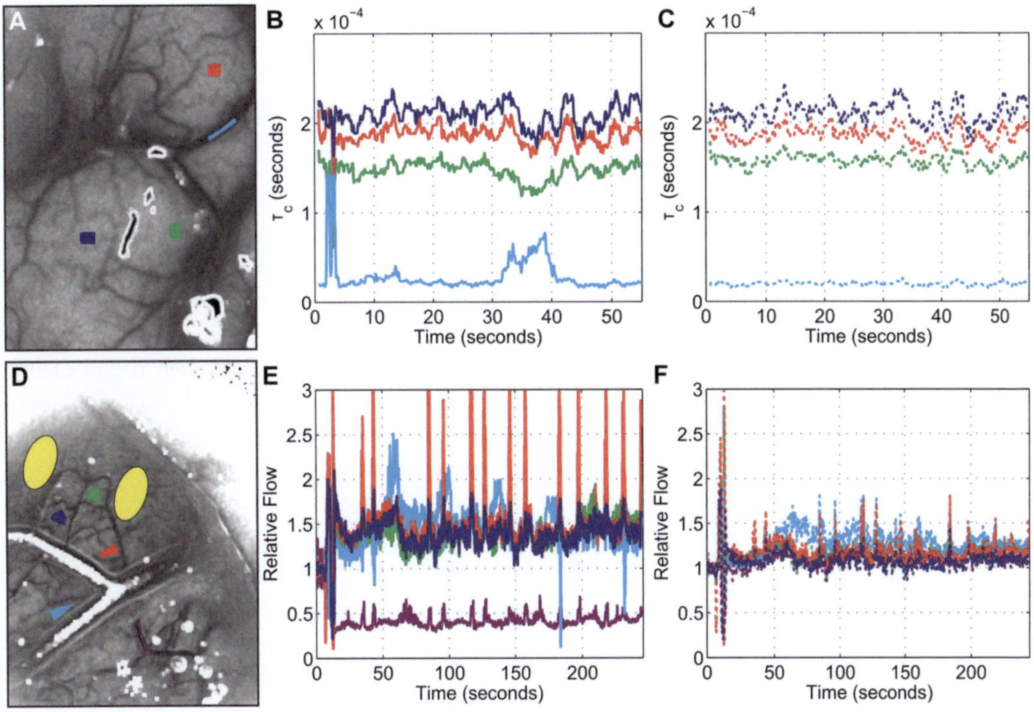

Fig. 9 A comparison between the time courses of regions of interest (ROIs) before and after translation-based registration for Case 1 (baseline, **a–c**) and Case 6 (stimulation, **d–f**). The left column (**a**, **d**) displays the speckle contrast images at the start of acquisition averaged over 10 frames with a field of view of ~2 × 1.5 cm. The ROI locations used for analysis are color-coded and match the colors used for the plots. The middle column of plots (**b**, **e**) displays the time course of the original recorded correlation time τ_c during the course of the intraoperative acquisition. The right column of plots (**c**, **f**) displays the time course of the correlation time after translation-based image registration of the image set. All plots have been ECG filtered to remove the cardiac motion artifact. Plots for the same patient were plotted with the same axes for easier comparison before and after registration

than the closer regions. This suggests that a cortical spreading depression may have occurred after stimulation [39], which was not clearly observable in the original time course (Fig. 9e). The magenta ROI in the original time course is also very deceiving, as it implies that the vessel is partially occluded to <50 % of baseline flow after the stimulation. Registration confirms that this was simply from the ROI moving into the nearby parenchyma tissue, as the magenta ROI follows the trend of the blue, red, and green ROIs in the registered result. The results from the stimulation case confirm the importance of removing the tissue motion artifact from the image sets, as it is crucial for proper visualization of the changes in blood flow that are occurring in the surgical field. The time courses seen in Fig. 9 clearly show the benefits of combining both the ECG filtering and the image registration, as artificial changes in flow from the cardiac cycle and tissue motion are removed and we are left with a more accurate view of the physiological changes in flow occurring in the cortex.

5 Conclusions

LSCI is an emerging technique for high resolution, real-time imaging of blood flow. The relatively simple instrumentation makes adoption of speckle imaging straightforward. However, basic single exposure speckle imaging is limited to measurements of relative blood flow changes within a single subject during a single experiment. Multi-exposure speckle imaging extends the capabilities of single exposure speckle imaging by capitalizing on the strong dependence of the measured speckle contrast on camera exposure time. When these measurements are combined with the appropriate mathematical model, the quantitative accuracy of the blood flow measures is improved and flow values can be tracked chronically. Speckle imaging is also an extremely useful tool for intraoperative blood flow imaging during neurosurgical procedures in humans due to its ability to provide real-time images.

References

1. Briers JD (2001) Laser Doppler, speckle and related techniques for blood perfusion mapping and imaging. Physiol Meas 22(4):R35–R66
2. Tuchin VV (2007) Tissue optics: light scattering methods and instruments for medical diagnosis. SPIE Press, Bellingham
3. Fercher A, Briers J (1981) Flow visualization by means of single-exposure speckle photography. Opt Commun 37:326–329
4. Briers JD, Webster S (1996) Laser speckle contrast analysis (LASCA): a nonscanning, full-field technique for monitoring capillary blood flow. J Biomed Opt 1:174–179
5. Kirkpatrick SJ, Duncan DD, Wells-Gray EM (2008) Detrimental effects of speckle-pixel size matching in laser speckle contrast imaging. Opt Lett 33(24):2886–2888
6. Boas DA, Dunn AK (2010) Laser speckle contrast imaging in biomedical optics. J Biomed Opt 15(1):011109
7. Richards LM, Kazmi SMS, Davis JL, Olin KE, Dunn AK (2013) Low-cost laser speckle contrast imaging of blood flow using a webcam. Biomed Opt Express 4(10):2269–2283
8. Bolay H, Reuter U, Dunn AK, Huang Z, Boas DA, Moskowitz MA (2002) Intrinsic brain activity triggers trigeminal meningeal afferents in a migraine model. Nat Med 8(2): 136–142
9. Dunn AK, Devor A, Dale AM, Boas DA (2005) Spatial extent of oxygen metabolism and hemodynamic changes during functional activation of the rat somatosensory cortex. Neuroimage 27(2):279–290
10. Forrester KR, Tulip J, Leonard C, Stewart C, Bray RC (2004) A laser speckle imaging technique for measuring tissue perfusion. IEEE Trans Biomed Eng 51(11):2074–2084
11. Huang Y-C, Ringold TL, Nelson JS, Choi B (2008) Noninvasive blood flow imaging for real-time feedback during laser therapy of port wine stain birthmarks. Lasers Surg Med 40(3): 167–173
12. Durduran T, Burnett MG, Yu G, Zhou C, Furuya D, Yodh AG, Detre JA, Greenberg JH (2004) Spatiotemporal quantification of cerebral blood flow during functional activation in rat somatosensory cortex using laser-speckle flowmetry. J Cereb Blood Flow Metab 24(5):518–525
13. Ayata C, Dunn AK, Gursoy-OZdemir Y, Huang Z, Boas DA, Moskowitz MA (2004) Laser speckle flowmetry for the study of cerebrovascular physiology in normal and ischemic mouse cortex. J Cereb Blood Flow Metab 24(7):744–755
14. Parthasarathy AB, Kazmi SMS, Dunn AK (2010) Quantitative imaging of ischemic stroke through thinned skull in mice with Multi Exposure Speckle Imaging. Biomed Opt Express 1(1):246–259
15. Bandyopadhay R, Gittings AS, Suh SS, Dixon PK, Durian DJ (2005) Speckle-visibility spectroscopy: a tool to study time-varying dynamics. Rev Sci Instrum 76(9):93110
16. Briers JD, Fercher AF (1982) Retinal blood-flow visualization by means of laser speckle photography. Invest Ophthalmol Vis Sci 2: 255–259

17. Briers JD, Richards G, He XW (1999) Capillary blood flow monitoring using laser speckle contrast analysis (LASCA). J Biomed Opt 4(1):164
18. Duncan DD, Kirkpatrick SJ (2008) Can laser speckle flowmetry be made a quantitative tool? J Opt Soc Am A Opt Image Sci Vis 25(8):2088–2094
19. Parthasarathy AB, Tom WJ, Gopal A, Zhang X, Dunn AK (2008) Robust flow measurement with multi-exposure speckle imaging. Opt Express 16(3):1975–1989
20. Kazmi SMS, Parthasarthy AB, Song NE, Jones TA, Dunn AK (2013) Chronic imaging of cortical blood flow using Multi-Exposure Speckle Imaging. J Cereb Blood Flow Metab 33(6): 798–808
21. Boas DA, Yodh A (1997) Spatially varying dynamical properties of turbid media probed with diffusing temporal light correlation. J Opt Soc Am A 14:192–215
22. Lemieux PA, Durian DJ (1999) Investigating non-Gaussian scattering processes by using nth-order intensity correlation functions. J Opt Soc Am A 16(7):1651–1664
23. Wu X-L, Pine DJ, Chaikin PM, Huang JS, Weitz DA (1990) Diffusing-wave spectroscopy in a shear flow. J Opt Soc Am B 7(1):15–20
24. Pine DJ, Weitz DA, Chaikin PM, Herbolzheimer E (1988) Diffusing wave spectroscopy. Phys Rev Lett 60(12):1134
25. Dunn AK (2012) Laser speckle contrast imaging of cerebral blood flow. Ann Biomed Eng 40(2):367–377
26. Armitage GA, Todd KG, Shuaib A, Winship IR (2010) Laser speckle contrast imaging of collateral blood flow during acute ischemic stroke. J Cereb Blood Flow Metab 30(8): 1432–1436
27. Atchia Y, Levy H, Dufour S, Levi O (2013) Rapid multiexposure in vivo brain imaging system using vertical cavity surface emitting lasers as a light source. Appl Optics 52(7):C64–C71
28. Rege A, Murari K, Seifert A, Pathak AP, Thakor NV (2011) Multiexposure laser speckle contrast imaging of the angiogenic microenvironment. J Biomed Opt 16(5):056006
29. Thompson OB, Andrews MK (2010) Tissue perfusion measurements: multiple-exposure laser speckle analysis generates laser Doppler-like spectra. J Biomed Opt 15(2):027015
30. Duncan DD, Lemaillet P, Ibrahim M, Nguyen QD, Hiller M, Ramella-Roman J (2012) Absolute blood velocity measured with a modified fundus camera. J Biomed Opt 15(5):056014
31. Ishikawa M, Sekizuka E, Shimizu K, Yamaguchi N, Kawase T (1998) Measurement of RBC velocities in the rat pial arteries with an image-intensified high-speed video camera system. Microvasc Res 56(3):166–172
32. Parthasarathy AB, Weber EL, Richards LM, Fox DJ, Dunn AK (2010) Laser speckle contrast imaging of cerebral blood flow in humans during neurosurgery: a pilot clinical study. J Biomed Opt 15(6):066030
33. Richards LM, Towle EL, Fox DJ, Dunn AK (2013) Laser speckle imaging of cerebral blood flow. In: Madsen SJ (ed) Optical methods and instrumentation in brain imaging and therapy. Springer, New York, pp 117–136
34. Richards LM, Weber EL, Parthasarathy AB, Kappeler KL, Fox DJ, Dunn AK (2012) Intraoperative laser speckle contrast imaging for monitoring cerebral blood flow: results from a 10-patient pilot study. In: Photonic Therapeutics and Diagnostics Viii, Pts 1 and 2, pp 1–12
35. American National Standard for the Safe Use of Lasers (2007). In: ANSI Z136.1-2007. Laser Institute of America
36. Raabe A, Van De Ville D, Leutenegger M, Szelényi A, Hattingen E, Gerlach R, Seifert V, Hauger C, Lopez A, Leitgeb R, Unser M, Martin-Williams EJ, Lasser T (2009) Laser Doppler imaging for intraoperative human brain mapping. Neuroimage 44(4):1284–1289
37. Deckers RHR, van Gelderen P, Ries M, Barret O, Duyn JH, Ikonomidou VN, Fukunaga M, Glover GH, de Zwart JA (2006) An adaptive filter for suppression of cardiac and respiratory noise in MRI time series data. Neuroimage 33(4):1072–1081
38. Klein S, Staring M, Murphy K, Viergever MA, Pluim JPW (2010) Elastix: a toolbox for intensity-based medical image registration. IEEE Trans Med Imaging 29(1):196–205
39. Woitzik J, Hecht N, Pinczolits A, Sandow N, Major S, Winkler MKL, Weber-Carstens S, Dohmen C, Graf R, Strong AJ, Dreier JP, Vajkoczy P (2013) Propagation of cortical spreading depolarization in the human cortex after malignant stroke. Neurology 80(12): 1095–1102

Chapter 16

Cerebral Monitoring and Surveillance Using High-Resolution Functional Optical Imaging

Anthony D. Uglialoro, Douglas S. Pfeil, Tigran Gevorgyan, Harry L. Graber, Yong Xu, Sundeep Mangla, Frank C. Barone, Jenny Libien, Jean Charchaflieh, John G. Kral, Sergio A. Ramirez, LeRone Simpson, Daniel C. Lee, and Randall L. Barbour

Abstract

Near-infrared (NIR) optical techniques, using large-area sensing arrays that provide for transcranial measures of cortical function, are finding increasing use for functional assessment of brain activity. In this chapter, we review current capabilities of these methodologies and describe their application to an in vivo large animal model, Bonnet macaque, for the purpose of imaging hemodynamic responses to clinically significant events, including detection of cerebral ischemia and hemorrhage. The described methodology outlines the technology and method of high-density diffuse optical tomography (DOT), as explored using recently developed analysis resources whose accuracy is independently validated. Also discussed are solutions to commonly encountered problems related to data collection and analysis.

Key words Diffuse optics, Diffuse optical imaging, Near-infrared spectroscopy (NIRS), Topography, Diffuse optical tomography (DOT), Cerebral monitoring and surveillance

1 Introduction

Disturbances in brain function can have lifelong consequences. Tools available for assessment of brain function vary in their sensitivity to underlying physiology or pathology, as well as their technology footprints. Together these factors can significantly influence the clinical usability of a given technology. Ideally, in addition to providing clinically relevant information (e.g., presence, position, and spatial extent of pathologies), the considered technology would have favorable attributes that support its usability, including but not limited to, system size, cost, noninvasiveness, safety, and easy-of-use.

Anthony D. Uglialoro and Douglas S. Pfeil have contributed equally to this chapter.

Mingrui Zhao et al. (eds.), *Neurovascular Coupling Methods*, Neuromethods, vol. 88,
DOI 10.1007/978-1-4939-0724-3_16, © Springer Science+Business Media New York 2014

The noninvasive methods available for assessing brain function include various imaging technologies that are well-suited for defining brain anatomy, the impact of acute events, and the brain's principal functional properties. Chief among the anatomical imaging technologies are X-ray computed tomography (CT) and magnetic resonance imaging (MRI). In cases of trauma or cerebral vascular accidents, non-contrast CT imaging is the preferred modality [1]. Owing to its enhanced specificity, MRI is often used in non-acute situations for further investigation of cerebral pathologies, including the detection of tumors and monitoring the progression of multiple sclerosis or encephalopathy [2, 3].

Several direct and indirect clinical neuromonitoring techniques are currently available for examining functional brain activity. These technologies have documented that brain activation and many pathologies (e.g., Alzheimer's [4], Parkinson's diseases [5], and cognitive impairment due to traumatic brain injury [6]) produce spatially distributed and temporally varying responses. Techniques that provide direct information of neuron-derived brain activity include electroencephalography (EEG) [7, 8] and its magnetic equivalent, magnetoencephalography (MEG) [9]. These methods provide a high degree of temporal resolution but have markedly reduced spatial resolution compared to the principal anatomical imaging techniques (MRI, CT). Indirect methods for monitoring brain activity have been derived from measures of the hemodynamic response coupled to neuroactivation [10], as well as from associated metabolic transformations that can be assessed by the method of positron emission tomography (PET) [11]. Among the former is the method of functional Magnetic Resonance Imaging (fMRI) [12, 13], which is based on neuroactivation-induced variations in the concentration of (paramagnetic) deoxyhemoglobin [14].

At present, while each of the above techniques has its distinct advantages for monitoring brain function, the direct methods (EEG, MEG) tend to have limited spatial resolution, whereas the indirect methods are restricted to inferring neuronal activity after it has been filtered through a complex neurovascular coupling mechanism. As discussed subsequently, over the past decade a new approach for indirect monitoring of brain function has emerged that offers the potential for wide use in many clinical specialties and subspecialties, owing to its favorable technology footprint and its sensitivity to important clinical disturbances.

1.1 Optical Imaging: Technology and Capabilities

Extensive clinical experience with pulse oximetry has demonstrated that the optical method is well-suited for measuring important features of the hemoglobin signal. Favoring its use is compact and low-cost sensing hardware, and its ability to noninvasively detect temporal variations in the content and oxygenation level of hemoglobin (Hb) with high sensitivity. In turn, the Hb content and oxygenation are closely tied to tissue perfusion, and this has prompted many to exploit the inherent flexibility of optical methods for potential clinical applications. Two distinct approaches have emerged. First, low-density sensing arrays (i.e., delivering light into and detecting it from the head at only a few—typically 1–4—locations) have been adopted as a tool for tissue oximetry [15–19]. Second, the extension of these methods to include high-density arrays (i.e., involving larger numbers of light delivery and detection points, distributed over a larger area), together with use of more advanced mathematical treatment of the data, has allowed for the development of new classes of functional imaging techniques.

Intrinsic optical imaging is an example of a method having high spatial resolution. This technique employs broad-area illumination of an exposed brain preparation, combined with use of high-resolution digital cameras. While this technique is mainly used as an experimental tool, it has been shown capable of real-time measures of both the intrinsic hemodynamic signal and membrane potential changes due to the neuronal response, especially when it is used in combination with voltage-sensitive dyes [20].

The use of near-infrared (NIR) spectroscopic methods to explore the temporal dynamics of the Hb signal in a closed cranium are built upon an observation made by Jöbsis in 1977 [21]. More than a decade later, this observation motivated two distinct technology development efforts that shared the goal of noninvasive functional imaging.

Beginning in the late 1980s, Barbour and coworkers showed that model-based methods for analyzing light transport in diffusing media can be adopted as a strategy for spatially resolving local variations in the oxygenation states of heme-containing proteins, principally Hb [22, 23]. The notable advance here was the recognition that low-resolution images of spatial features of subsurface structures can be recovered from measurements of the emission profile of *diffusely scattered* light. Prior to these demonstrations, it had been assumed that these features were irretrievably obscured due to the intense scattering of light by tissue. The spirit of this technique, known as *diffuse optical tomography*, is similar to other methods (i.e., X-ray CT), in that image formation is based on the treatment of multi-view measures. A principal difference is that scattering causes light to assume probabilistic paths (unlike the straight-line paths taken by X-rays) whose particulars are dependent on features of the medium and details of illumination and detection [24]. These particulars, however, can be explicitly considered using mathematical models [25]. These early studies coincided with groundswell of related activities throughout the 1990s by other groups [26–31] that built on the many ways optical techniques can be employed to examine features of dense scattering media. Among these was the evidence that use of arrays of emitters and detectors can detect local variations in the hemodynamic response to neuroactivation (see Ferrari and Quaresima for review of the historical development of functional near-infrared spectroscopy (fNIRS) [32]).

These considerations have led to development of a series of instrumentation platforms (see Note 1) that ultimately rely on more simplified mathematical treatment of the data than that used for tomography studies to explore spatiotemporal features of the Hb signal. This technique, known as *optical topography* [33], maps weighted volume-averaged measures of the Hb signal to the surface of the scalp or, by linear projection, to the cortex (see Fig. 3). Similar to the tomography approach, the detected signal is diffusely scattered. Both techniques employ grossly similar instrumentation, but differ in the density of sensing elements deployed, with higher-density arrays used for tomography and, as indicated, differ in the approaches used for data analysis. For both approaches, light delivery can be accomplished either through use of intervening optical fibers or by placement of miniaturized illumination and detection elements directly on the scalp. Typical for topography studies is the use of a rectilinear array having ~3 cm spacing between the elements; tomography arrays detect re-emitted light at multiple distances from any one source (see Note 1 for a description of trade-offs between technology complexity and derived information, for topographic and tomographic measures).

Underscoring the attractiveness of the optical method is its use of non-damaging energy sources and instrumentation having form factors that can be made compact, even wearable [34]. Also, because Hb is the dominant time-varying optical attenuator in tissue, many applications that are based on examination of this signal can be explored. For instance, a frequent concern in clinical medicine is whether perfusion of the brain has been compromised. While there have been numerous studies of the use of low-density tissue oximetry as a monitoring tool, the use of large-area arrays for assessment of brain function in clinical settings has been thus far underutilized, owing to the need for more refined system designs that are appropriate for these settings.

As a first step in bridging this gap, in this chapter we describe a general methodology for using the tomography approach to investigate acute vascular events having clinical significance, and we demonstrate its application to a large animal model. Specifically examined here were the responses to acute cerebral vascular maneuvers, such as injection of a contrast agent, and the response to clinically relevant events, such as induced cerebral ischemia or subarachnoid hemorrhage (SAH).

Key elements of the method are the use of a high-density sensing array, inversion of measured signals based on a group-averaged anatomical atlas, and use of an integrated analysis environment to produce a three-dimensional (3D) image time series of various components of the Hb signal. Also presented are the results from independent studies that validate the optical findings based on use of radiological and postmortem histological techniques.

2 Methods

2.1 Animal Preparation

2.1.1 Bonnet Macaque

The arterial vasculature of the Bonnet macaque brain is anatomically similar to that of the human brain. (The most apparent difference is that the macaque has a single anterior cerebral artery (ACA), while the human has distinct left and right ACAs.) This has led a number of investigators to employ the macaque as an animal model for stroke and other cardiovascular research.

For the current studies, subjects investigated were adult male macaques weighting approximately 7 kg, aged 5–6 years. A total of five animals were studied under an IACUC-approved protocol. For purposes of documenting the method of high-resolution functional optical imaging, results from a single animal are reported.

2.1.2 Anesthesia and Setup

Using standard animal anesthesia protocols, the animal was sedated and intubated [35]. Once this was established, hemodynamic monitors were applied that recorded blood pressure, oxygen saturation (SaO_2), and electrocardiography (ECG) throughout experimentation. An intra-venous (IV) catheter was inserted for administration of fluids and medications, along with an arterial line to monitor the blood pressure more accurately.

In order to gain access into the arterial system for cerebral angiography and for temporary occlusion of the cerebral vessels, a femoral artery catheter sheath [5 Fr (i.e., 5/3 mm)] was inserted in the groin of the animal and secured using standard cut-down surgical technique.

2.1.3 Exposure of Scalp

In order to apply the optical interface, the overlying tissues including the skin and head muscles were removed using standard surgical techniques. A 4 × 5 cm area in the mid-frontal–parietal region of the scalp of the animal was reflected, exposing the skull. This was performed to achieve optimal optical penetration to subsurface tissues. Scalp reflection was necessary in this macaque model because the very thick (1–2 cm) frontalis muscle would otherwise absorb most of the light. In human studies, significant optical penetration is achieved without any need for equivalent procedures, because the overlying tissues are much thinner. Please see Note 2 for a description of expected influence of superficial tissues to applied optical measures.

2.2 Equipment Setup for Functional Optical Imaging

2.2.1 Preparation for Optode Interface and Array

The optode–tissue interface was custom-designed for the animals. The physical support structure for the optodes was a 7 cm diameter, 9 mm thick sheet of black Delrin® that was machined to provide a pattern of hexagonal segments to allow close fitting between the optical hardware and the skull. Shown in Fig. 1 is a photograph of the optode holder secured to the skull of the anesthetized animal. The optode interfaces containing the optical fibers were placed

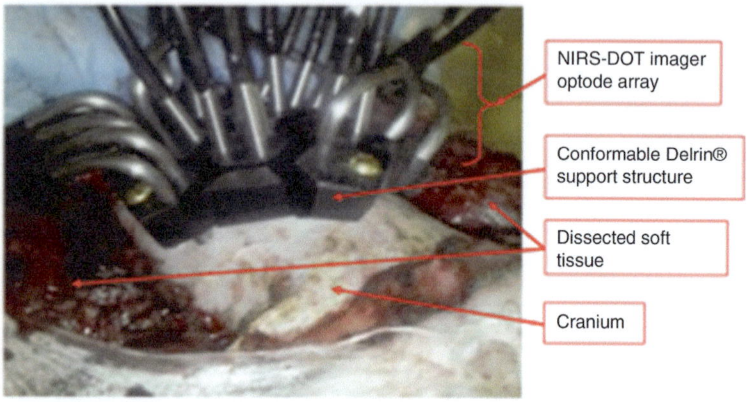

Fig. 1 Photograph of fNIRS optode holder secured to the Bonnet macaque's skull after reflection of the overlying soft tissues

over the exposed 4 × 5 cm skull surface and secured using small screws. The optodes were positioned 10 mm apart. Shown in Fig. 2 is a schematic layout of the optode array. There were 9 illumination sites, which also served as detection sites, and 21 detection-only sites, for a total of 30 detectors. Thus there were 9 × 30 = 270 data acquisition channels in total. The co-located design was implemented by combining source fibers with a halo of detector fibers to produce a common end.

2.2.2 Instrumentation and Data Collection

Data collection was achieved using a continuous wave (CW)-NIRS imaging system (DYNOT, NIRx Medical Technologies). After the optode array within its holder was secured to the skull using small screws, the amplification settings for each source–detector pair were established using an automated procedure that sweeps the amplifier gain over the full sensing range available (nine decades) and identifies the gain that yields an optimal signal without saturation of the detection circuitry. These settings are stored in instrument memory and successively accessed for each illumination site, thereby achieving real-time adaptive gain control.

Time-multiplexed illumination (i.e., each source is sequentially illuminated) was performed by optically combining light from two diode lasers, operating at 760 and 830 nm, into small-diameter fiber bundles (~3 mm) that terminate on the skull. Light received into the fiber bundles was measured by parallel readout from the light sensing elements (i.e., silicon photodiodes). The sensing array was read out at the rate of 8 Hz. System control, real-time data viewing, and data storage are performed using a LabView (National Instruments Corporation, www.ni.com) interface. See Schmitz et al. [36, 37] for details of system design and operation. See Note 3 for a discussion of illumination-detection strategies.

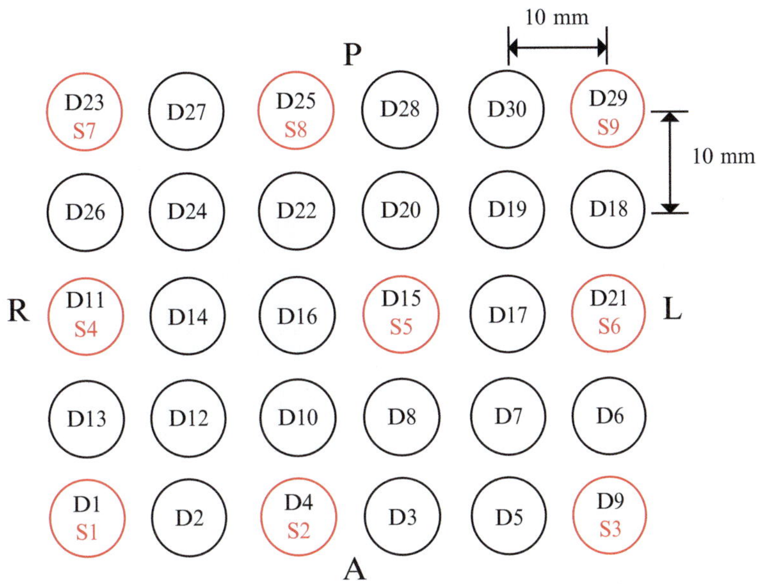

Fig. 2 A schematic layout of the optode array. *Black numbers* represent detectors and *red numbers* represent light sources. Letters denote orientation: (A): anterior; (L): left; (P): posterior, (R): right side

2.3 Angiography and Cerebral Vascular Manipulation

After a baseline recording of approximately 10 min, intra-arterial digital subtraction angiography (DSA) of the brain was performed to identify the correct cerebral vessel in which to induce temporary ischemia. In this case, the middle cerebral artery (MCA), which is a continuation of the internal carotid artery (ICA), was located using a 5-Fr diagnostic catheter with the use of iodinated contrast (Iohexol 300 mgI/mL) to assist in the navigation to the MCA. For the purposes of establishing regional acute cerebral ischemia, an occlusive angioplasty microcatheter [2.7 Fr (i.e., 0.9 mm outer diameter)] was threaded through the femoral artery sheath under radiological guidance and then wedged at the origin of the MCA. Before the occlusion attempts, verapamil, a drug known for its vasodilatory effects, was locally injected into the right ICA/MCA. Several contrast injections were performed to assist the guidance of the catheter to the point of occlusion. The microcatheter was left in place for 3 h, then removed to restore blood flow to the occluded area. The specific protocol implemented here was modified in response to initial difficulties in guiding the microcatheter to the initial planned cerebral vessel (right MCA) as a consequence of vasospasm. As a result, the contralateral (left) ICA was occluded instead.

2.4 MR Validation and Postmortem Studies

After 1 h of arterial reperfusion, the NIRS recording was stopped and the animal was transported to the imaging suite. Standard stroke protocol sequences including diffusion, T2, T1, and fluid-attenuated inversion recovery (FLAIR) were performed.

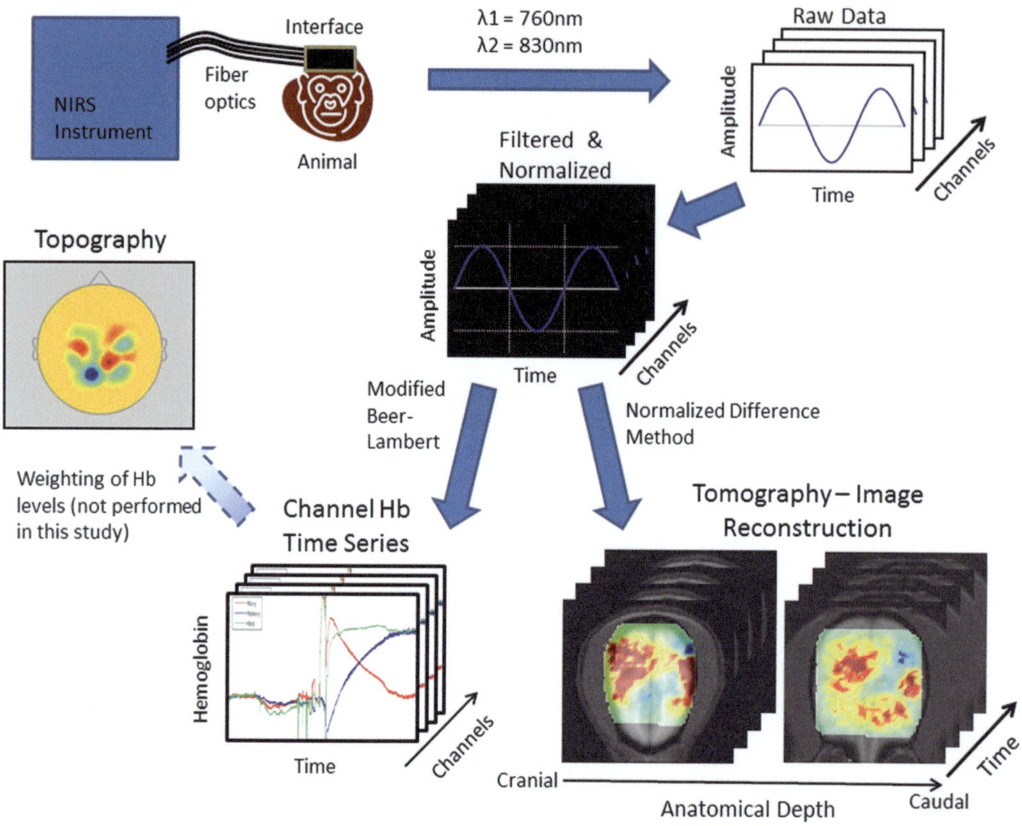

Fig. 3 A schematic depicting the data processing steps involved in NIRS measurements

The animal was then transported to the autopsy room, where euthanasia was performed using pentobarbital, 120 mg/kg, IV. At the conclusion of euthanasia, the brain was removed and fixed in 10 % buffered formalin for at least 2 weeks, after which coronal sections were cut at approximately 5 mm intervals. Brain regions corresponding to the center of the infarct, as determined by the postsurgical radiological data and from the corresponding location on the contralateral side, were submitted for routine tissue processing and paraffin embedding. Hematoxylin and eosin-stained sections, each 8–10 μm in thickness, were then examined by light microscopy.

2.5 Analysis of NIRS Measures and Image Reconstruction

Wavelength-dependent time-series measures obtained from the sensing array may be processed in two principal ways: either through the Modified Beer–Lambert Law (MBLL) [39] to yield channel-based hemoglobin time series (the first step in topography), or through image reconstruction algorithms to yield 3D Hb time series (tomography). See Fig. 3 for a schematic of data processing steps.

The Beer-Lambert Law (Absorbance = [ε] × [Concentration] × [Length]) is commonly used in spectrophotometric determinations of the concentrations of substances suspended in clear media (e.g., protein in an aqueous solution or the concentration of DNA). In optical monitoring applications that are based on low-density arrays or on optical topography, Hb concentrations are estimated by using a modified Beer-Lambert Law that takes account of the increased light propagation distance resulting from photon scattering in tissue. This correction is called the differential pathlength factor (DPF) [38]. Thus, the MBLL can be written as:

$$\log \frac{I(j)}{I_{bl}} = \varepsilon \times \Delta c(j) \times (\text{DPF} \times l),$$

where $I(j)$ is the detected signal value in the jth measurement frame and I_{bl} is the average value of the signal detected over a selected "baseline" time interval. ε is the extinction coefficient of the chromophore [oxy- and deoxyhemoglobin (Hb_{oxy}, Hb_{deoxy})] and l is the linear distance between light source and detector. Because it can be difficult to measure the precise value of light intensity entering the tissue, it is convenient to report Hb concentrations as changes with respect to the baseline value (ΔHb).

Image reconstruction, on the other hand, uses a different method to calculate ΔHb values. The algorithm we used for reconstruction is called the normalized difference method (NDM) [67]. In this method, the differences between predicted (forward model) and measured channel data are related to changes in the interior optical properties (e.g., absorption) of the investigated medium, with respect to a defined reference medium. The Jacobian operator [67] and reference detector values were computed prior to data analysis, from forward-model solutions to the diffusion equation, using a 3D tessellated and segmented group atlas. See Note 4 for discussion of forward modeling techniques.

For this experiment, image reconstruction was performed via the following three-step process:

Step 1: Raw data was subjected to a low-pass filter (0.2 Hz cutoff) and normalized to its temporal mean value, separately for each data acquisition channel and each measurement wavelength.

Step 2: Using a 3D tessellated and segmented group monkey atlas, whose details are described below, the Jacobian operator and reference detector values were computed.

Step 3: The quantities computed in Steps 1 and 2 were subsequently processed by applying the NDM to generate a wavelength-dependent 3D image time series of the recovered absorption coefficients. Conversion of these spatial maps to corresponding maps of the hemoglobin components was accomplished using methods described by Schmitz et al. [39]. See Note 5 for discussion of limitations of linear inverse solvers applied to recovery of image time series.

Steps 1 and 3 were accomplished using routines available in the NAVI (*N*ear-infrared, *A*nalysis, *V*isualization, and *I*maging) software package [40]. This MATLAB (The MathWorks, Inc., Natick, MA, USA)-based software is freely available and can be downloaded from the NITRC website at http://www.nitrc.org/projects/fnirs_downstate. Forward solutions to the diffusion equation (Step 2) were obtained using a previously described finite element method (FEM)-based solver written in Fortran [41]. To accomplish this, first a tetrahedral FEM mesh was applied to the atlas, followed by assignment of optical coefficients to the segmented anatomy as described in references [42, 43]. To more accurately reflect the conditions of surgery, numerical solutions to the forward problem were obtained by carefully locating the sensor positions onto the skull of the atlas, rather than to the scalp.

2.5.1 Generation of a Hybrid Group Atlas

Figure 4 illustrates the principal steps involved in generating the hybrid group atlas used in these studies. Briefly, by carefully measuring the dimensions along the principal axes from MRI images of a single-animal brain and from a group-averaged brain obtained from the literature [44], and then executing an affine transformation [45], we substituted the group brain for the individual brain to

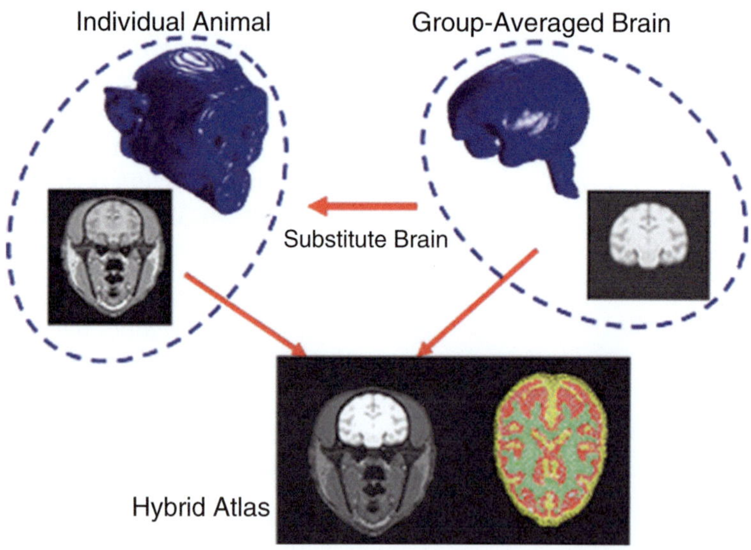

Fig. 4 A schematic showing the process used to create the macaque atlas. Information from a group-averaged brain is superimposed onto an individual MRI scan to yield the hybrid atlas. The image in the *bottom right* shows one of the segmented and tessellated slices from the atlas

yield a hybrid atlas. As reported by McLaren et al. [44], the group-averaged brain was derived from 80 adult male and 32 adult female Rhesus macaque monkeys, which are anatomically similar to Bonnet macaques. Group atlases are frequently adopted in order to minimize biases inherent to specific anatomical features of an individual, as well as to provide for comparisons across platforms, sensing modalities, and other atlases.

3 Results of the Applied Methodology

Shown in Figs. 5 and 6 are cross-sectional views of reconstructed 3D images of the Hb response to the vascular events described above. Figure 5a shows an axial view of the amplitude of the ΔHb_{oxy} image obtained during the baseline period (sampled at 20 min). Figure 5b is the coronal view obtained from the transect indicated in Fig. 5a. Both images reveal a relatively low-amplitude spatial variation in signal contrast. Shown in Fig. 5c, d are corresponding

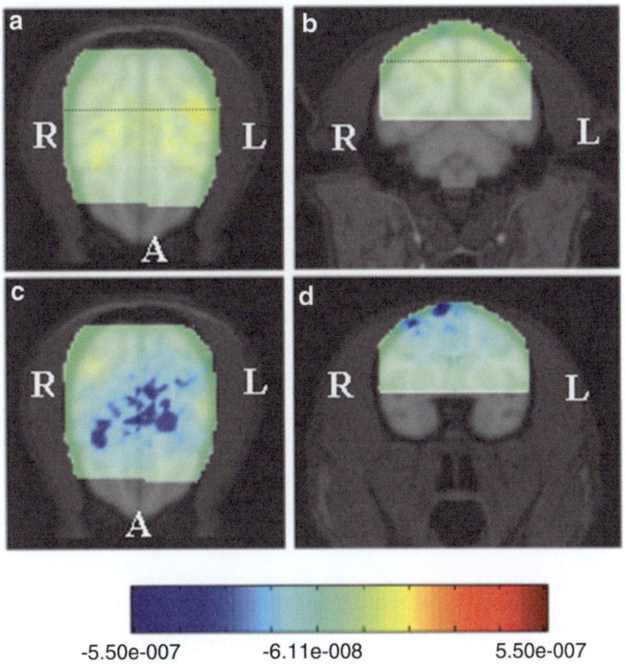

Fig. 5 Axial and corresponding coronal cross-sectional image views of ΔHb_{oxy} (mol/L) during (**a**) and (**b**): baseline and (**c**) and (**d**): post-injection of contrast agent (approximately 10 s post-injection). (**a**, **b**) reveal a relatively low-amplitude spatial variation in signal contrast, while (**c**, **d**) reveal displacement of hemoglobin from the field leading to a reduced ΔHb_{oxy} signal (*blue* contrast feature). *Dotted lines* in (**a**, **b**) indicate the intersections among the plotted two-dimensional (2D) sections

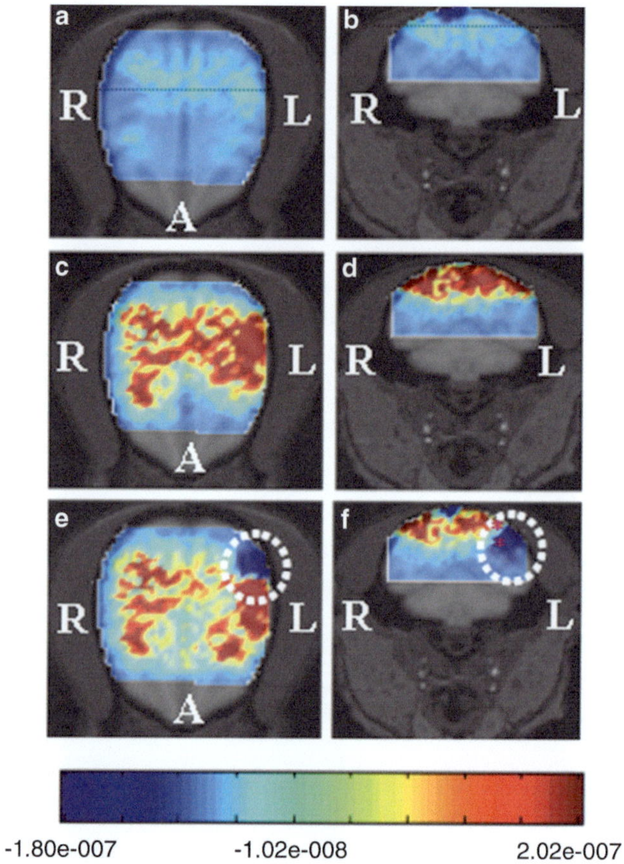

Fig. 6 Axial and corresponding coronal cross-sectional image views of ΔHb_{total} (mol/L) obtained during the (**a**, **b**): baseline (sampled at 20 min), (**c**, **d**): post-SAH (sampled at 125 min), and (**e**, **f**): post-occlusion (sampled at 166 min, approximately 40 min post-occlusion) of the cerebral vessel. Subarachnoid hemorrhage is revealed by the presence of elevated ΔHb_{total} signal (*red* contrast feature). A region of reduced ΔHb_{total} signal is highlighted by the *dotted circle* (*blue* contrast feature). *Dotted lines* in (**a**, **b**) indicate the intersections among the plotted 2D sections. The *asterisks* in (**f**) mark the pixels represented in Fig. 7

images obtained during periods of injection of the contrast agent (sampled at 90 min) that is expected to cause a bolus displacement of Hb from the field, leading to a reduced signal, as indicated by the areas of dark blue.

Shown in Fig. 6 are similar views of images of the ΔHb_{total} ($\Delta Hb_{total} = \Delta Hb_{oxy} + \Delta Hb_{deoxy}$) obtained during the baseline period (sampled at 20 min, Fig. 6a, b), post-SAH (sampled at 125 min, Fig. 6c, d), and post arterial occlusion (sampled at 166 min, Fig. 6e, f) periods. Comparison of the image features reveals three principal findings. First, owing to a large rapid change in the recorded

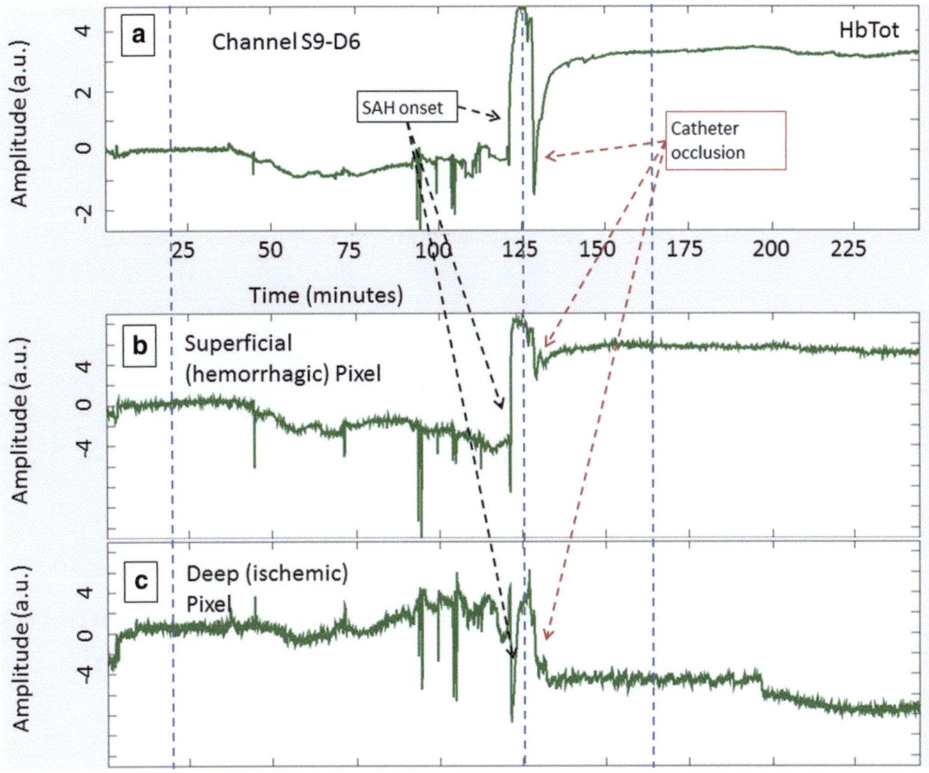

Fig. 7 (**a**) S9-D6 channel data for ΔHb_{total} (see Fig. 2 for channel location). Both the onset of SAH and catheter occlusion are visible. (**b**) Time-series data for a pixel (from the reconstructed image) just under the skull. Note the SAH is prevalent but catheter occlusion only produces a small change. (**c**) Time-series data for a pixel in the ischemic region. Note there is very little change in ΔHb_{total} at the time of the SAH, but a decrease corresponding to catheter occlusion is prevalent. The *blue dashed lines* correspond to the times sampled in Fig. 6

optical signals during manipulation of the catheter in attempts to occlude the MCA (see Fig. 7), a rupture of one of the cerebral vessels was suspected and was concurrently confirmed by DSA, and later by postsurgical radiology and postmortem evaluation. This is revealed by the presence of elevated ΔHb_{total} signal in the recovered images (red contrast feature, Fig. 6c, d). Second, a region of reduced ΔHb_{total} contrast in the vicinity of the posterior tributaries of the ICA is seen after approximately 40 min of ischemic occlusion [Fig. 6e, f (affected region is indicated by a dotted circle)]. The third feature is clear evidence of depth discrimination between the more superficial confinement of hemorrhage and areas of ischemia at greater depths in the brain. Additionally, it is instructive to compare reconstructed time series for selected pixels in these affected regions (denoted by asterisks in Fig. 6f) to the response of a source–detector pair positioned near the left frontal–parietal region (S9-D6, Fig. 2). The MBLL was used to compute the ΔHb_{total} time series for this channel. These results are shown in Fig. 7.

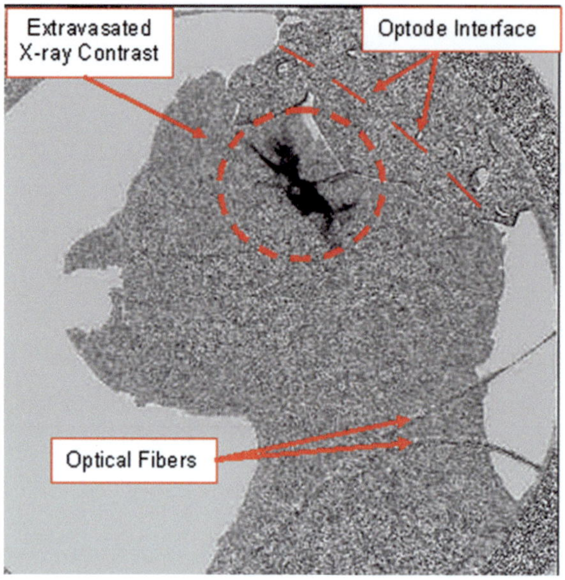

Fig. 8 Lateral DSA demonstrating right ICA perforation and SAH prior to subsequent occlusion. Dark material is extravasated X-ray contrast agent

Figure 7a reveals that at approximately 120 min, the ΔHb_{total} signal rapidly increases, which is indicative of SAH. Additionally, a drop at 130 min marks the time of occlusion by the microcatheter. However, interpretation of the data from Fig. 7a would be difficult without prior knowledge of the events of the experiment. Even after catheter occlusion, ΔHb_{total} values remain above baseline. The extent of the ischemia in the deeper tissue is not fully revealed. Figure 7b, c show the reconstructed ΔHb_{total} times series for a pixel just under the skull and a pixel about 1 cm deeper in the brain, respectively. These two time series demonstrate how the different regions of the brain tissue are affected during this experiment. Figure 7b clearly shows the increased ΔHb_{total} signal corresponding to the SAH, while very little change is evident during catheter occlusion. Figure 7c shows the opposite. This is a clear demonstration of the utility provided by image reconstruction.

Corroborating evidence in support of the optical findings, obtained by DSA and MRI, is shown in Figs. 8 and 9. Figure 8 reveals a high-contrast feature showing extravasation of X-ray contrast agent, indicating cerebral vessel perforation. Figure 9 reveals a region of hyperintensity as seen by FLAIR MRI, confirming that ischemic damage has occurred subsequent to occlusion. Also seen in this image is the evidence of hemorrhage.

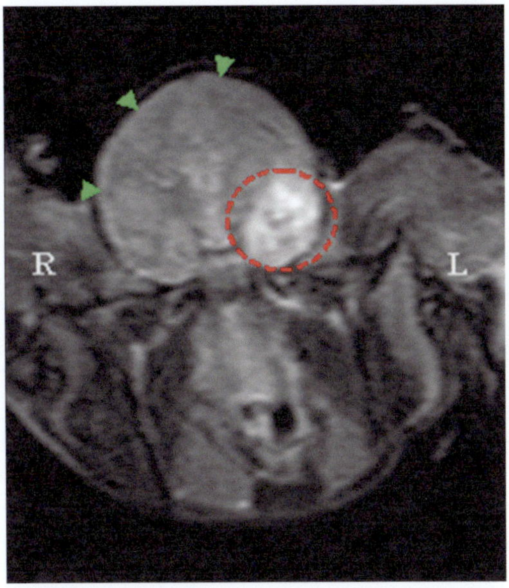

Fig. 9 Postsurgical coronal FLAIR MRI sequences (slice locations: left—L, right—R) confirm onset of ischemia (*dotted circle*) in the left temporal and inferior frontal lobe following the production of a period of left MCA occlusion. Images also confirm the presence of subarachnoid hemorrhage (indicated by *green arrowheads*)

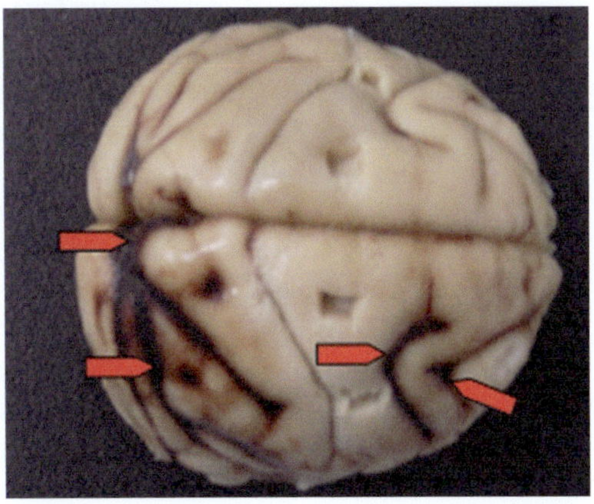

Fig. 10 Gross pathology (indicated by *red arrows*) of the dorsal surface of the brain, illustrating SAH

Figure 10 shows gross examination results confirming that a SAH is present bilaterally in the depths of the sulci over the dorsal aspect of the brain. An ill-defined area of pallor and blurring of the gray–white junction, consistent with edema subsequent to ischemia, were identified in the superior temporal gyrus and posterior hippocampus (results not shown).

4 Conclusions

Disturbances in brain function can have immediate, severe, and lifelong consequences. Therefore, monitoring activity of the brain holds considerable significance in a broad range of clinical fields such as neonatology, psychiatry, gerontology, emergency care, neurology, physical rehabilitation, and surgery. Diffuse optical imaging, deployed to yield topographic or tomographic information, has emerged as a promising, versatile method. We hope, by means of the review of the methodology presented in this chapter, to encourage and spur others to adopt it for different applications. As outlined here, the tomography method can resolve different forms of hemodynamic events having clinical significance, and yields information that is strongly corroborated by standard independent measures.

The distillation of methods for clinical application requires consideration of many factors beyond what is necessary to support investigative studies. World-wide, there are hundreds of laboratories using NIRS tomography and topography systems in support of clinical investigative studies. At present, the method remains experimental. However, the favorable attributes of a compact and versatile technology footprint, and its use of non-damaging energy sources, together with continued development in advanced processing environments [see Notes 4 and 5], strongly favor development of useful clinical applications. It is also recognized that, similar to other neuroimaging methods, it can be desirable to merge technologies in order to extend their overall sensing capabilities. Of special interest is the capacity to merge NIRS studies with EEG. Time-series measures, in general, are rich in information and expected improvements to system design and integration should serve to significantly extend their utility. Direct evidence of this is a recent report that showed improved performance of brain–computer interface applications [46] from concurrent NIRS-EEG measures. Additionally, advances in data analysis [47] and modeling [48] have shown promise for extending the information content accessible from NIRS measures.

To date, most multichannel NIRS topography and tomography studies focus primarily on brain-function applications [49–53]. Many multichannel NIRS topography applications have been developed [54–56], and the method has already been shown to detect brain ischemia and hemorrhage [57–60]. For example, this has been accomplished by using topographic (2D) absorption images in a newborn piglet model (16 sources, 2 detectors) [61]. While NIRS topography clinical applications may have illustrated the capability and importance of detecting such phenomena, the 3D localization of such entities is still lacking. In our study, we have illustrated that the temporal and spatial features seen in the reconstructed 3D image time series faithfully coincide with the observed events.

5 Author's Notes on Optical Imaging: Solutions to Common Problems

In this section we discuss solutions to some of the problems one may encounter when implementing the methodology described in this chapter.

Note 1: *Topography* vs. *Tomography*: This is a frequently considered issue. Favoring the use of the former is a sparser sensing arrangement and consequently the capacity for greater surface coverage. A notable concern for topography studies, however, is the evidence that scalp contributions can significantly bias estimates of true cortical responses, and that in some cases these biases dominate [62]. Having a larger dynamic range, tomography systems allow for a denser sensor arrangement. Shorter distance source–detector arrangements preferentially probe near-surface tissue volumes. The use of this information to account for scalp contributions is directly considered by model-based image reconstruction methods. More simplified empirical strategies (e.g., simple regression) can be employed to remove or account for near-surface tissue information, but because the sensitivity of the measurement to underlying features is strongly nonlinear, such strategies can distort information. As discussed more fully in Notes 4 and 5, implementation of a tomography approach requires access to thoughtfully considered analysis software. At present, these are not commercially available, although university-based freeware resources can be found at www.nitrc.org.

Note 2: *Optode–Tissue Interface*: Note 1 mentions biases originating from vascular responses occurring in the scalp. In the study described in this chapter, this concern was largely avoided because the scalp and underlying muscle tissue was surgically resected. However, two additional considerations remain. First is the need to mechanically stabilize the sensing array. This is important because even small variations in mechanical stability can cause artifacts. Here, stability was achieved by introducing screws to lock down the array. Second, while removal of the overlying tissues significantly improves sensitivity to deeper-lying structures, it affects the boundary conditions that apply to the forward and inverse modeling computations. To overcome this concern, when performing these computations using the developed hybrid atlas, the position of the sensing array was assigned to the surface of the skull rather than to the scalp.

In noninvasive studies (such as those performed on humans), consideration of the optode–tissue interface has added importance when the measurements are performed at sites where hair is present. Darkly pigmented hair, in particular, is a strong optical attenuator and must be displaced to ensure high-quality measures. One solution is to use small-diameter optodes that can fit between the hair follicles. While effective, this raises two other concerns. First, signal levels scale with the sensor aperture. Second, extended contact with small-diameter optodes can become notably uncomfortable

Fig. 11 View of the modified EEG cap showing LED sources and detection fibers, and a portion of the over-cap

due to the induction of scalp ischemia. A simple solution is to use optodes having a larger contact surface area. This can pose added efforts for hair displacement, but we have found the use of hair gel prevents the displaced hair from reverting to its original location. In our experience, optodes having a diameter of 3 mm can be worn for extended periods of time (at least 2 h).

A related issue is the profile of the optodes. Commonly considered is the use of spring-load optode housings. These are effective in stabilizing the sensing array against small-amplitude head movements. Unfortunately, these designs impose a higher profile on the optodes (i.e., they protrude farther from the scalp), rendering the array more susceptible to inertial forces produced by head movements. One solution that we have found effective is to employ a double-layered cap design consisting of an elastomeric under-cap (of the type commonly used in EEG studies) into which optode holders and optodes are introduced. This array is then covered by an over-cap which also has elastomeric properties (see Fig. 11). This approach confers three advantages. First is that it supports the use of low-profile optode designs, thereby reducing susceptibility to inertial forces and helping to ensure constant optode contact with the scalp. Second, the over-cap produces a more uniform distribution of force onto the head, improving subject comfort. Third, for optical measures that are sensitive to ambient light, the use of darkly colored caps serves as an effective light shield.

Note 3: *Illumination-Detection Strategies*: Mainly, the considerations discussed in this note are factors hard-wired into system design and consequently are not easily changed. We include this material because it is our experience that those who are not closely aligned with the biomedical optics community frequently are unsure of the trade-offs inherently associated with the many approaches available to system designers.

1. Optical Detectors: Three widely available sensors for optical measurements are silicon photodiodes, avalanche photodiodes, and photo-multiplier tubes. The second and third have greater sensitivity to low light signals, but add to system cost, complexity, and form-factor size. Silicon photodiodes are inexpensive, small, have a large linear dynamic range, and are increasingly the sensor of choice, especially where the aim is to deploy wearable devices [63].
2. Optical Sources: Measures performed in the NIR region are frequently accomplished using either LEDs (light-emitting diodes) or laser diodes. Laser diodes offer a significantly narrower emission profile (typically 2–3 nm full-width at half-maximum) relative to LEDs, especially at wavelengths >900 nm, but are more sensitive to damage from voltage instabilities and can require active temperature control to achieve acceptable stability. For studies involving measures of the Hb signal (690–850 nm range), the two source types can yield nearly equivalent information. However, in cases where resolution of endogenous chromophores in addition to Hb is desired (e.g., detection of lipid, water signal in the 900–1,000 nm range), the use of laser diodes are preferred owing to the notably narrower emission profile.
3. Combined Illumination-Detection Strategies: There are two principal considerations that frequently arise. One of these concerns the type of information that is being sought. Frequency- and time-domain measures are methods sensitive to the absolute background concentration of Hb and other constituents that absorb radiation at the employed wavelengths, as well as to the scattering properties of tissue [32]. Compared to the more simplified instrumentation employed with continuous wave (CW) methods (the type that is used here), these methods do have the potential to yield additional information beyond that easily obtained with CW methods. Nevertheless, the instrumentation required for such measures is notably more costly than CW, and not easily adapted to wearable systems. Some investigators have asserted that it is possible to obtain "absolute" measures of the Hb signal in tissue, particularly with respect to frequency-domain measures. However, the fact is that the complexity of tissue (e.g., the differing optical properties of each tissue layer) assures that, without considerable additional computational effort, the Hb quantities obtained are not actually absolute, but are convolved in a complex manner whose relationship to the true value is

not easily discerned. Nevertheless, these methods do offer the advantage of providing measurement signals (time-domain: temporal point spread function; frequency-domain: frequency-dependent amplitude and phase [24]), while CW measurements are restricted to the overall light intensity only. The second consideration related to combined source–detector measures is the arrangement of the optode array in relation to the expected location of the target area. A simple approach often used for topographic studies is to arrange the sources and detectors in an alternating pattern, usually separated by ~3 cm. This separation distance allows for at least a portion of the light to interact with cortical structures. Evidence of this interaction is gained from the presence of a cardiac signal in the data. In fact, the latter signal, while ordinarily taken as evidence of light interacting with arterial structures, in the brain is mainly evidence of brain motion due to mechanical forces resulting from ventricular contraction of the heart. In cases where multi-distance measures are desired (particularly for tomography), the decision on how best to arrange the array can depend on other factors, one of which is the influence that other optical sources may have on the detected signal (e.g., optical cross talk; see report by Al abdi et al. [64] for helpful suggestions).

Note 4: *Approaches to Forward Modelling*: The need for forward modelling arises primarily in cases where the aim is to perform tomography. In fields such as EEG, this problem is often dealt with by using analytic or approximate analytic solvers that are computationally efficient. Similar approaches can also be considered for the NIRS problem, but the many tissue layers (e.g., scalp, muscle, bone, meninges, cortex, sub-cortical white matter) of the head and the frequent influence of local events on the detected signal tend to argue for use of more accurate but computationally intensive methods. The approach we and others [65] have adopted for forward modeling is to employ a 3D finite element solver to compute numerical solutions to the diffusion equation. This and the related topic of inverse methods (see Note 5) can be nontrivial to implement. Fortunately, teams, including ourselves, have made efforts to create packaged solutions that are straightforward to use and are readily available (for example, visit www.nitrc.org).

A straightforward simplification is to pre-compute required information for a large number of possible illumination-detection conditions from a head model, thereby establishing a large look-up table. We have implemented this approach for most locations on the adult head (see Sect. 2.5) and for the hybrid atlas used here. Still, we are aware that more flexible solutions are desirable (e.g., efficient evaluation of subject-based DICOM images, precomputed tables for standard atlases, etc.). To this end, the implementation of fast and efficient computing strategies based on GPU (Graphical Processing Units) processing strategies is becoming increasingly feasible [66].

Note 5: *Limitations of Linear Inverse Solvers Applied to Time-Series Measures.* A standard approach for solving inverse problems is to employ recursive methods that repeatedly update intermediate forward and inverse solutions [67]. While this can lead to accurate findings, it remains thoroughly impractical when applied to a time series, especially if analytic solvers are not employed. Fortunately, at least for situations where the goal is to define relative changes, first-order linear solvers (which are less computationally intensive than recursive methods) provide for accurate recovery of the temporal information and usefully accurate spatial information, even for complex media [68, 69]. This is the approach adopted here, where the qualitative accuracy of the optical inverse-problem results is corroborated by independent validation methods. The computational methods used are very efficient, with inversion of a dataset consisting of several thousand time frames requiring only a few minutes using a I7-processor. Nevertheless, details of these types of solutions are beyond what most would care to consider. This emphasizes the need for a well-constructed and documented advanced computing environment, such as the one that we continue to develop.

Acknowledgements

This research was supported in part by the National Institutes of Health (NIH) under Grants nos. R21NS067278, R42NS050007, and 5R44NS049734; by Defense Advanced Research Projects Agency project N66001-10-C-2008; and by the Empire Clinical Research Investigator Program (ECRIP) of the New York State Department of Health.

References

1. Updated mild traumatic brain injury guideline for adults: a part of CDC's "Heads Up" series. http://www.cdc.gov/concussion/pdf/TBI_Clinicians_Factsheet-a.pdf
2. Saunders D, Jäger HR, Murray AD, Stevens JM (2008) Skull and brain: methods of examination and anatomy. In: Grainger RC, Allison D, Adam, Dixon AK (eds) Diagnostic radiology: a textbook of medical imaging, 5th edn. Churchill Livingstone, New York, Chapter 55
3. Wilkinson ID, Paley MNJ (2008) Magnetic resonance imaging: basic principles. In: Grainger RC, Allison D, Adam A, Dixon AK (eds) Diagnostic radiology: a textbook of medical imaging, 5th edn. Churchill Livingstone, New York, Chapter 5
4. Klunk WE, Engler H et al (2004) Imaging brain amyloid in Alzheimer's disease with Pittsburgh Compound-B. Ann Neurol 55(3): 306–319
5. Stanzione P, Marciani MG, Maschio M, Bassetti MA, Spanedda F, Pierantozzi M, Semprini R, Bernardi G (1996) Quantitative EEG changes in non-demented Parkinson's disease patients before and during L-dopa therapy. Eur J Neurol 3:354–362
6. Castellanos NP, Paúl N, Ordóñez VE, Demuynck O, Bajo R, Campo P, Bilbao A,

Ortiz T, del-Pozo F, Maestú F (2010) Reorganization of functional connectivity as a correlate of cognitive recovery in acquired brain injury. Brain 133:2365–2381

7. Niedermeyer E, da Silva FL (2004) Electroencephalography: basic principles, clinical applications, and related fields. Lippincot Williams & Wilkins, New York. ISBN 0-7817-5126-8
8. Swartz BE (1998) The advantages of digital over analog recording techniques. Electroencephalogr Clin Neurophysiol 106(2):113–117
9. Hansen PC, Kringelbach ML, Salmelin R (eds) (2010) MEG: an introduction to methods. Oxford University Press Inc., New York
10. Lindauer U, Gethmann J, Kühl M, Kohl-Bareis M, Dirnagl U (2003) Neuronal activity-induced changes of local cerebral microvascular blood oxygenation in the rat: effect of systemic hyperoxia or hypoxia. Brain Res 975(1–2): 135–140
11. Bailey DL, Townsend DW, Valk PE, Maisey MN (2005) Positron emission tomography: basic sciences. Springer, Secaucus, NJ. ISBN 1-85233-798-2
12. Buxton RB (2002) Introduction to functional magnetic resonance imaging: principles and techniques. Cambridge University Press, Cambridge. ISBN 0-521-58113-3
13. Cabeza R, Kingstone A (eds) (2006) Handbook of functional neuroimaging of cognition, 2nd edn. MIT Press, Cambridge, MA. ISBN 0-262-03344-5
14. Moonen CTW, Bandettini PA (eds) (2000) Functional MRI. Springer, New York, Chapters 9–11
15. Blas ML, Lobato EB, Martin T (1999) Noninvasive infrared spectroscopy as a monitor of retrograde cerebral perfusion during deep hypothermia. J Cardiothorac Vasc Anesth 13(2):244–245
16. Prabhune A, Sehic A, Spence PA, Church T, Edmonds HL Jr (2002) Cerebral oximetry provides early warning of oxygen delivery failure during cardiopulmonary bypass. J Cardiothorac Vasc Anesth 16(2):204–206
17. Davie SN, Grocott HP (2012) Impact of extracranial contamination on regional cerebral oxygen saturation: a comparison of three cerebral oximetry technologies. Anesthesiology 116(4):834–840
18. Slater JP, Guarino T, Stack J, Vinod K, Bustami RT, Brown JM III, Rodriguez AL, Magovern CJ, Zaubler T, Freundlich K, Parr GV (2009) Cerebral oxygen desaturation predicts cognitive decline and longer hospital stay after cardiac surgery. Ann Thorac Surg 87(1):36–44, discussion 44–5
19. Edmonds HL Jr, Ganzel BL, Austin EH III (2004) Cerebral oximetry for cardiac and vascular surgery. Semin Cardiothorac Vasc Anesth 8(2):147–166
20. Ma H, Zhao M, Schwartz TH (2013) Dynamic neurovascular coupling and uncoupling during ictal onset, propagation, and termination revealed by simultaneous in vivo optical imaging of neural activity and local blood volume. Cereb Cortex 23(4):885–899
21. Jobsis FF (1977) Noninvasive, infrared monitoring of cerebral and myocardial oxygen sufficiency and circulatory parameters. Science 198:1264–1267
22. Barbour RL, Graber HL, Aronson R, Lubowsky J (1990) Model for 3-D optical imaging of tissue. International geoscience and remote sensing symposium (IGARSS), vol 2, pp 1395–1399
23. Barbour RL, Lubowsky J, Graber HL (1988) Use of reflectance spectrophotometry as a possible 3-dimensional spectroscopic imaging technique. FASEB J 2:a1772
24. Graber HL, Chang J, Aronson R, Barbour RL (1993) A perturbation model for imaging in dense scattering media: derivation and evaluation of imaging operators. In: Mueller GJ, Chance B, Alfano RR, Arridge SR, Beuthan J, Gratton E, Kaschke M, Masters BR, Svanberg S, van der Zee P (eds) Medical optical tomography: functional imaging and monitoring, vol IS11. Society of Photo-Optical Instrumentation Engineers, Bellingham, WA, pp 121–143
25. Arridge SR (1999) Optical tomography in medical imaging. Inverse Probl 15(2):R41–R93
26. Colier W, Quaresima V, Barattelli G, Cavallari P, van der Sluijs M, Ferrari M (1997) Detailed evidence of cerebral hemoglobin oxygenation in response to motor cortical activation revealed by a continuous wave spectrometer with 10 Hz temporal resolution. Proc SPIE 2979:390–396
27. Delpy DT, Cope M (1997) Quantification in tissue near-infrared spectroscopy. Philos Trans R Soc Lond B Biol Sci 352:649–659
28. Eda H, Oda I, Ito Y, Wada Y, Oikawa Y, Tsunazawa Y, Takada M, Tsuchiya Y, Yamashita Y, Oda M, Sassaroli A, Yamada Y, Tamura M (1999) Multichannel time-resolved optical tomographic imaging system. Rev Sci Instrum 70:3595–3602
29. Gratton E, Mantulin WW, vandeVen MJ, Fishkin JB, Maris MB, Chance B (1993) A novel approach to laser tomography. Bioimaging 1:40–46
30. Hock C, Villringer K, Müller-Spahn F, Wenzel R, Heekeren H, Schuh-Hofer S, Hofmann M,

Minoshima S, Schwaiger M, Dirnagl U, Villringer A (1997) Decrease in parietal cerebral hemoglobin oxygenation during performance of a verbal fluency task in patients with Alzheimer's disease monitored by means of near-infrared spectroscopy (NIRS)-correlation with simultaneous rCBF-PET measurements. Brain Res 755:293–303

31. Villringer A, Chance B (1997) Non-invasive optical spectroscopy and imaging of human brain function. Trends Neurosci 20:435–442
32. Ferrari M, Quaresima V (2012) A brief review on the history of human functional near-infrared spectroscopy (fNIRS) development and fields of application. Neuroimage 63:921–935
33. Koizumi H, Yamamoto T, Maki A, Yamashita Y, Sato H, Kawaguchi H, Ichikawa N (2003) Optical topography: practical problems and new applications. Appl Opt 42:3054–3062
34. Piper SK, Krueger A, Koch SP, Mehnert J, Habermehl C, Steinbrink J, Obrig H, Schmitz CH (2014) A wearable multi-channel fNIRS system for brain imaging in freely moving subjects. Neuroimage 85(pt 1):64–71
35. Fish RE, Brown MJ, Danneman PJ, Karas AZ (2008) Anesthesia and analgesia in laboratory animals. Elsevier Inc., San Diego, CA. ISBN 978-0-12-373898-1
36. Schmitz CH, Löcker M, Lasker JM, Hielscher AH, Barbour RL (2002) Instrumentation for fast functional optical tomography. Rev Sci Instrum 73:429–439
37. Schmitz CH, Graber HL, Pei Y, Farber MB, Stewart M, Levina RD, Levin MB, Xu Y, Barbour RL (2005) Dynamic studies of small animals with a four-color DOT imager. Rev Sci Instrum 76:094302
38. Scholkmann F, Wolf M (2013) General equation for the differential pathlength factor of the frontal human head depending on wavelength and age. J Biomed Opt 18(10):105004
39. Schmitz CH, Klemer DP, Hardin RE, Katz MS, Pei Y, Graber HL, Levin MB, Levina RD, Franco NA, Solomon WB, Barbour RL (2005) Design and implementation of dynamic near-infrared optical tomographic imaging instrumentation for simultaneous dual-breast measurements. Appl Optics 44:2140–2153
40. Xu Y, Ewald A, Graber HL, Nichols JD, Pflieger ME, Ossadtchi A, Schmitz CH, Barbour RL (2012) A computing environment for multimodal integration of EEG and fNIRS. Poster 46 at functional near infrared spectroscopy, London, UK, 26–28 October 2012
41. Graber HL, Xu Y, Pei Y, Barbour RL (2005) Spatial deconvolution technique to improve the accuracy of reconstructed three-dimensional diffuse optical tomographic images. Appl Optics 44:941–953
42. Xu Y, Graber HL, Barbour RL (2007) An image correction algorithm for functional 3D DOT brain imaging. Appl Optics 46:1693–1704
43. Barbour RL, Graber HL, Xu Y, Pei Y, Schmitz CH, Pfeil DS, Tyagi A, Andronica R, Lee DC, Barbour S-LS, Nichols JD, Pflieger ME (2012) A programmable laboratory testbed in support of evaluation of functional brain activation and connectivity. IEEE Trans Neural Syst Rehabil Eng 20:170–183
44. McLaren DG, Kosmatka KJ, Oakes TR, Kroenke CD, Kohama SG, Matochik JA, Ingram DK, Johnson SC (2009) A population-average MRI-based atlas collection of the rhesus macaque. Neuroimage 45:52–59
45. Okamoto M et al (2005) Spatial registration of multichannel multi-subject fNIRS data to MNI space without MRI. NeuroImage 27: 842–851
46. Fazil S, Mehnert J, Steinbrink J, Curio G, Villringer A, Muller KR, Blankertz B (2012) Enhanced performance by a hybrid NIRS-EEG brain computer interface. Neuroimage 59: 519–529
47. Wylie GW, Graber HL, Voelbel GT, Kohl AD, DeLuca J, Pei Y, Xu Y, Barbour RL (2009) Using co-variations in the Hb signal to detect visual activation: a near infrared spectroscopic imaging study. Neuroimage 47:473–481
48. Fantini S et al (2013) Dynamic model for the tissue concentration and oxygen saturation of hemoglobin in relation to blood volume, flow velocity and oxygen consumption: implications for functional neuroimaging and coherent hemodynamic spectroscopy (CHS). Neuroimage S1053–8199(13):00315–00317
49. Chen S et al (2002) Auditory-evoked cerebral oxygenation changes in hypoxic-ischemic encephalopathy of newborn infants monitored by near infrared spectroscopy. Early Hum Dev 67:113–121
50. Okamoto M et al (2006) Prefrontal activity during taste encoding: an fNIRS study. Neuroimage 47(2):220–232
51. Holper L, Muehlemann T, Scholkmann F, Eng K, Kiper D, Wolf M (2010) Testing the potential of a virtual reality neurorehabilitation system during performance of observation, imagery and imitation of motor actions recorded by wireless functional near-infrared spectroscopy (fNIRS). J Neuroeng Rehabil 7:57
52. White BR, Culver JP (2010) Quantitative evaluation of high-density diffuse optical tomography: in vivo resolution and mapping performance. J Biomed Opt 15:026006

53. Habermehl C, Holtze S, Steinbrink J, Koch SP, Obrig H, Mehnert J, Schmitz CH (2012) Somatosensory activation of two fingers can be discriminated with ultrahigh-density diffuse optical tomography. Neuroimage 59: 3201–3211
54. Hyde DC, Boas DA, Blair C, Carey S (2010) Near-infrared spectroscopy shows right parietal specialization for number in pre-verbal infants. Neuroimage 53:647–652
55. Nakahachi T, Ishii R, Iwase M, Canuet L, Takahashi H, Kurimoto R, Ikezawa K, Azechi M, Kajimoto O, Takeda M (2010) Frontal cortex activation associated with speeded processing of visuospatial working memory revealed by multichannel near-infrared spectroscopy during Advanced Trail Making Test performance. Behav Brain Res 215:21–27
56. Kameyama M, Fukuda M, Yamagishi Y, Sato T, Uehara T, Ito M, Suto T, Mikuni M (2006) Frontal lobe function in bipolar disorder: a multichannel near-infrared spectroscopy study. Neuroimage 29:172–184
57. Stankovic MR, Maulik D, Rosenfeld W, Stubblefield PG, Kofinas AD, Gratton E, Franceschini MA, Fantini S, Hueber DM (2000) Role of frequency domain optical spectroscopy in the detection of neonatal brain hemorrhage—a newborn piglet study. J Matern Fetal Med 9(2):142–149
58. Leistner S et al (2011) Non-invasive simultaneous recording of neuronal and vascular signals in subacute ischemic stroke. Biomed Tech (Berl) 56:85–90
59. Zweifel C et al (2010) Continuous assessment of cerebral autoregulation with near-infrared spectroscopy in adults after subarachnoid hemorrhage. Stroke 41:1963–1968
60. Salonia R, Bell MJ, Kochanek PM, Berger RP (2012) Cerebral hemorrhage / ischemia in DOT the utility of near infrared spectroscopy in detecting intracranial hemorrhage in children. J Neurotrauma 29(6):1047–1053
61. Stankovic MR, Maulik D, Rosenfeld W, Stubblefield PG, Kofinas AD, Drexler S, Nair R, Franceschini MA, Hueber D, Gratton E, Fantini S (1999) Real-time optical imaging of experimental brain ischemia and hemorrhage in neonatal piglets. J Perinat Med 27: 279–286
62. Takahashi T, Takikawa Y, Kawagoe R, Shibuya S, Iwano T, Kitazawa S (2011) Influence of skin blood flow on near-infrared spectroscopy signals measured on the forehead during a verbal fluency task. Neuroimage 57: 991–1002
63. Strangman G, Boas DA, Sutton JP (2002) Non-invasive neuroimaging using near-infrared light. Biol Psychiatry 52:679–693
64. Al abdi R, Graber HL, Xu Y, Barbour RL (2011) Optomechanical imaging system for breast cancer detection. J Opt Soc Am A Opt Image Sci Vis 28:2473–2493
65. Kak AC, Slaney M (1999) Principle of computerized tomographic imaging, classics in applied mathematics. SIAM, Philadelphia, PA
66. Fang Q, Boas DA (2009) Monte Carlo simulation of photon migration in 3D turbid media accelerated by graphics processing units. Opt Express 17(22):20178–20190
67. Dehghani H, Eames ME, Yalavarthy PK, Davis SC, Srinivasan S, Carpenter DM, Pogue BW, Paulsen KD (2008) Near infrared optical tomography using NIRSFAST: algorithm for numerical model and image reconstruction. Comm Numer Meth Eng 25:711–732
68. Pei Y, Graber HL, Barbour RL (2001) Influence of systematic errors in reference states on image quality and on stability of derived information for DC optical imaging. Appl Optics 40:5755–5769
69. Graber HL, Pei Y, Barbour RL (2002) Imaging of spatiotemporal coincident states by DC optical tomography. IEEE Trans Med Imaging 21:852–866

Index

Mingrui Zhao et al. (eds.), *Neurovascular Coupling Methods*, Neuromethods, vol. 88,
DOI 10.1007/978-1-4939-0724-3, © Springer Science+Business Media New York 2014

MIX
Papier aus verantwortungsvollen Quellen
Paper from responsible sources
FSC® C105338

If you have any concerns about our products,
you can contact us on
ProductSafety@springernature.com

In case Publisher is established outside the EU,
the EU authorized representative is:
Springer Nature Customer Service Center GmbH
Europaplatz 3, 69115 Heidelberg, Germany

Printed by Libri Plureos GmbH
in Hamburg, Germany